VOLCANOES

WHAT'S HOT & WHAT'S NOT

on Earth and in Our Solar System

by Ian M. Lange

This book is dedicated to the
Late Professor Richard E. Stoiber (1911-2001),
Dartmouth College, who inspired so many
to become geologists in industry,
public service, and teaching.

ISBN: 978-1-59152-168-6

Published by Ian M. Lange

Front cover: One of the most active volcanoes in the world is the hot spot volcano
Piton de la Fournaise, La Reunion Island, Indian Ocean. (Julien Grondin photo/Hemera
collection/Thinkstock).
Back cover: Astronaut photo of ash cloud from Mount Cleveland, Alaska (NASA photo).

For more information, contact Ian M. Lange at ianlange@rockisland.com.

You may order extra copies of this book by calling Farcountry Press toll free at (800) 821-3874.

sweetgrassbooks
an imprint of Farcountry Press
Produced by Sweetgrass Books.
PO Box 5630, Helena, MT 59604; (800) 821-3874; www.sweetgrassbooks.com.

Produced in the United States of America.
Printed in China.

20 19 18 17 16 1 2 3 4 5

Acknowledgments

I thank Dr. Fred Bodholt, Missoula, MT; Dr. Arthur Bookstrom with the U.S. Geological Survey, Spokane, WA; the late Dr. Eric Braun, Missoula, MT; Dr. Warren Nokleberg with the U.S. Geological Survey, Menlo Park, CA; Professor Steven Sheriff, Geosciences Department, University of Montana, Missoula, MT; and Dr. John Whetten, Lopez, WA, for their careful, thoughtful, and informative reviews of selected book chapters. Their insights and perspectives greatly improved the manuscript. I also am very appreciative of Director of Publications Kathy Springmeyer, Senior Editor Will Harmon, and Senior Designer Shirley Machonis, all with Farcountry Press, for their careful editing, book design, and suggestions which greatly improved the manuscript.

I also thank Chuck Blay and Robert Siemers for the generous use of several figures from their book entitled "Kauai's Geologic History," and Ludovic Ferriere, Curator at the Natural History Museum, Vienna, Austria, for use of his map showing the distribution of confirmed earthly meteorite impact craters. Finally, credit is due to Wikipedia for being such a useful, free, online resource.

Pavlof volcano on the Alaskan peninsula (U.S. Fish and Wildlife Service photo).

Table of Contents

Legends and Volcanoes

Most legends are based upon some event in prerecorded history. The stories were told from generation to generation. Prior to written language, or the advent of television, the oral transmission of stories tied societies together. "Facts" changed with time, and stories commonly grew more exciting. Many of us have played the game of sitting in a circle and relaying a story from person to person. The last one to hear the story usually tells a far different tale than the original yarn.

All prehistoric peoples, whether they lived in the highlands of New Guinea or the Sahara Desert of North Africa, had legends with origins that dated back centuries or more. In many cases, gods were central to the legends: they were born of them or were part of them.

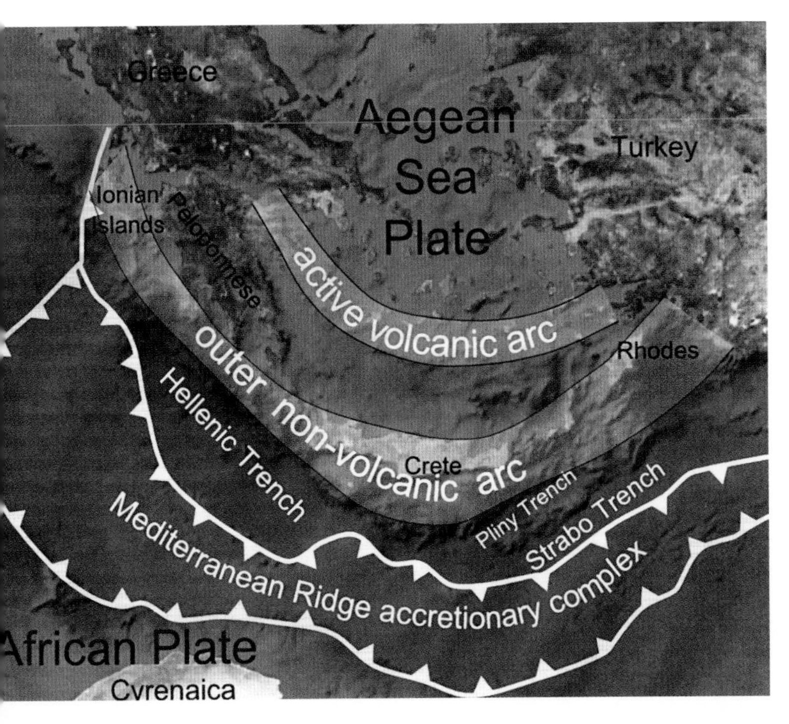

FIGURE INTRO 1: *Tectonic-geographical map of the Aegean Sea (Hellenic arc) including Greece and Turkey, showing major faults in white and the Santorini volcanic arc in green. The southern white-toothed line depicts the edge of the oceanic crustal portion of the African continental plate that is subducting, or being driven, under the Hellenic continental plate, resulting in earthquakes and the volcanic arc (from Wikipedia based on NASA World Wind Software).*

Eventually these legends, while interesting to hear, were debunked as nothing more than superstition. Over time, however, the findings of archeologists and others have shown that many legends were sparked by actual events. One of the most famous, documented by Plato in the 4th Century B.C., concerns the lost continent of Atlantis.

Within the Critias, the second in series of three dialogues by Plato, is a description of Atlantis. Here, Critias is talking with Socrates and Hermocrates. We join them as Critias says *". . . let us remind ourselves that it is in all 9,000 years since the general war, of which we are now to relate the course, was declared between those who dwelt without and those that dwelt within the Pillars of Hercules. The command of the latter was taken, and the war conducted throughout, as the story ran, by our own city [Athens]; the leaders of the other party were kings of the* ***island of Atlantis. Atlantis, as you will recall, was once, we said, an island larger than Libya and Asia together; it now has been engulfed by earthquakes and is the source of the impassable mud which prevents navigators from this quarter from advancing through the straits to the open ocean. . . .*** *As we have said before, when we were speaking of the "lots," the gods divided the whole earth into lots, some larger, some smaller, and established their temples and sacrifices in them. Poseidon, then, receiving as his lot the isle of Atlantis, settled his sons by a mortal woman in a district of it which must now be described. By the sea, in the center of the island there was a plain, said to have been the most beauteous of all such plains and very fertile, and again, near the center of this plain, at a distance of some 50 stadia, a mountain which nowhere of any great altitude . . . it was the island itself which furnished the main provision for all purposes of life. In the first place it yielded all products of the miner's industry. . . . It also bore in its forests a generous supply of all timbers . . . and maintained a sufficiency of animals wild and domesticated; even elephants were plentiful. There was ample* *pasture for this largest and most voracious of brutes. . . . Besides all this, the soil bore all aromatic substances still to be found on earth. So the kings employed all these gifts of the soil to construct and beautify their temples, royal residences, harbors, docks, and domain in general. . . ."* And so Atlantis was described by Plato, through Critias, who lived between 427 and 347 B.C.

Now the people of Atlantis grew wealthy and powerful and attempted to conquer their neighbors. This did not go over well with the Greeks to north, however, and so they bested them on the battlefield. Adding insult to injury, the gods were also so unhappy that they sent forth earthquakes and floods[1] to destroy the island civilization: Atlantis sunk into the sea in less than a day.

The effects of the island's destruction were felt well beyond the Mediterranean Sea. Historians of China's earliest recorded dynasty, the Xia, noted that "at the time of King Chieh the sun was dimmed. Three suns appeared and winter and summer came irregularly." China witnessed July frosts and heavy rainfall resulting in crop failures and famine. This was followed by a seven-year drought.[1]

To understand the significance of this legend and others requires knowledge of what the ancient world was like then. Thirty-five hundred years ago, the western world was the greater Mediterranean Sea. Yes, some learned people knew that "barbarians" lived to the north, east, and west. So some continents in those days were probably large Mediterranean islands like Thera.

These island people lived in one of the most advanced civilizations. Therans traded with the Minoans on Crete, the Egyptians south of Crete, and the Greeks to the north. Those living to the east in what are now Turkey and the Holy Land, and Italy to the west, no doubt also traded with the Minoans.

Unknown to those living on lush, vegetated Thera was the volcanic origin of their island. By 1420 B.C. the people living on the volcano, which had lain dormant for as much as 15,000 years, were prospering. Soils were deep and rich, yielding bountiful harvests. Little did they know that their world would be greatly affected by the eruption of the dormant volcano Santorini.

So what happened to Thera? The island, roughly circular in shape, was formed by at least 10 volcanoes, much like the big island of Hawaii is today. Santorini was not tall as volcanoes go, perhaps reaching 2,000 feet above sea level.

FIGURE INTRO 2: *Minoan-age mural of ladies discovered on a wall in the buried Bronze-age village of Akrotiri on Thera (See Figures 3 and 6 for location of Akrotiri). The mural was moved to the National Museum in Athens (Lange photo).*

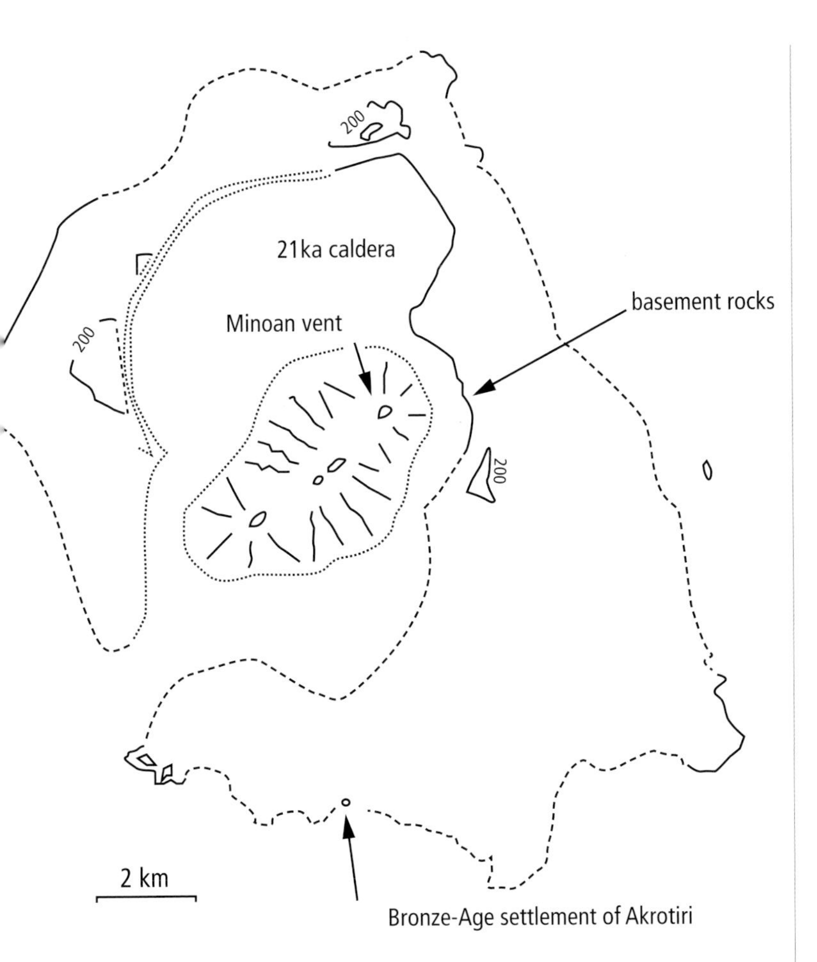

FIGURE INTRO 3: *Thera as it might have been before the great Minoan eruption (3,600 years ago) with known, inferred, and possible shorelines (solid, dashed, and dotted lines, respectively) (modified from Druitt et al., 1999).*

FIGURE INTRO 4: *Minoan-age air fall. Note the basal thick layer of essentially uniform grain size versus the thinner upper air fall units containing different grain sizes (Lange photo).*

The sequence of events leading to Thera's destruction began with a major earthquake below the island, apparently due to rising magma. Minoan settlements were either damaged or destroyed. Months, perhaps up to two years, passed before volcanism commenced. During this time, some buildings were repaired.

The first eruptive phase consisted of a strong ejection of gas, ash, pumice, and volcanic rock fragments ripped from the vent walls. We know this because the erupted material rests on ancient soil. The volcanic layer varies from 0.5 to 6 m thick, and is silica rich (rhyodacite).

Following this phase, seawater apparently entered the volcano at depth and contacted magma. Water then expanded greatly as it flashed to steam, creating powerful phreatomagmatic explosions.

Huge blocks were hurled outward, with some destroying houses in the village of Akrotiri located near the south end of the island. This eruptive phase, lasting perhaps days to weeks, deposited material up to 12 m (40 feet) deep. The eruption

FIGURE INTRO 5: *Minoan-age ash flow (light gray containing dark rock clasts ripped from vent wall during the eruption) overlain by younger dark gray outwash derived from the slopes above (Lange photo).*

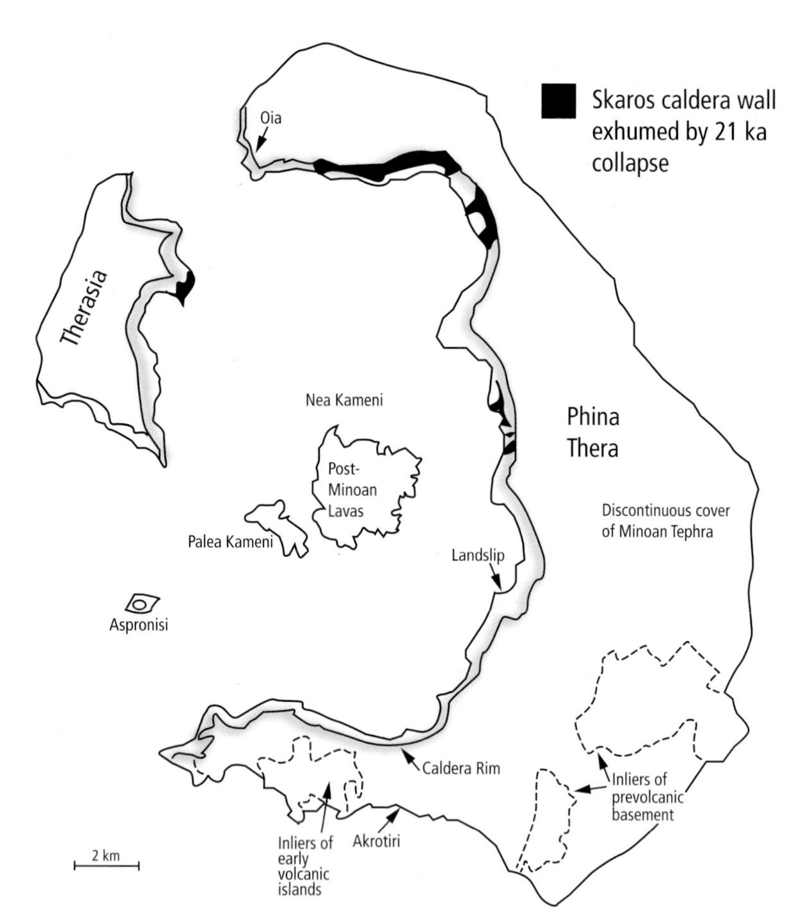

FIGURE INTRO 6: *The Island of Thera today showing areas of prevolcanic basement rocks and locations of Minoan tephra on caldera walls (modified from Druitt and Francaviglia, 1992).*

then evolved into a plinian-type eruption,[2] jetting a continuous stream of very hot gas and entrained particles (ash to boulder size) high into the stratosphere. Gravity prevailed at the top of the erupting column, so while the inner, lower part of the column rose, very hot (up to 850°C) and ash-rich gas descended along the outer margin of the column. Upon reaching the ground, this mixture flowed away at speeds of up to 500 km (300 mi) per hour. Anything flammable burned, and the particles, being heavier than air, settled out and formed a relatively low-density volcanic rock called tuff.

Some ash flow material was deposited on the island, but most landed in the surrounding sea. Island ash flow deposits range from 56 m (180 feet) thick to nonexistent in places.

A huge quantity of lithic (preexisting volcanic rock) fragments, comprising 20 to 30 percent of the total ejecta with some pieces weighing several tons, were ripped from older lava flows during the eruption. Fresh, vesiculated (full of gas escape holes) magma from the chamber constituted the rest of the ejecta. When over, between 28 and 34 km^3 of ejecta, and the central part and bulk of Thera was gone. It was replaced by a huge depression, a caldera 83 km^2 (32 mi^2) and up to 390 m (1,280 feet) deep, filled with seawater. Only three pieces remained of the former island. They are today's Thera, Therasia, and tiny Aspronisi.

The sea within the caldera probably steamed for years, and then a small volcano developed on the floor of the caldera. It breached the sea surface, forming the island Kameni, in 197 B.C. This volcano has erupted intermittently ever since, with the last outburst ending in 1950 A.D.

Will the volcano erupt again? Probably. The oldest known rocks date from 645,000 +/- 90,000 years ago, with Thera having had several episodes of growth followed by violent, partial destruction. In fact, the rich soils that supported the islanders 3,500 years ago had developed on deposits from the last major eruptive phase dated at 21,000 years ago.

This Minoan-age catastrophic eruption may have been responsible for a distant event that morphed into an equally famous legend. According to the Bible, airborne plagues arrived in Egypt, some 850 km (530 mi) south of Thera, during the time of Moses. The plagues persuaded the Pharaoh to let the Israelites leave. The Pharaoh evidently had second thoughts as the Israelites approached the north end of the Red Sea, and he sent his troops after them. The sea parted for the Israelites but closed back on the troops in hot pursuit. Does a relationship exist between the eruption and the Moses story? Perhaps. Was the "plague" really ash from Santorini? Did an earthquake briefly displace the waters of the Red Sea?

Santorini is one of Earth's many volcanoes of various kinds and sizes. Between 1,300 and 1,500 terrestrial volcanoes have erupted during the past 10,000 years. These volcanoes, fed from at least 1,000 major magmatic systems, produce 50 to 70 eruptions each year. Remember that 70 percent of the Earth is covered with water. Beneath the oceans there may be more than 1 million additional volcanoes. They form a 70,000-km (43,500-mi), sinuous, mostly submerged mountain range that runs like a baseball seam from the Atlantic to the Pacific. This active, ocean crust-forming system (discussed in Chapter 2) sees more eruptions than occur on land each year, and produces more lava.

Most of the land-based volcanoes occur in linear and arcuate belts totaling 32,000 km (19,900 mi) in length. Due to differences in magma composition, land-based volcanoes are generally more explosive, and so more dangerous, than marine volcanoes. Fortunately for humankind, the larger the eruption, as with earthquakes, the less frequently they occur.

The following chapters examine where volcanoes are located and why. We also look at the different types of volcanoes and the various materials and landforms they produce. We'll also explore the different kinds of volcanoes on other planets in our solar system.

A fuller understanding of volcanism also requires looking below the surface at the roots of volcanic systems—the huge igneous systems that feed volcanoes. These systems also develop many types of mineral deposits, vital to our world's ever-growing need for resources.

Modern society benefits from volcanic processes, but is sometimes harmed by them, as described above in the Minoan example of Thera. Scientists have learned how huge past eruptions affected climate. Some eruptions have resulted in mass destruction, civil unrest, famine, and death. As world population grows, and people increasingly live close to and even on potentially dangerous volcanoes, scientists are redoubling their efforts to accurately forecast eruptions. We'll examine the challenges of this still inexact science.

INTRODUCTION – NOTES

1. Heavy rainfall is a common occurrence during and after large volcanic eruptions.
2. A plinian-type eruption is named for Pliny the Younger, who witnessed the 79 A.D. eruption of Mount Vesuvius in Italy that buried the Roman resort towns of Pompeii and Herculeum. The eruption killed his uncle, Pliny the Elder, who was trying to save Romans.

INTRODUCTION – REFERENCES

Abbott, 2004; Critias, 1961; Druitt and Francaviglia, 1992; Druitt et al., 1999; McCoy and Heiken, 2000; Pichler and Friedrich, 1980; Simkin and Siebert, 2000; USGS Bull. 1850.

FIGURE INTRO 7: *Photograph of the northeast caldera wall showing the complex, violent, volcanic history of Thera. White Minoan deposits (3,600 years old) appear on the skyline, whereas Lower Pumices 1 (203,000 +/- 24,000 years old) and overlying Lower Pumice 2 (180,000 years old), light gray, thick air fall and ash flow units lie prominently above dark lava flows. The brown layer between Pumices 1 and 2 is a paleosol or soil. An abandoned pumice loading facility is at the waters edge (Lange photo).*

CHAPTER ONE

Magma, Fire, and Brimstone —an Earth Primer

Introduction

No matter how the Earth formed, whether from accretion of hot or cold particles, material ripped from our Sun by another passing star, or in some other way, we have learned much about its structure, composition, and history in the last 100 years. These data have been obtained by examining surface materials and features, and by indirect means. With increasingly more sophisticated technology, scientists are discovering more information about our incredible, living blue planet, the third from the Sun of the four small, rocky inner planets in our solar system.

Directly observable features include the various rock types, fossils, and structures now more or less completely mapped on the land surface except where covered with thick soils, glaciers, and glacial deposits. Other data come from pieces of rocks brought to the surface in volcanic eruptions in places like Hawaii, and in diamond-bearing kimberlite pipes,[1] whose kimberlite source is in the mantle. Also invaluable are the data found in the slices of seafloor, including upper mantle rocks exposed on land by tectonic forces and core samples from deep wells drilled in both continental and seafloor rocks.

We have also learned much about the Earth by examining and mapping its magnetic, electrical, and gravitational properties. Further studies of seismic (earthquake) waves and volcanic and human-caused explosions have helped us understand the Earth's composition and structure.

This information, together with various geochemical, isotopic, and age-dating techniques, have not only allowed scientists to determine ages of rocks, fossils, and events, but also to determine source regions of magmas and understand complex rock-forming processes. Furthermore, studies of magnetism locked in ancient rocks—called paleomagnetism—have enabled scientists to know where our present continents once were, and who their continental rocky neighbors might have been long ago.

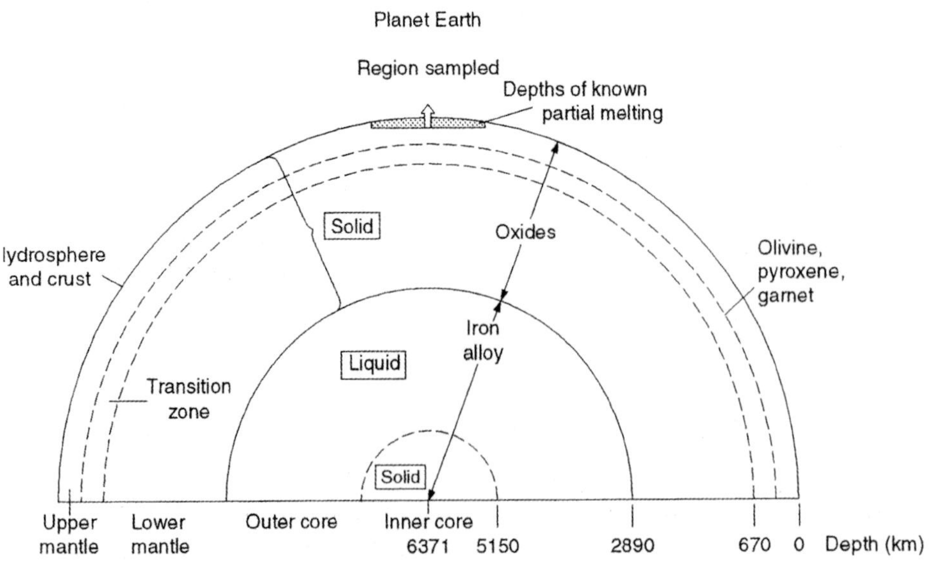

FIGURE 1.1: *Earth cross section of zones, whether they are solid or liquid, and thickness in kilometers. Note that the line thickness of the "Hydrosphere and crust" is to scale (modified from Jeanloz, 2000).*

What We Know about the Physical Earth

The Earth, a sphere with an average diameter of 12,740 km (7,916 mi), is divided into three major zones: a crust, a mantle, and the core. The crust, very thin in relation to the diameter of the Earth, varies not only in thickness, but in composition as well. Continental, granitic crust averages 50 km (30 mi) in thickness, while oceanic crust is basaltic and averages 6 km (3.6 mi) thick.

The thick zone below the crust, the mantle, is a solid, but

Radioactivity and Absolute Age Dating

Absolute age dating is based upon the decay (break down or disintegration) of unstable or radioactive isotopes of elements such as uranium and thorium. An atom of an element has an inner nucleus containing positively charged particles with mass called protons, and particles with mass but no charge called neutrons. Electrons with negative charges and essentially no mass equal the number of protons and circle the nucleus. However, the number of neutrons an element has varies. These species of an element with different numbers of neutrons, then, are call isotopes.

The standard way of designating isotopes, using uranium as an example, goes like this: ${}_{92}^{238}U$, where 92 is the number of protons and the atomic number of the element, and 238 is the sum of protons plus neutrons. Therefore, this isotope of uranium has 238 - 92 = 146 neutrons.

Some elements such as uranium, however, have unstable arrangements of these subatomic particles and so decay in one or more steps to a stable, or non-radioactive element called a "daughter" element. For example, isotope ${}^{238}U$ decays to ${}^{206}Pb$ (lead) in 14 steps, and the other U isotope, ${}^{235}U$ decays to ${}^{207}Pb$ (lead) in 15 steps. With each isotopic decay step, not only is another isotope or daughter isotope formed, but heat is produced in the process.

The rate of decay is constant for any particular radioactive isotope, so in effect we have a radioactive clock. Geochronologists, then, can date the formation age of minerals and rocks if they contain an element such as uranium.

For example, it takes approximately 700 million years for one pound of ${}^{235}U$ to decay to one-half pound, producing ${}^{207}Pb$ in the process, and 4.6 billion years for one pound of ${}^{238}U$ to decay to one-half pound, thereby resulting in the daughter product ${}^{206}Pb$. In another 700 million years and 4.6 billion years, there will only be one-quarter pounds of ${}^{235}U$ and ${}^{238}U$ left, respectively. These time spans are called half lives of the respective radioactive elements. Each radioactive isotope has a different half-life which varies from microseconds to billions of years, depending on the relative stability of the isotope. The greater the isotope's stability, the longer its half-life.

The Earth, then, receives a tremendous amount of heat from these long-lived radioactive isotopes. In its earliest years, the Earth also received much heat from short-lived isotopes such as aluminum (${}^{26}Al$), which, with a half-life of only 720,000 years, decayed to the stable isotope magnesium (${}^{26}Mg$).

Some isotopic schemes can also be used for determining geologic processes such as magma generation and source regions in the earth.

heat and massive pressures cause it to behave like a plastic or putty over time. The mantle is thus capable of slow flow. The interface of the mantle and outer core is 2,890 km (1,800 mi) below the surface. From that depth down to 5,150 km is the liquid outer core. The center of the Earth, located within the inner core at a depth of 6,370 km (3,820 mi), is believed to be as much as 7,000°C—nearly as hot as the surface of our Sun—and solid due to the great pressure at that depth.

These parts of the Earth and their respective thicknesses were determined seismically (measuring earthquake and explosion energy waves), as was the discovery that at least the outer core is liquid. (The inner core may be solid due to intense pressure at that depth.) More recently, other Earthly attributes were discovered, including the existence of 100-km (60-mi) thick lithospheric plates[2] composed of the crust and upper mantle. These and other facets of the theory of plate tectonics are described in Chapter 2.

Something else we know about the Earth, and have since prehistoric times, is that the Earth emits heat. We swim in hot springs, see it coming from volcanoes, feel the hot walls in deep mines, and actually measure heat rising from ordinary, everyday crustal rocks. Heat is not only necessary to fire volcanoes, it also helps fuel the movement of 100-km (60-mi) thick lithospheric plates composed of the crust and upper mantle. And, as you

	Density	Thickness or depth to	Mass Fraction of the Earth	Composition
Continental Crust	2.8	50 km (30 miles)		Granite
Oceanic Crust	3.1	6 km (3.7 miles)	0.5%*	Basalt
Mantle	4.5	Extends to (1790 miles)	67.0%	Ultramafic***
Outer Core	10.7	3470 km (2160 miles)	32.5%**	Iron & Nickel
Inner Core	10.7	1220 km (760 miles)		Iron & Nickel
Whole Earth	5.5	12740 km (7920 miles)		

* = includes continental and oceanic crust
** = for entire core
*** = includes the minerals olivine (FeSiO4) and Pyroxene (FeMgSiO3)

TABLE 1.1: *Earth physical characteristics including density, zone thicknesses, as fraction of the Earth, and composition.*

probably already knew, temperature increases the deeper we probe into the Earth. The rate of temperature rise—the geothermal gradient—varies according to geological environments. (See Figure 8.1, page 118, for the crustal-upper mantle geothermal gradient.) In general, the base of the crust ranges from 500°C to 900°C, and the base of the mantle is at least 4,000°C. Temperatures at the inner core may exceed 7,000°C. By comparison, the temperature of the Sun's photosphere—the region we perceive as visible sunlight—is 5,500°C.

Why is the Earth Hot?

So why is the Earth, and especially the core, hot? There are three primary sources for the heat: radioactive decay of unstable elements, early gravitational contraction, and residual heat from meteoric impacts. Other heat sources include accretion as the Earth formed, frictional heat from tidal forces, and frictional heat generated by mantle convection.

Radioactive decay is the major cause of Earth's heat today. It so happens that certain elements are unstable or radioactive due to the nature of and arrangement of their subatomic particles. Because of this characteristic, these unstable atoms disintegrate, decay, or break down (all three words mean the same thing) in steps to stable or non-radioactive "daughter" elements. For example, the radioactive element uranium (U) decays in many steps to the stable element lead (Pb). During the process, heat is emitted, which warms the material around it, mainly continental crust where these elements are now concentrated. And because the Earth is a relatively well-insulated body, the escape of this heat to the atmosphere is slow.

Another major mechanism that produced heat was the accretion of particles during the early years of Earth's formation. As material at depth became more compressed due to the weight of the overlying rock, its temperature increased.

The Earth has also retained heat from the impacts of meteorites during its early history. When these huge chunks of metal and rock, traveling at velocities measured in kilometers per second, crashed into the Earth, they produced incredible amounts of heat. The peak intensity of lunar bombardment was about 3.8 billion years ago; there is little doubt the Earth was also subjected to this intense punishment.

Much of this heat was in turn captured by the victim of these celestial assaults, Mother Earth. And because the Earth is a poor conductor of heat, much of this heat has been stored ever since.

The 1.6-km-wide, 50,000-year-old Arizona Meteor Crater created by an iron-nickel meteorite provides an example of impact forces and heat generation. Upon impact, this meteorite, only about 30 m (100 feet) in diameter, released energy equivalent to approximately 4 million tonnes of TNT. (The atomic bomb dropped Hiroshima during World War II, in comparison, released about 20,000 tonnes of equivalent energy.)

Another mechanism that also produces heat is tidal force. We are familiar with ocean tides, caused by the gravitational attraction of the Moon and Sun on our water envelope. Tidal distortions, interestingly, also affect our mantle by as much as 11 cm vertically. In addition, these tides have slowed the Earth's rate of rotation with the braking action producing heat roughly equivalent to the heat produced by the radioactive decay of thorium (^{232}Th).

This is not an insignificant amount. Elsewhere in the solar system, the tidal force between Jupiter and its moon Io is so great that enough heat is produced to keep Io in a state of perpetual volcanic eruption.

Composition of the Earth

The study of seismic waves has allowed us to delineate the structural zones of the Earth, but what are these zones composed of? Once again, direct as well as indirect investigations have enhanced our understanding. For example, we are able to examine chunks of the mantle that arrive at the surface in lava flows, kimberlite pipes,[1] and tectonically uplifted seafloor. We have examined the rocks in these samples to determine their density and other physical properties. As seismic waves travel, they move faster and slower—and also shorten or lengthen—depending on the elasticity, density, and composition of the rock they move through. This allows us to deduce what types of material are (and are not) present in the different zones. We have learned, for example, that mantle rocks are much richer in magnesium and lower in silica, and so are denser and heavier than basalt, the most common volcanic rock. Basalt forms the ocean crust and covers large swaths of continents. The Earth's core is composed of much heavier, denser material than found in the mantle. The best estimates suggest the core is rich in iron, with lesser amounts of nickel and sulfur, plus other elements.

Surface Shaping Forces

We have been fortunate during the last several decades of manned and unmanned space probes to see wonderful, clear, color photographs of our Moon, Mars, Venus, Mercury, and the gaseous giants beyond Mars, plus some of their moons. What stands out is how different the surface of the Earth looks in comparison.

One of the most noticeable differences, certainly between the planets, is the watery envelope that we presently believe only Earth possesses. But if we look at the land surface of the Earth, it is strikingly devoid of impact craters compared with Mars, and especially the Moon and Mercury. In addition, our land and seafloor surfaces have long, sinuous mountain ranges.

Until recently, scientists didn't think much about the difference in density of impact craters found on the Earth's surface in comparison with those found on Mars, the Moon, and Mercury. But our thinking changed dramatically in the 1980s when mounting evidence gave more credence to the "killer impact" theory for the extinction of the dinosaurs

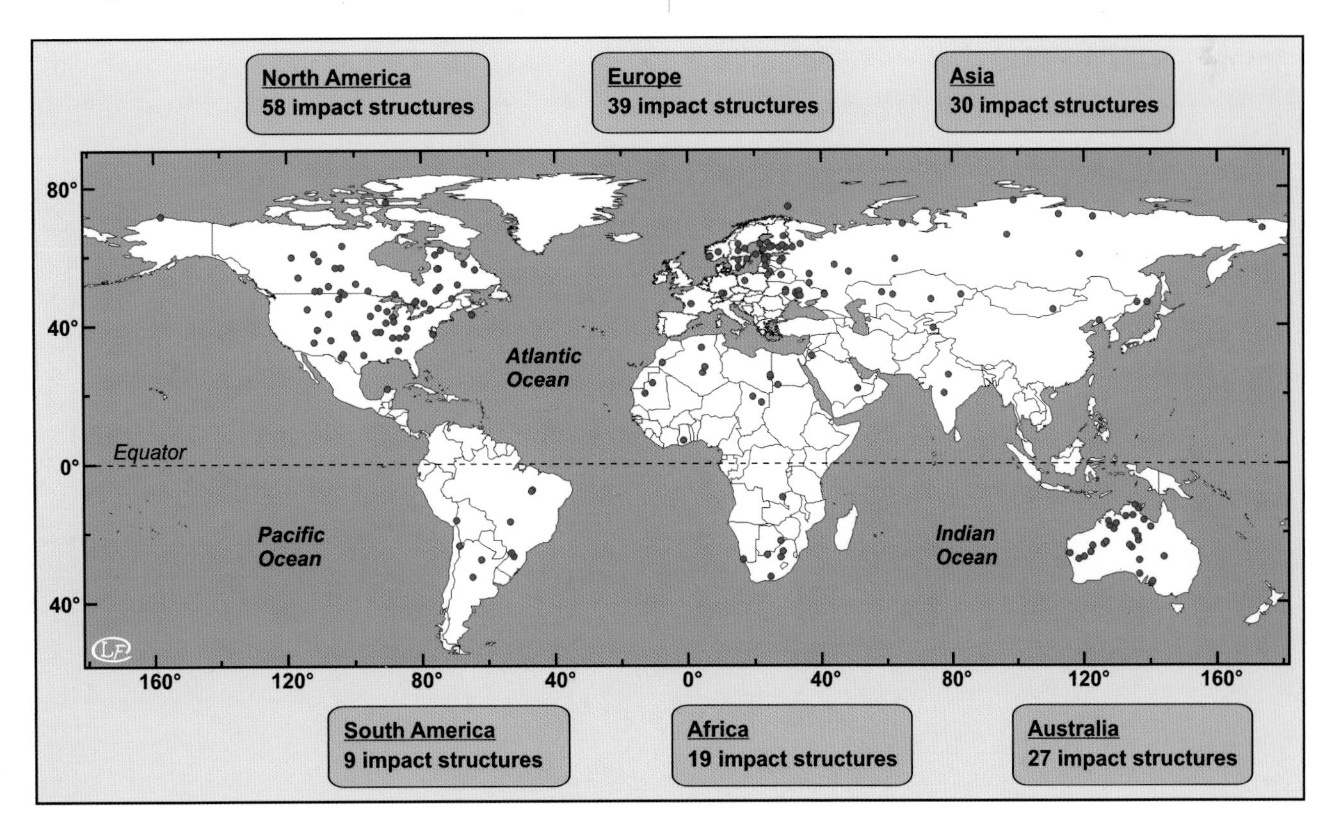

FIGURE 1.2: *Known Earth Impact Craters (from Earth Impact Database, 2012). I thank Ludovic Ferriere, Natural History Museum in Vienna, Austria, for permission to use this diagram. www.MeteorImpactOnEarth.com*

65 million years ago. Scientists realized that the Earth must have been hit repeatedly by meteorites during the early history of our solar system, likely to a greater extent than the Moon because of Earth's greater mass and gravitational attraction. So why don't we find as many craters on Earth?

Once scientists started looking carefully at the Earth's surface, impact craters, or evidence for their former existence, became apparent both on continents and in ocean basins. To date, at least 186 impact craters have been found on the Earth. But why aren't they more obvious and exposed, as on the Moon? The answer is twofold: Earth's atmosphere produces rain, wind, and chemical factors that lead to rock weathering and erosion, and active plate tectonics continue to shape the planet's surface. The latter explanation includes surface-masking volcanism.

Our atmosphere, with its oxygen, water, and carbon dioxide, is responsible for mineral and rock weathering or disintegration, and erosion or movement of the particles by running water and glaciers. The presence of water also facilitates land sliding and other types of ground failure that is apparently less common on the other rocky planets and moons.

Plate tectonics, which involves the movement of the surface plates, also destroys craters by burying them under slabs of over-thrusted rock and subducting ocean basin crusts containing craters down into the mantle. None of the other planets, we believe, have or have had plate tectonics like ours. Venus, on the other hand, has an even more corrosive and hotter atmosphere than Earth's, so rock weathering most probably proceeds at a much greater rate there.

The Earth geologically, then, is a very special place. Both internal and external processes rapidly (in geologic time) erase or bury surface features such as mountains, valleys, and impact craters. In addition, it appears that the oldest preserved rocks from which we can obtain ancient geological information are on the continents. Due to ongoing subduction, the ocean floors are no older than about 180 to 200 million years old. So we can't look there to find ancient rocks or evidence of older meteorite impacts.

CHAPTER 1 – NOTES

1 Kimberlite pipes—kimberlite is a heavy, dark rock, due to great amounts of contained iron and magnesium, that originated in the mantle and rose to the Earth's surface. Kimberlite pipes have also brought diamonds to the surface and have been mined for these hardest of all minerals. Diamonds are composed solely of carbon and form under extremely high pressure achieved only in the mantle. (See also Chapter 10, including endnote 10.)

2 Lithosphere - lithos means rock or rigid. The lithosphere is the rigid slab, about 100 km (60 mi) thick, comprising the crust and upper mantle. The lithosphere slides over a more plastic zone about 250 km (150 mi) thick of mantle below called the asthenosphere.

CHAPTER 1 – REFERENCES

Grieve, 1990; Jeanloz, 2000.

Plate Tectonics and Volcanoes

Introduction

Ever wonder why volcanoes occur in some parts of the world and not others? Why are there volcanoes in California, Oregon, Washington, Alaska, Hawaii, and Wyoming, but not in the other 44 states? Volcanoes occur in particular geologic—or more precisely, geotectonic—environments. But it wasn't long ago that volcanologists knew only that most volcanoes were associated with zones of earthquakes and young mountain ranges (Williams, 1951). For example, in the early nineteenth century, Alexander von Humboldt, while exploring South America, noted that volcanoes were not distributed randomly on the Earth's surface. He speculated that there was some deep-seated reason for their alignment.

Our understanding of volcanoes—why they exist where they do—is clearer thanks to insights gained during the 1960s into plate tectonics. Let's begin by looking at Earth's geologic history.

The oldest Earthly rocks are igneous, from western Australia. Zircon minerals ($ZrSiO_4$) within these rocks are 4.03 billion years old. Some zircons derived from even older igneous rocks but now found in sedimentary rocks are 4.374 billion years old. Most earth scientists, however, believe the Earth is even older, about 4.6 billion years old, based on the ages of meteorites. Almost all meteorites we've sampled, using radioactive decay absolute age-dating techniques, are 4.6 billion years old. (Meteorites younger than 4.6 billion years are from Mars and the Moon; see Martian Meteorites sidebar, page 149.)

For at least 4.0 billion years, rain and running water have been eroding the landscape and depositing sediments, and volcanoes have erupted lava, ash, and gas. In places, slabs of the Earth's rigid crust and upper mantle also subducted, sliding down into the mantle below the crust.

These rigid slabs, called **lithospheric plates,**[1] come in various sizes and shapes, and vary between 70 and 120 km (42 and 86 mi) thick. Three types of plates exist—continental, oceanic, and those composed of both continental and oceanic crust.

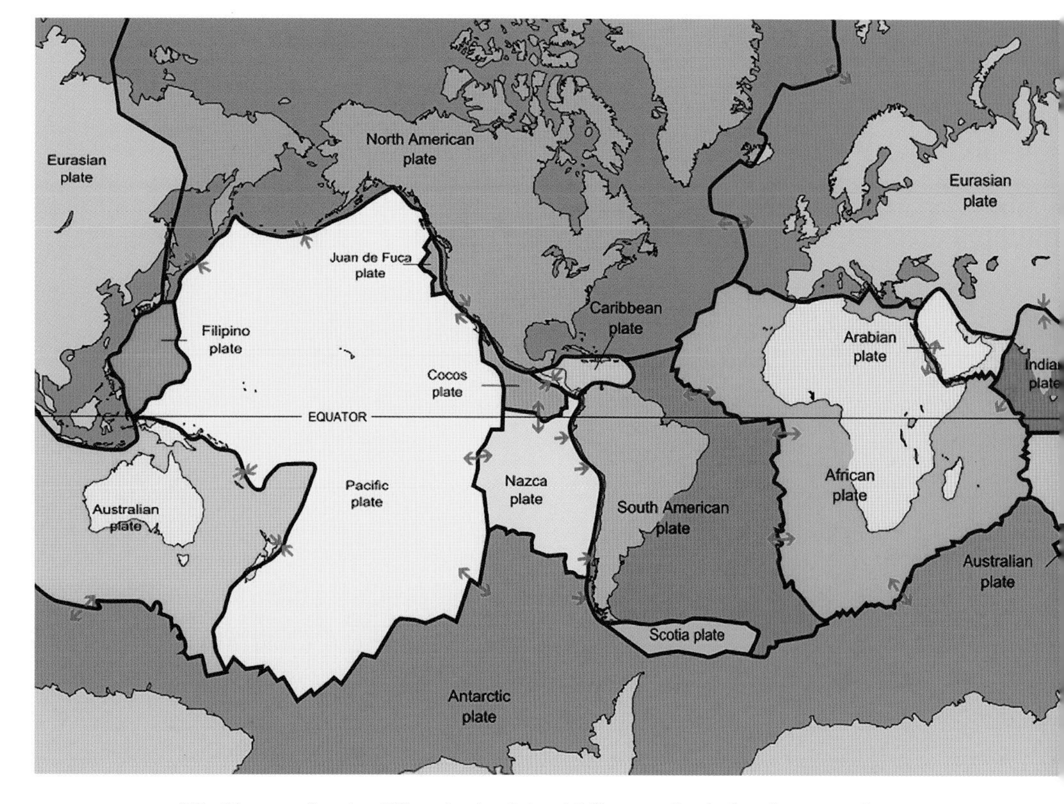

FIGURE 2.1: *World map of major lithospheric plates (different colors) showing spreading centers (constructive or divergent plate boundaries) with opposed motion denoted by red arrows (e.g. Mid-Atlantic Ridge), and subduction (destructive plate boundaries) with red arrows facing each other (e.g. Aleutian Islands of Alaska). Transform lithospheric plate boundaries, for example the San Andreas fault in California, are denoted with red arrows "sliding" past each other. Rates of spreading greatly vary and are centimeters per year (see Figure 2.4) (USGS map).*

The seven largest of the 30 major plates—the Pacific, North American, African, Eurasian, South American, Australian, and Antarctic—cover 94 percent of the Earth's surface. Smaller plates include the Indian, Philippine, Caribbean, and Arabian plates. All of these rigid plates slide over a more plastic upper zone of the mantle called the **asthenosphere.**

Lithospheric plates also move in different directions. At **spreading centers,** such as the Mid-Atlantic Ridge, volcanic material is added, and the plates move apart. In subduction zones, the plates dive into and are consumed by the mantle. A classic example of subduction is the Nazca Plate along the west coast of South America, which is diving under the westward-advancing South American continental plate. In some places, plates slide past one another along steep or vertical **transform faults.** A classic example of a transform fault is the San Andreas in California, which is over 1,280 km (800 mi) long, is steeply angled, and runs about 16 km (10 mi) deep. These faults are zones of crushed rock that may be nearly 2 km (1 mi) wide and kilometers deep: they cut through entire lithospheric plates.

The rate of plate spreading varies, as do plate velocities. The African Plate, for example, is moving very slowly north as it collides with the European plate. This process will eventually close the Mediterranean Sea. Also note the "notches" along the axes of the spreading centers seen on Figure 2.1. These ridge axes offsets result from transform faulting, or horizontal fault motion.

The newest part of a lithospheric plate develops in, and then moves away from long, linear volcanic rift systems primarily located on the ocean floor. These systems sit atop wide ridges. Most of these volcanic systems lie well below sea level and are where the Earth loses the bulk of its internal heat.

We know that lithospheric plates form at the ridges and then spread away from the crests based on the discovery of magnetic field "stripes" in the basaltic seafloor rocks. These stripes exhibit variable widths and mirror-image patterns and ages on each side of the ridge crests, indicating that they had separated as the rift grew. (see SB Figure 2 in the Paleomagnetism sidebar, page 13). The lithospheric plates, then, grow at these constructive plate boundaries, with the oldest sea floor rocks located farthest from the crests.[2]

Folding, faulting, igneous activity, crushing, and metamorphism (the latter term literally meaning changing shape due to heat, pressure, and the interaction of fluids on the rocks) all affect lithospheric rocks during subduction and transform faulting. In turn, these forces move and erode land masses, build mountains, uplift and tear down, change ocean-land configurations, and recycle (through subduction) lithospheric rocks into the mantle. Plate tectonic processes have apparently existed for billions of years and show no sign of stopping. These processes eventually will stop when most of the heat-producing radioactive elements have decayed to their stable, non heat-producing daughter products. The interior of the Earth will then be too cool to generate plate tectonic forces.

But why do these rigid lithospheric plates move over the less rigid, 250-km (155-mi) thick asthenospheric mantle below? The answer involves the Earth's hot core, and the plastic behavior of the cooler mantle above it. Parts of the lower mantle are considerably hotter than areas around them. (The mantle receives heat from the core below that may be 7,000°C: some scientists believe there are anomalously hotter portions of the core as well.) These unusually hot parts of the lower mantle must expand relative to the cooler mantle around them, as most materials do when warmed. Expansion results in less dense, more buoyant material, which then rises. The heating and rising process is called convection. The mantle, a solid at any instant, is plastic over time and therefore very slowly flows or

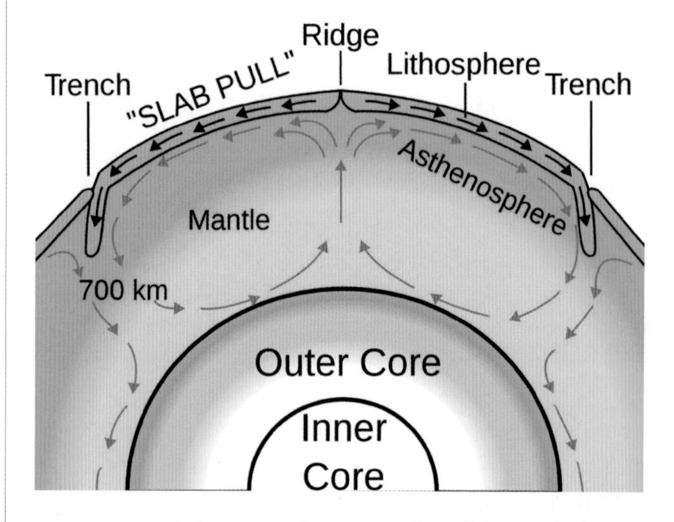

FIGURE 2.2: *Diagrammatic cross section of the Earth showing the rigid 100-km-(62-mi)-thick lithosphere being formed at spreading centers and then being subducted or forced back into the mantle at destructive plate boundaries (subduction zones). Note the convective circulation of mantle with upwelling occurring at ridge locations (USGS diagram).*

Paleomagnetism

Navigation was greatly facilitated some 1,500 years ago when Chinese sailors discovered that a linear piece of lodestone suspended from a thread would rotate until one end pointed north. Lodestone contains the magnetic mineral magnetite (Fe_3O_4), and with time this crude navigational magnet evolved into the modern compass. In 1600, Sir William Gilbert postulated that the Earth is a huge magnet with its magnetic poles situated near the rotational or geographic poles. But why this magnetic field existed wasn't understood until 1954.

The theory by two geophysicists, Englishman E. C. Bullard and American W. M. Elsasser, attributes the magnetic field to fluid motion in the Earth's liquid outer core that is composed mostly of iron. It is theorized that molten iron, a non-magnetic but electrical conducting material, generates the magnetic field as the Earth rotates. The molten iron moves in the direction of the spin, producing an electrical dynamo and resulting magnetic force fields.

The Earth's magnetic field is analogous to having a bar magnet running through the Earth approximately, but not exactly, parallel to the axis of Earth's rotation. Those of us who played with bar magnets as kids know they have polarity: a north end and a south end. We also discovered that oppositely marked ends attract each other while like-labeled ends repel each other.

The Earth, then, has north and south magnetic poles (it is dipolar). Each pole is more than 800 km (500 miles) from the geographic north and south poles. Furthermore, because we are dealing with an active, dynamic system, the geographic locations of the magnetic poles are always changing, though they never move far from their respective geographic poles. Navigation and topographical maps for any particular location show true north together with the magnetic declination from true north. This declination must be accounted for during navigation.

To further complicate the Earth's magnetic picture but also increase its usefulness, the polarity of the magnetic poles switches or reverses about every 800,000 years: some durations last more than a million years and some less than 200,000 years. Ironically, this magnetic pole reversal, a phenomenon we don't yet understand, has greatly helped us understand plate tectonics.

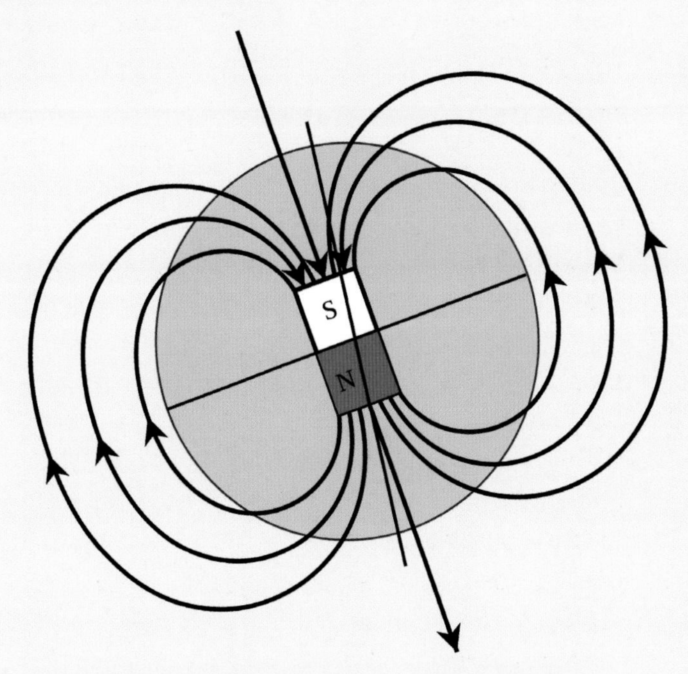

SB FIGURE 1: A diagrammatic view of the Earth's magnetic field. Note that the magnetic and geographic poles of Earth's rotation are close but do not coincide. Magnetic needle inclinations vary from vertical to horizontal at latitudes from the North Pole south to the Equator, respectively. These inclinations are a function of the Earth's magnetic field flow lines (Google).

Also important to the magnetic story was the discovery in 1895 by French scientist Pierre Curie that all magnetic substances lose their magnetism at some particular elevated temperature below 800°C. This temperature, called the Curie point, is 578°C for magnetite.

Applications of the Earth's Magnetic Field Properties

These magnetic attributes have proved valuable in three ways in deciphering Earth's history. Magnetic properties in igneous, sedimentary, and metamorphic rocks may determine (1) the direction to the magnetic pole at the time of sedimentation or igneous rock formation, (2) the latitude at which rock formation took place, and (3) the magnetic polarity at the time of rock formation.

Why? Magnetite-bearing rock locks in these magnetic properties when formed. The magnetic signature preserved in rock is known as paleomagnetism. And the paleomagnetic properties are valid unless the Curie point was subsequently attained by such things as deep, high-temperature burial or reheating by an igneous event.

A good way to see how paleomagnetic information is obtained is to visualize a series of lava flows that were erupted long ago in the northern hemisphere. When the lowest flow in the sequence was emplaced, tiny magnetite crystals recorded the Earth's magnetic field—the direction to the magnetic pole, magnetic polarity, and magnetic inclination (explained below) the moment the flow cooled below the Curie point. Because of the bipolar nature of magnets, the tiny mineral grains will either have their "north" or "south" ends pointing north, depending on the polarity of the Earth's magnetic field.

The magnetic pole direction indicated by the mineral grains in the ancient rocks may or may not point towards the present day north magnetic pole. And the older the lava flow, the less likely the directions will coincide with the present magnetic pole location because of tectonic crustal changes including lithospheric plate movement and plate rotation.

Because most rocks can now be absolute age dated, a magnetic-reversal chronology has been established. Magnetic polarity, therefore, can also be used to date rocks.

Finally, the latitude at which the lava flow erupted can be determined by examining the magnetic inclination of the magnetic minerals. Note the magnetic field lines in space around the Earth (SB Figure 1). If a lava flow issues from a volcano near the equator, the particles align horizontally (parallel to the earth's surface) or close to zero inclination to the horizontal plane. But as we proceed towards the north (or south) pole, the inclination, or dip of magnetic minerals into the Earth, steepens.[1]

With these magnetic tools, geophysicists have been able to reconstruct the migration paths of ancient land masses, including the break up of the super continent Pangaea starting about 200 million years ago. Also, the reversal in magnetic polarity formed the mirror image magnetic "stripes" seen in seafloor basalt on both sides of spreading centers such as the Mid-Atlantic Ridge and the East Pacific Rise. In fact, these stripes are conclusive proof that lithospheric plates form at spreading centers and move away, that is, that seafloor spreading does indeed occur.

NOTES

1 Critical to paleomagnetic studies is ascertaining the original depositional orientation, including the horizontal plane of samples used for analysis. This is easier said than done, so several samples at any location are analyzed to obtain statistical validity of the data.

REFERENCES

Carrigan and Gubbins,1979: Gilluly et al., 1968; Hamblin and Christiansen, 2004; Marshak, 2004.

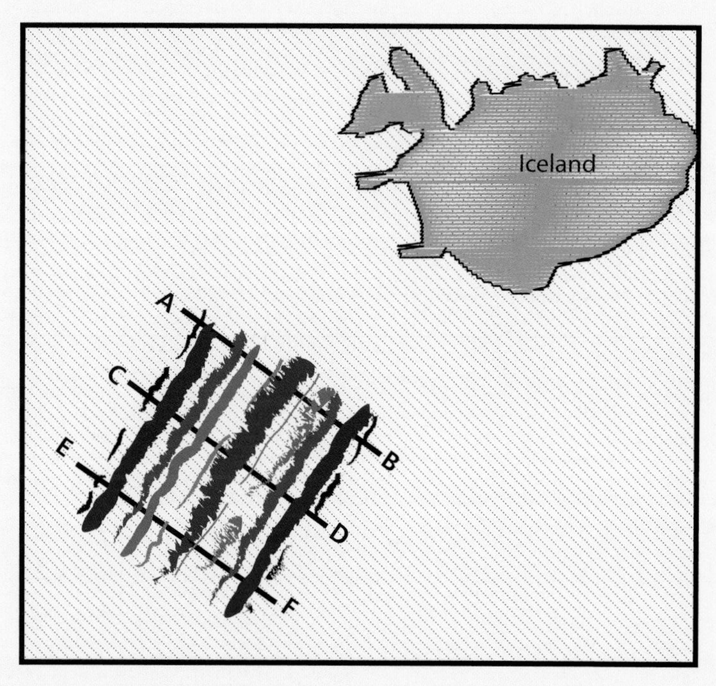

SB FIGURE 2: Magnetic "stripes" in oceanic basalt located south of Iceland on the Mid-Atlantic Ridge. Lines A – B, C – D, and E – F show a mirror image of the magnetic intensity in gammas (magnetic intensity) and polarity (normal polarity) on each side of the center of the spreading center. The stripes increase in age away from the center (modified from Hamblin and Christiansen, 2004).

convects in a circular motion. In turn, cooler, upper mantle rocks sink, only to be heated and rise again.

The convecting mantle moves at a velocity of just centimeters per year. That may not seem like much until you consider the enormous mass—trillions of tonnes—that's in motion. It's believed that the circular motion of these mantle convection cells rarely stops but may vary in speed and location.

The mantle rock under convection is hot (up to 1,400°C) and rich in magnesium and iron but poor in silica. As it reaches the upper mantle (anywhere from 200 km (124 mi) to less than 10 km (6 mi) beneath the surface), the pressure on this material is reduced, and as much as 10 to 40 percent of this rock melts. (See Chapter 8 for Magma Generation). This produces heavy, black lava called **basalt.**[3]

How the basaltic liquid moves through the mantle and crust is unknown, but researchers suspect that it flows through cracks. As it migrates upward, the liquid incorporates small amounts of other components while also interacting with mantle rocks (peridotite). Additives include carbon dioxide (230 to 550 ppm), water (80 to 950 ppm), and trace amounts of methane. The chemistries of the basalts produced by this process vary slightly from one place to the next.

Basalt also covers central Washington and Oregon, especially along the Columbia River, through southern Idaho, and into northeastern California. Basalt forms the Giant's Causeway in Ireland, the Hawaiian Islands, most of the seafloor, and also the lunar Maria, or the dark, flat areas of the Moon easily observed from Earth. Basalt, then, was born in, and is the result of partially melted mantle rock. By volume, it is by far the most abundant igneous rock on the surface of the Earth. The intrusive, chemical equivalent of basalt is gabbro. Gabbro, also abundant, is distinguished from basalt by its totally crystalline (non-glassy) texture.

Two main types of basalt exist: tholeiitic and alkaline. They differ slightly in chemical composition. Tholeiitic basalt, most abundant, forms the bulk of the ocean crust and most of the flood basalt provinces both on land and beneath the sea (see Chapter 5 – Hawaii description). Tholeiitic basalt also forms the bulk of the intra-oceanic plate volcanic islands. It is poorer in several elements (potassium, phosphorous, titanium, rubidium, zinc, niobium, uranium, and thorium) than alkaline basalt.

Alkaline basalt contains less silica but more potassium and sodium. These basalts form small volcanoes and constitute part of numerous oceanic islands, including the Hawaiians. Alkaline basalt can be distinguished visually from tholeiitic basalt by the presence of green crystals of the mineral olivine $(Fe,Mg)_2SiO_4$) (see Chapters 5 and 8 for a discussion of the origin of these basalts).

Constructional Plate Boundaries

Basalt covers more than 60 percent of the Earth's surface, mostly in ocean basins. It erupts within very long, mostly submerged, and sinuous rifts or cracks formed under tension.[4] These rifts have been called Earth wounds that never heal because of their almost continual state of rifting and lava production. The rifts form in the central axis of a mostly submerged ridge, or mountain range, which varies in height above the ocean basin floor.

This mid-oceanic ridge system is the longest (75,000 km, 46,600 mi), continuous mountain range in the world. But because most of the igneous activity occurs below sea level, we rarely observe eruptions. Ridge eruptions average between 2 km^3 and 3 km^3 (0.5 to 0.7 mi^3) per year. However, seven times as much intrusive activity, or 18 km^3 (4.3 mi^3) per year, occurs here.

Oxide	Tholeiite (MORB)	Alkali Olivine (OIB)	Mantle*	Lunar Mare**
SiO2	48.77 to 50.5	47.52	43.15	43.6
Al2O3	15.9	15.95	2.5	7.9
Fe2O3	1.33	3.16	–	–
FeO	8.62	8.91	9.02	21.7
MnO	0.17	0.19	0.13	0.3
MgO	7.5 to 9.67	5.18	33.35	14.9
CaO	11.16	8.96	2.25	8.3
Na2O	2.43 to 2.6	3.56	0.21	0.2
K2O	0.08 to 0.2	1.29	0.21	0.1
TiO2	1.15	3.29	0.21	2.6
P2O5	0.09	0.64	0.07	
Selected Trace Elements				
V	262	350	70	
Cr	528	421	3,200	
Ni	214	153	2,300	
S	approx. 1,000	up to 5,000		

Mantle*=Composition of inclusions found in Hawaiian lavas / basalt lavas are formed from partial melting of the mantle
Lunar Mare**=Apollo 12 olivine basalt
After Hyndman, 1985; Rogers and Hawkesworth, 2000

TABLE 2.1: *Main mantle-derived basalt types, lunar mare.*

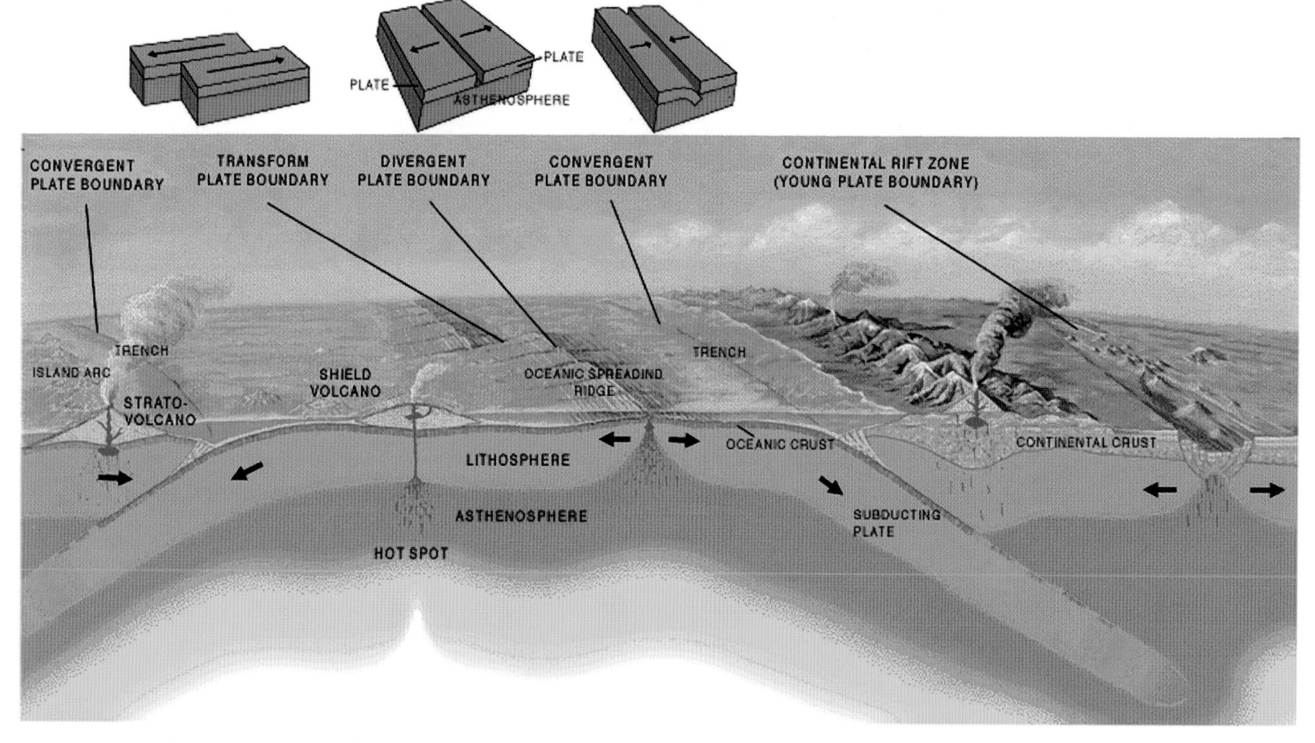

FIGURE 2.3: *Schematic cross section of a spreading center or constructional plate boundary showing the ridge crest offset in places by transform faults. An Andean-type destructional boundary (subduction zone) is on the right and an inland arc-type subduction system is on the left of the diagram. Note the formation of magmas (red) above the descending slabs in the subduction zones (Jose F. Vigil – USGS diagram).*

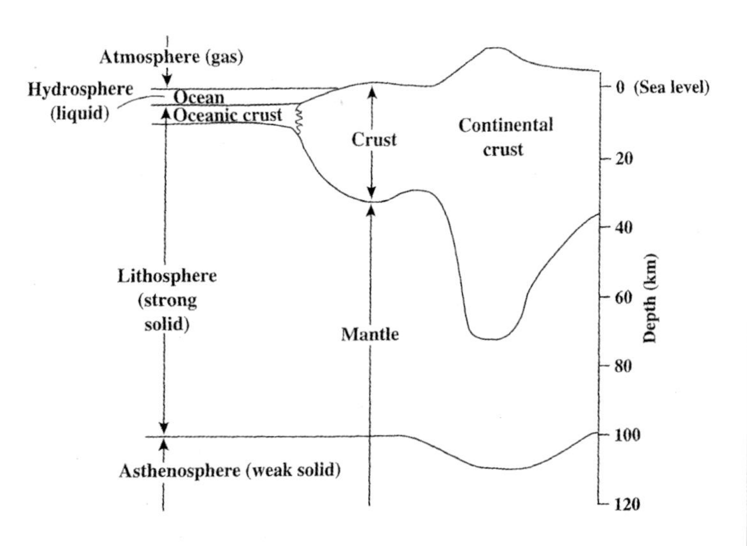

FIGURE 2.4: *Crust-lithosphere-mantle cross section (from Abbott, 2004).*

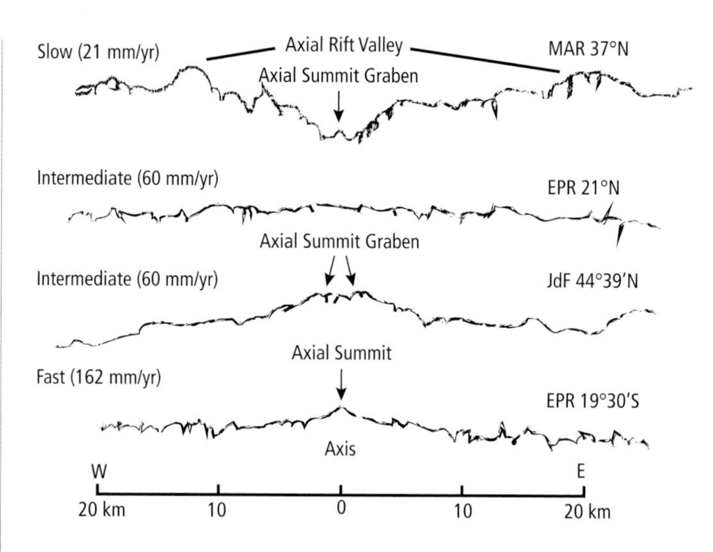

FIGURE 2.5: *Cross section profiles top to bottom of the slow-spreading northern Mid-Atlantic Ridge (MAR 37°N) south of Iceland, intermediate spreading East Pacific Rise (EPR 21°N) south of Baja, Mexico, and the Juan de Fuca Ridge (jdF 44° 39'N) and the fast spreading southern East Pacific Rise Ridge (EPR 19° 30'). Note the change in topography from slow to rapid spreading. Please refer to Figure 2.1 for approximate locations (modified from Perfit and Davidson, 2000).*

And as eruptions take place along the constructional plate boundaries, new sea floor forms, which then moves away from the ridges due to the convection process. These ridges are spreading centers and constructional lithospheric plate boundaries, where lithospheric plates grow. While oceanic eruptions above the water surface are rare, the formation of the volcanic island Surtsey in 1963 off the coast of Iceland provided a glimpse of the process.

Divergent or constructional plate boundaries are divided into slow-, intermediate-, and fast-spreading systems. Examples of each, respectively, are the northern Mid-Atlantic Ridge, Juan de Fuca Ridge off the northwestern U.S. coast, and the East Pacific Rise off South America. The major physical differences between them are a function of the rates of magma supply, spreading speeds, and effectiveness of hydrothermal cooling of the magmatic systems by ingested seawater. Observable differences include

	FAST	INTERMEDIATE	SLOW
Spreading speed*	8–16 cm/yr.	4–8 cm/yr.	1-4 cm/yr.
Ridge Crest** height	100m	–	4 to 5 km
Axial Valley (Width) Axial Valley (Depth)	<2km; commonly 40–250 5 – 40m	1 – 5 km 50 – 1000 m	8 – 20 km 1–2 km
Loci of Magmatism	Neovolcanic zone***	same	same
Lava Type(s)	<4 cm to 10's cm thick smooth to ropy	Smooth and pillows	Dominantly pillows in Hommocky ridges
Eruption Rates	Yearly to 10 year intervals	Intermediate intervals	100 to 1,000 year intervals
Off Axis Volcanism	Cones, flows up to 6km out board of axis	Cones, flows	?
Examples	EPR off of South America****	Juan de Fuca, Galapagos	SE Northern Atlantic Ridge

*Full rate of spreading
**Ridge Crest above surrounding sea floor
***Neovolcanic zone = zone of magnetism from dikes and hydrothermal activity of ridges located within axial valley
****EPR=East Pacific Rise
References: Kelley et al., 2002; Perfit and Davidson, 2000

TABLE 2.2: *Characteristics of mid-ocean ridge spreading centers.*

ridge topography, lava flow morphology, and width of the axial ridge crest valleys where most igneous activity takes place.

Rates of spreading vary from about 1.8 cm (0.7 inch) per year on the slow-spreading North Atlantic Ocean to more than 16 cm (>6 inches) per year on the fast spreading East Pacific Rise. While displacement is not much in human terms, it becomes significant over millions of years.[5]

Rapidly spreading ridges have smooth topography, low relief, an almost continuous central rift up to tens of meters wide, and more or less continuous volcanism. Basalt erupts from **dikes**, or vertical feeder zones, only a few meters wide. Dikes tap magma chambers that are one to four kilometers below and that may extend tens of kilometers laterally. Modeling suggests the chambers are lense-like, 1 km (0.6 mi) wide and at least 40 km (25 mi) long, and typically parallel to the ridge axes.

Dike eruptions may last for several weeks. The intense earthquake activity accompanying eruptions demonstrates that dikes may propagate laterally along strike (parallel to the ridge) up to 60 km (37 mi) away in two days through the seafloor crust. The resulting underwater erupted lavas form ropy and lobate pillows[6] and sheet flows.

Slow-spreading centers have rough, fault-controlled topography, great relief, and low extrusion rates due to magma supplied intermittently from small intrusions. The central, down-dropped valleys can be several kilometers wide and contain eruptive centers composed of pillowed basalt flows forming small hills and hummocky ridges up to hundreds of meters tall. Together magma-driven and tectonic processes cause gaps in the ridge structure, varying from less than 1 km to more than 30 km.

Off-axis and seamount volcanism varies according to spreading rates. Lava flows form low hills up to 6 km (3.7 mi) away, while seamounts generally less than 500 m (1,312 feet) tall have been observed up to 15 km (9.3 mi) away from axial ridges in fast-spreading systems. These phenomena appear restricted to the axial valleys in the slowest spreading systems. Volcanic cones aligned in chains essentially parallel to plate motion (roughly perpendicular to ridge axes) occur in both fast- and intermediate-spreading speed systems. Seismic data suggest that the tops of the magma chambers feeding these submarine flows and axial valley extrusions may be only 1.2 km (0.75 mi) below the surface.

FAULTED RIDGE CRESTS

Axes of spreading center ridges are segmented, and offset by major faults. These are transform faults with primarily horizontal crust displacements, and they cut through the lithosphere. They may offset ridge axis segments by 50 km (31 mi) or more. (Note Figure 2.1 that shows major offsets especially on the East Pacific Rise).

Strange Life Forms! Life Without Light at the Bottom of the Sea!

Life, as we all knew until 1977, was based upon photosynthesis. Plants, utilizing solar energy that powers photosynthesis, produce food plus oxygen as a by-product. This process changed the composition of the atmosphere from carbon dioxide (CO_2) rich and oxygen (O_2) deficient to one with about 20 percent oxygen and, today, +400 ppm carbon dioxide. This dramatic atmospheric change eventually allowed animals, including humans, to develop. Many species of animals, then, showed their gratitude by eating plants (and other animals). And so life has gone on like this for a long time, or for at least 3.5 billion years.

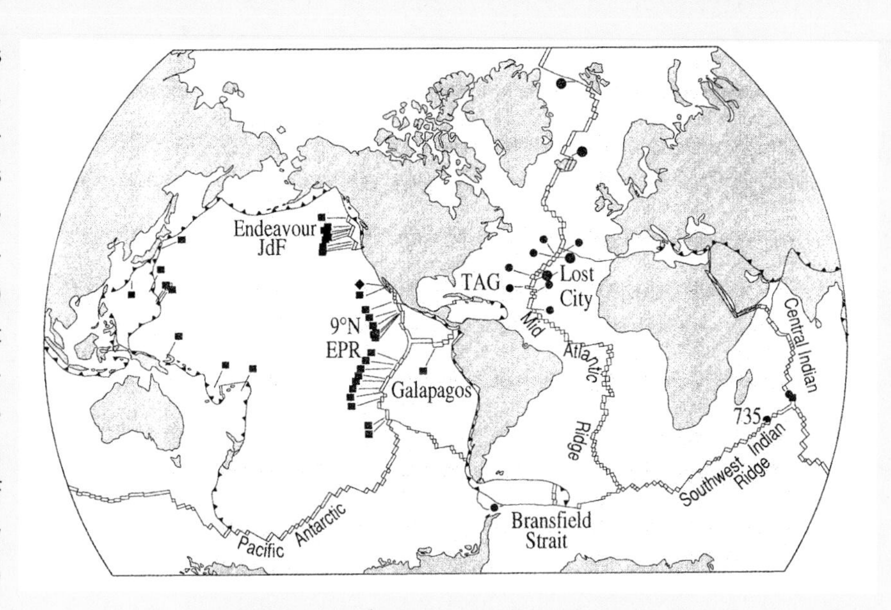

SB FIGURE 1: Location map of seafloor vents, mostly located in axial valleys of spreading centers (from Kelly et al., 2002).

However, along came inventions in the latter part of the 20th century resulting in such wonderful things as manned and unmanned space ships that went to the Moon and beyond, and iPods, and iPhones. Submersibles—small craft capable of going into the deepest parts of our oceans—were also built.

Most people are aware of the manned lunar trips (Apollo 11 through 17), but few know about our adventures into the deep abysses of the sea. Where did the submarine explorers go and what did they find? They made really astounding geological and biological discoveries in some of the deepest and most fascinating parts of the oceans, the trenches, and especially along the constructive, spreading plate boundaries. It was here that they discovered life-forms based not on sunlight but on chemicals.

Before we look at these ocean discoveries, let's first examine how photosynthesis works. Through the process of photosynthesis, plants absorb carbon dioxide (CO_2) from the air through leaves and, utilizing sunlight as an energy source, produce carbohydrate (CH_2O) for food. Oxygen, a by-product of the process, is released into the atmosphere. Plants have been processing CO_2 at least as far back as 2.7 billion years. In doing so, plants converted our early carbon dioxide-rich and essentially oxygen-free atmosphere into one with relatively little carbon dioxide (about 280 ppm prior to the industrial revolution) but about 20 percent oxygen. (Nitrogen makes up about 80 percent of Earth's atmosphere.) Below is the photosynthesis process in equation form.

$CO_2 + H_2O$ —(energy from sunlight)⟶ CH_2O (food) $+ O_2$

Our understanding of biology changed dramatically in 1977 when scientists aboard the submersible *Alvin* discovered hydrothermal, or hot water, vents, along the Galapagos spreading center off the coast of Ecuador. The water from the vents was only a few degrees centigrade (°C) warmer than surrounding seawater, but some strange macro and micro life-forms (together with sulfide minerals) were found around the vents' warm waters in the otherwise deep, cold, lightless seawater.

In 1979, a return trip was made to the deep, this time to the axial spreading rift valley south of the southern tip of the Baja Peninsula of Mexico. Researchers found not only similar and bizarre life-forms, such as giant 3-m (10-foot) long red and white colored tube worms, but also large clams, fish, snails, limpets, and crabs. All were living in total darkness and evidently quite happily around very hot seafloor hydrothermal vents. Some of the micro life-forms even thrive in temperatures above 100°C, and some possibly in water as warm as 200 to 300 degrees C.

The submersible illuminated what appeared to be black smoke jetting from tall, thin, dark tubes or chimneys built of iron-, zinc- and copper-bearing sulfide minerals. Minerals here include pyrite (FeS_2), sphalerite (ZnS), and chalcopyrite ($CuFeS_2$). Some of the sulfide mineral vent tubes were more than 10 m (+30 feet) high, having grown by the continuous precipitation of sulfide minerals on and around the chimneys and vents. The metals incorporated in sulfide mineral chimneys apparently are leached from crustal basalt below by the very hot and salty seawater heated by magma located a few kilometers below the vents. The "smoke" is, in reality, microscopic particles of iron sulfide that form as the hot (up to 400°C), acid, salty, reduced (oxygen deficient), metal- and hydrogen sulfide (H_2S)-rich solution encounters cold (4°C), alkaline, and oxygenated seawater[1] on the seafloor.

In addition to this amazing geological discovery of hot sulfide- and metal- bearing axial valley vents, biologists were astounded to learn that the life-forms in the deep, lightless water are based not on photosynthesis, but chemosynthesis! In chemosynthesis, tube worms and clams living near the vents take in vent-issuing hydrogen sulfide (H_2S). Bacteria living within the clams and worms then process the H_2S into carbohydrate (CH_2O)—food—for the animals that house the bacteria, as shown below.

$$H_2S + 2CO_2 + 2H_2O \longrightarrow 2(CH_2O) + H_2SO_4$$

Bacteria also use methane (CH_4), ammonia (NH_3), and hydrogen (H_2) released from the vents to produce carbohydrate as shown, respectively, below.

$$CH_4 + O_2 \longrightarrow CH_2O + H_2O$$
$$NH_3 + O_2 + CO_2 \longrightarrow CH_2O + HNO_3$$
$$4H_2 + O_2 + CO_2 \longrightarrow CH_2O + 3H_2O$$

Life down under, then, is built on chemical energy in the form of H_2S, CH_4, NH_3, and H_2 emanating from the volcanic crust! And because the worms and clams can't live without the bacteria and vice versa, an amazing symbiotic relationship exists between them, totally different from how plants function in the atmosphere.

Subsequent dives to spreading centers, including the Mid-Atlantic Ridge, have also encountered black sulfide smokers, with some up to 45 m (148 feet) tall. White smokers emitting sulfate, instead of H_2S, from seafloor vents also exist, producing white to light-gray colored anhydrite ($CaSO_4$) chimneys.

More than 440 species of animals, with greater than 90 percent endemic to vents, have been discovered in such geothermal areas. Most species were found in fast- and intermediate-rate spreading centers.

Hydrothermal venting is the result of cold, heavy, bottom seawater descending into fractured, porous seafloor basalt near spreading centers. Seawater is able to penetrate up to 8 km (5 mi), and while on this downward journey it may be heated to more than 500°C as its chemistry is changed, as is that of the rocks it flows through. This circulatory process is called convection: cold, heavy seawater sinks as hot, more buoyant water rises to the seafloor and vents, forming smokers. The resulting hydrothermal fluids are enriched, with respect to seawater, in lithium (Li), potassium (K), rubidium (Rb), cesium (Cs) sulfur dioxide (SO_2), copper (Cu), iron (Fe), lead (Pb) hydrogen sulfide (H_2S), carbon dioxide (CO_2), helium (He), molecular hydrogen (H_2), and methane (CH_4).

At temperatures as low as 150°C the elements potassium, rubidium, and lithium and ore metals iron, copper, zinc, and lead are leached from basalt as the warming, saline waters travel through the porous basalt. But during the watery passage, basaltic rocks, in turn, take up elements including magnesium, calcium, sodium, and water, resulting in a greening of the originally black basalt by the formation of minerals such as chlorite mica.

Some vent systems may last for tens of thousands of years, while others are quite ephemeral. While most systems probably lie on young spreading center axes, some have been discovered several kilometers away, while still others vent in crust up to 65 million years old.

These geological-biological discoveries have demonstrated that an important connection, previously unknown, exists between volcanism and life. In fact, Earth's earliest life may have developed in these deep, dark marine waters rather than in surface, sunlit environments. The important relationship between volcanism and life was further demonstrated in the 1980s with the discovery of massive microbial biomass emissions called "mega plumes" or "event plumes." These plumes emanated from porous seafloor basaltic rocks on the Juan de Fuca Ridge off the northwest U.S. coast.

These warm, buoyant, volatile-rich hydrothermal

water lenses may be hundreds of meters thick, rise 800 m (2,625 feet) above the seafloor, and extend for 20 km (12 mi) or more along their ocean ridge axes. The plumes, containing an enormous amount of heat, accompany dike or fissure eruptions of basalt on the seafloor.

Explanations for the huge plumes include (a) sudden emptying of a reservoir located within the porous, fractured, near surface volcanic crustal rocks that form due to dike eruptions; and (b) rapid heating of volcanic rock-hosted seawater by intruding dike or lava flows. These plumes demonstrate that abundant microbial life exists within the hot submarine crust associated with magmatic activity.

Biomass-rich mega plumes have also been detected above the Gorda Ridge also located off the west coast of North America, and in the North Fiji Basin in the western Pacific Ocean. This biomass includes archeae or hyperthermophile heat-loving (above 80°C), single-celled microorganisms that grow in an oxygen and sunlight-free environment.

The Lost City Field

Another startling but totally different type of deep sea life discovery was made in 2000 at 30° north in 700- to 800-m (2,300- to 2,625-foot) deep water along faults on the Mid-Atlantic spreading center. There, researchers found the Lost City (geothermal) Field consisting of light-gray to white spires and mounds, and pinnacles up to 60 m (197 feet) tall composed predominantly of calcium carbonate ($CaCO_3$) and magnesium hydroxide ($Mg(OH)_2$). These edifices were built from alkaline (pH 9 to 9.8), low silica-bearing hydrothermal fluids 40°C to 70°C. The fluids, low in hydrogen sulfide and metals compared with black smoker vent fluids, contain methane and hydrogen. This lightless environment also supports microbial communities, including light-gray to white filamentous strands centimeters in length and macrofaunal assemblages of crabs, sea urchins, sponges, and corals.

This chemosynthetic system, rather than relying on hydrogen sulfide and volcanically derived heat, is driven by the exothermically produced heat from the reaction of water encountering mantle rock, peridotite. Serpentine is produced in the reaction.[2] This scenario is also in contention for having produced the first Earthly life, perhaps as much as 4 billion years ago.

NOTES

1. While much or most of the sulfur found in the sulfide minerals such as pyrite (FeS_2) forming around the black smokers is derived from seawater (which contains 900 ppm within sulfate (SO_4)), some of the hydrogen sulfide (H_2S) emanating from the black smokers comes from magmas. The bulk of the hydrogen sulfide is made during reactions in hot basaltic seafloor rocks in which seawater SO_4 is reduced while reduced Fe++ is concurrently oxidized in reactions such as:

   SO_4 (in seawater) + $10H^+$ + $8Fe^{++}$(in minerals) $\longrightarrow$ $8Fe^{+++}$ + H_2S + $4H_2O$

   When metals such as iron, copper, and zinc and hydrogen sulfide in the hot waters start mixing with cooler seawater, the decreasing temperature causes the precipitation of sulfide minerals as illustrated below in the formation of sphalerite. H+ is released into sea water during the process.

   $H_2S + Zn^{++} \longrightarrow ZnS + 2H^+$

   Reactions of the evolving water within the hot oceanic crust are:

   path 1: $CO_2 + (H_2O + CO) \longrightarrow CO_2 + CH_4 + H_2O + C \longrightarrow CH_4 + H_2O + C$

   path 2: $CO_2 + (H_2O + CO) \longrightarrow H_2O$ +/- $CO_2 \longrightarrow H_2O$ +/- CO_2 + brine (brine is very salty water)

2. The exothermic reaction whereby heat is released (rather than in an endothermic reaction whereby heat is needed to make the reaction go) takes place when hot peridotite encounters water, as shown below. Products of the reaction include serpentine and the release of some elements such as iron and magnesium into solution. The resulting solution is alkaline or basic, which then facilitates the calcium (Ca) in seawater to precipitate.

   $5(Mg,Fe)_2SiO_4 + 10H_2O \longrightarrow 2Mg_3Si_2O_5(OH)_4 + 10Fe^{++} + 4Mg^{++} + 8(OH)^-$ + heat + alkaline solution

REFERENCES

Corliss et al., 1979; Kelley et al., 2002; Tunnicliffe, 1992.

Ridge axis segments between faults vary in length, with the longest, or first order segments, reaching 400 to 600 km (249 to 373 mi). The shortest, fourth-order segments, are only tens of kilometers long and occur at segment bends and offsets. Volcanism occurs in major transform faults on fast-spreading ridges.

While the yearly movement or offset on these faults is only a few centimeters, it is enough to generate most of the earthquakes associated with spreading centers. These fault segments, located between the axial ridge crests (note the opposite direction of motion only between axial crests) are lithospheric plate boundaries.

But there is more complexity associated with the ridges! For example, while the East Pacific Rise has nine major transform faults cutting it over a length of 5,000 km (3,107 mi), the axis is also disrupted by many smaller offsets. These axis segments vary from 10 to 200 km (6 to 124 mi) in length with axis offsets measured in a few kilometers. And the geometry along the offsets shows crustal crinkling and axis tip overlap.

The advent of manned and remotely operated submersibles with advanced technologies in the late 20th century has allowed for studies of these ridges and axial valleys thousands of meters below the ocean surface. In the late 1970s, scientists made startling discoveries while exploring the Galapagos spreading center off the coast of Ecuador and the East Pacific Rise just south of the southern tip of Baja, Mexico. For the first time, they observed amazing life forms based not on sunlight and photosynthesis, but on the chemical energy released from submarine hot springs (chemosynthesis) (see the sidebar on Strange Life Forms, page 18).

Destructional Plate Boundaries

Lithospheric plates grow larger at spreading centers, but because the Earth is not expanding, the plates must also be consumed elsewhere to compensate. Precisely where this occurred was long a puzzle to geologists, but we now know that collisional or destructive plate boundaries occur in the deepest parts of the ocean, in submarine trenches[7] found along such places as the west coast of South and Central America, east of Japan, and southeast of the Aleutian Islands of Alaska. These boundaries all feature landward belts of volcanoes and earthquakes zones (Figure 2.1). These convergent plate margins are where the oceanic lithosphere is subducted into the mantle, plunging beneath an adjacent continental or oceanic plate. Researchers calculate that during the last 180 million years, subduction zones have consumed an area equivalent to the seafloor of all of the Earth's oceans!

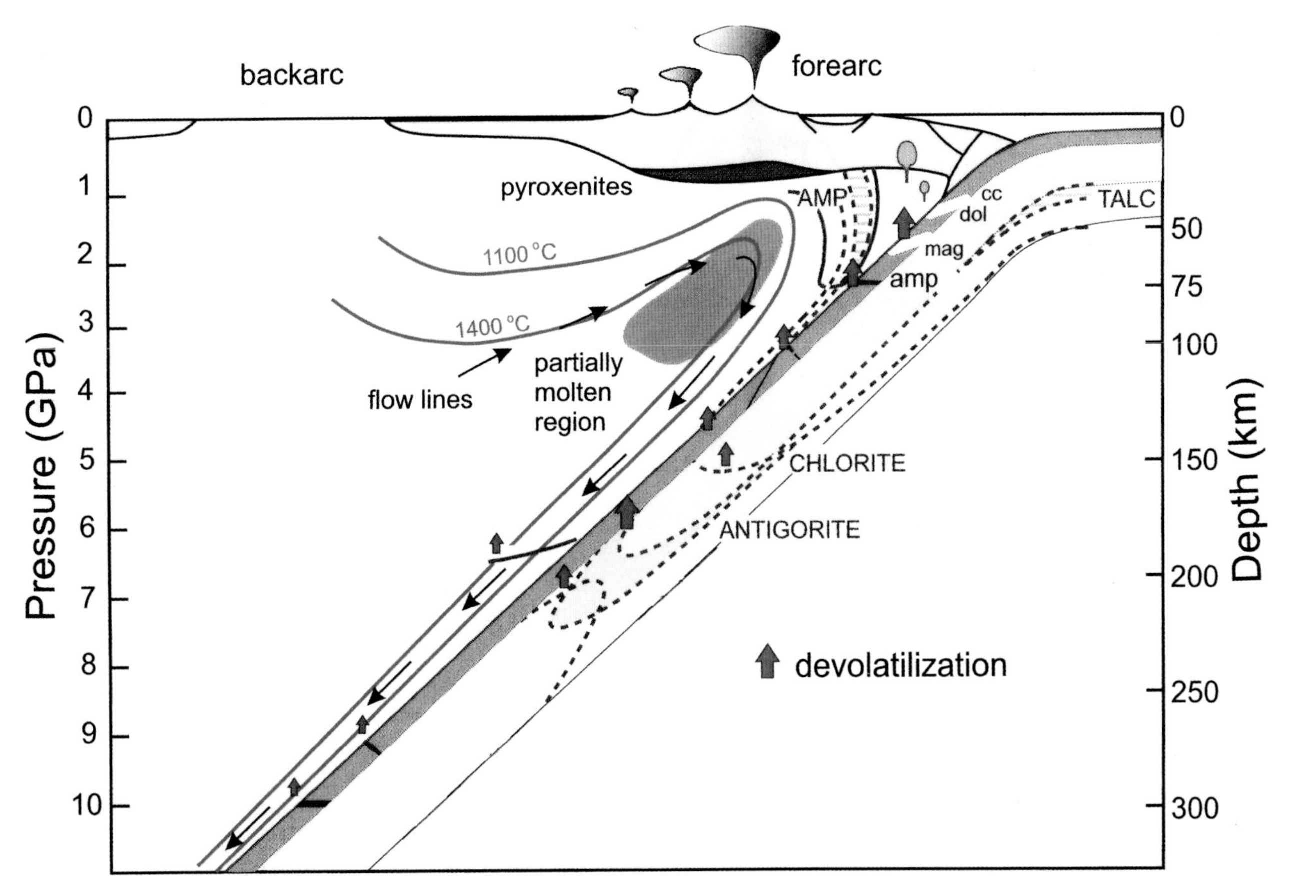

FIGURE 2.6: *Schematic diagram of an intra-oceanic or island arc subduction zone showing temperature isotherms (surfaces of equal temperatures including the crucial 1,100°C surface at and below which melting of mantle forming basalt takes place), mantle convection flow lines (black arrows), position of volcanoes located 100 km (60 miles) above the subducting slab, and some mineral formational boundaries and stability fields (heavy dashed lines) of volatile (primarily water and CO_2)-bearing phases that devolatilize during descent. Volatile-bearing (OH) minerals include: amphibole (amp) = $(Mg,Fe)_7(Si_8O_{22}(OH)_2$, talc = $Mg_3Si_4O_{10}(OH)_2$, Chlorite = $(Mg,Fe,Al)_6(Al, Si)_4O_{10}(OH)_8$, and antigorite = $Mg_6Si_4O_{10}(OH)_8$. Other minerals shown in Figure 2.5 that form during subduction of the slab include (proxene = $(Mg,Fe)SiO_3$, calcite (cc) = $CaCO_3$, dolomite (dol) = $Ca,Mg\ (CO_3)_2$, and magnetite (mag) = Fe_3O_4. The angle of slab descent may vary with depth (modified from Poli and Schmidt, 2002).*

This process generates earthquake epicenters and the arcuate chain of volcanoes that circles the northern Pacific Ocean from Japan, north to the Kamchatka Peninsula, and then northeastward to the Aleutians and into the Alaskan Peninsula eastward almost to Anchorage, Alaska. Other places in the world also have suduction-related volcanoes and earthquakes (see Figure 2.1 and Chapter 4).

But how does subduction operate? In collisions between plates, the less dense plate overrides the denser, less buoyant plate (see Figures 2.3 and 2.6). The two types of convergent plate boundaries are Andean, named for the Andes Mountains of South America, and island arc. In the Andean type, continental lithosphere overrides oceanic lithosphere. The classic example is the westward-moving, more buoyant South American plate in collision with the oceanic Nazca plate formed at the East Pacific Rise spreading center.

In island arc-type subduction, oceanic lithosphere overrides oceanic lithosphere. Examples of very tectonically active island arc-type subduction zones include the outer Aleutian Islands of Alaska, the Izu-Mariana, and the Tonga-Kermadec and New Hebrides systems in the southwestern Pacific Ocean (Figure 2.1). In these systems, magmas may erupt in three different structural locations: in front of the arc (fore arc), behind the arc (back arc), and within the island arc itself.

While these plate collisions typically compress the crust, in some back arc settings, the crust may pull apart, forming basins with associated basaltic volcanism.

OBDUCTION

Another convergent plate situation occurs when material on the subducting plate is scraped off or obducted onto the overriding plate, commonly a continental plate. **Obduction** also occurs in island arc-type settings.

The island of Cuba and the isthmus of Costa Rica and Panama formed through obduction. These land masses, rather than being underlain by continental crust, are obducted sea floor sediments, submarine volcanic plateaus, sea mounts, and ophiolites. Ophiolites are slices of the seafloor composed, top to bottom, of basalt and gabbro, in turn overlying dunite and peridotite of the upper mantle. Obducted ophiolites also occur in Oman, Cyprus, New Zealand, New Caledonia, Newfoundland, and the Klamath Mountains of northern California. In some volcanic arcs where obduction occurred, a new subduction zone formed seaward of the old subduction zone. This resulted in the superposition of a new volcanic arc on the older one.

SUTURING OF CONTINENTAL PLATES

Subduction zones may stop "swallowing" lithosphere when two continental plates collide. Instead, the continental plates attach or "suture" together. This causes folding, faulting, metamorphism, and uplift, in turn building high mountains. Contrast, for example, the relatively low volcanic mountains

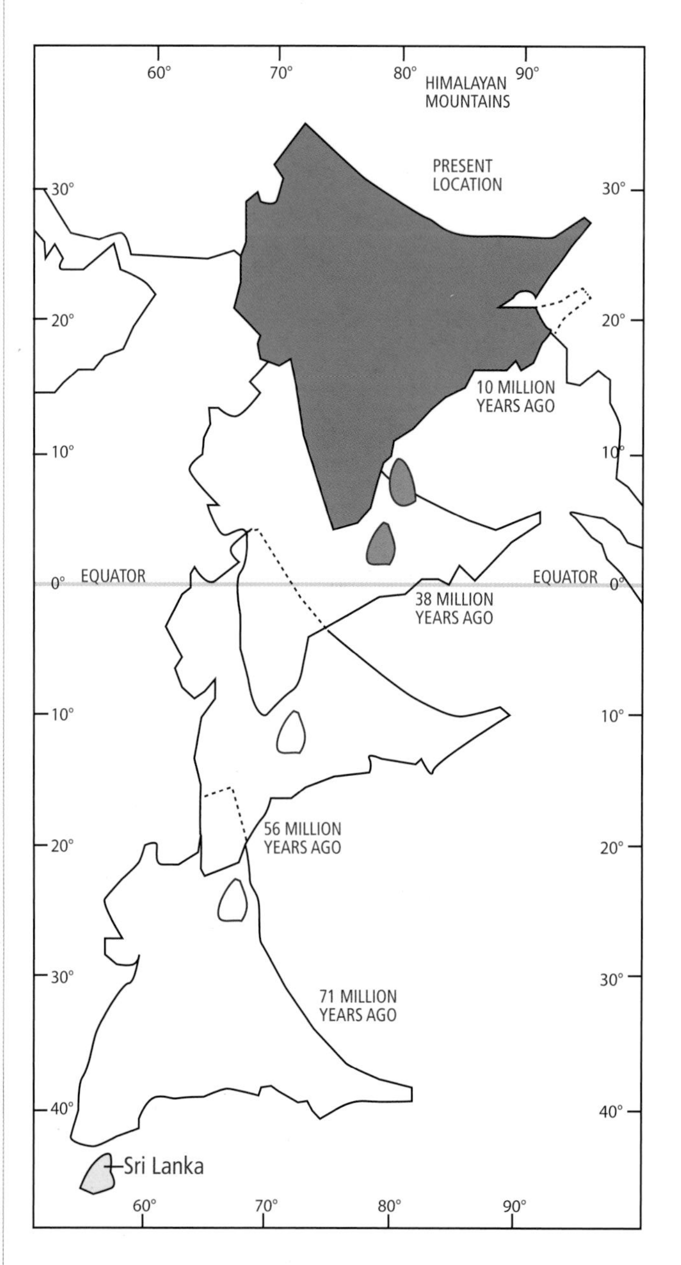

FIGURE 2.7: *Positions over time in millions of years of the Indian subcontinent drifting north and eventually suturing with Asia. Reconstructions indicate that India moved about 5,000 km (3,100 mi) to the north before colliding with Asia resulting in the formation of the Himalayan Mountains (modified from Molnar and Tapponnier, 1980).*

of Central America, where oceanic lithosphere is plunging beneath continental lithosphere, with the world's largest and highest mountain range, the Himalayas of Asia. The Himalayas were formed by the collision of northward-moving India with Asia starting about 50 million years ago (see Figure 2.7). A much older example of this phenomenon is the north-south-trending Ural Mountains of west-central Russia that mark a 385-million-year-old eroded suture zone between Europe and Asia.

Alternatively, when continental lithospheric pieces collide, subduction may continue, but its location may "jump" to the seaward side of the sutured continental land mass. An example is found along the northwestern coast of the United States from northern California to southern British Columbia. Here a collage of continental and oceanic fragments became attached to North America starting 80 million years ago. But a still-active subduction zone exists off the coast.

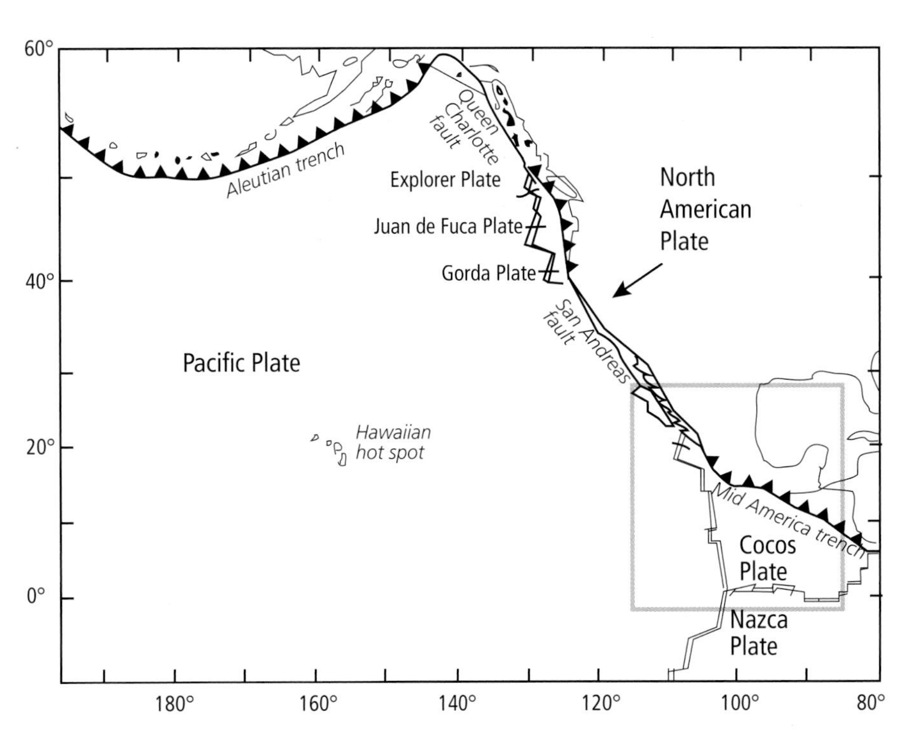

FIGURE 2.8: *Eastern Pacific coast map from northern Central America north to Alaska showing the types of Pacific-North America plate boundaries. The Mid East Pacific Rise spreading center is rifting Baja, California, away from Mexico but in turn is being overridden to the north and throughout California by the southwestward-moving North American plate. This has resulted in the San Andreas transform fault Pacific-North America plate boundary. A vestigial remnant of the East Pacific Rise remains off of northern California north to southern British Columbia. To the north, the Queen Charlotte transform fault system forms the Pacific-North American plate boundary. Eventually the Juan de Fuca Ridge system will also be overridden by the North American plate, and a transform fault system will mark the entire North American plate boundary (Figure 2.9 is within the outlined box) (modified from Abbott, 2004).*

Transform Fault Plate Boundaries

The third lithospheric plate-type boundary type, the transform fault, features lithospheric plates sliding past each other. These faults may extend 120 km (75 miles) or more deep. They consist of crushed rock and fault slivers up to 1.6 km or more wide and are hundreds to thousands of kilometers long. In places, mantle-derived basalt erupts from these faults. The geologically young volcanic rocks near Clear Lake in northern California represent a continental example.

The most famous transform fault is the 1,280-km (800-mi), northwest-trending San Andreas Fault located primarily in California. (Offset or displacement along the 30-million-year-old San Andreas fault is at least 560 km (350 miles).

The east-northeast side of the fault is part of the North American lithospheric plate, but the continental rocks on the west-southwestern side are now part of the huge Pacific plate, composed mostly of oceanic lithosphere. The Pacific plate is slipping northwestward at several centimeters per year toward northeast Asia and the Aleutian islands of Alaska. There the leading edge of the Pacific plate is being subducted below the overriding Asian and North American plates. The northeastern boundary of the Pacific plate with the North American plate in northwestern Canada and southeastern Alaska is a transform fault (the Queen Charlotte or Fairweather fault system). (See Figure 2.8.)

Plate Tectonic Complexities

Because the Earth is spherical and isn't expanding, the lithospheric plates interact in complex ways as they develop, move, and subduct. This is especially true in the western Pacific Ocean from Japan south to the Australian Plate, and in the Central America region (Figure 2.1).

One type of complex divergent plate boundary is the triple junction, with a classic example found in the Pacific Ocean off the coast of Central America. There, parts of three plates—the Pacific, Cocos, and

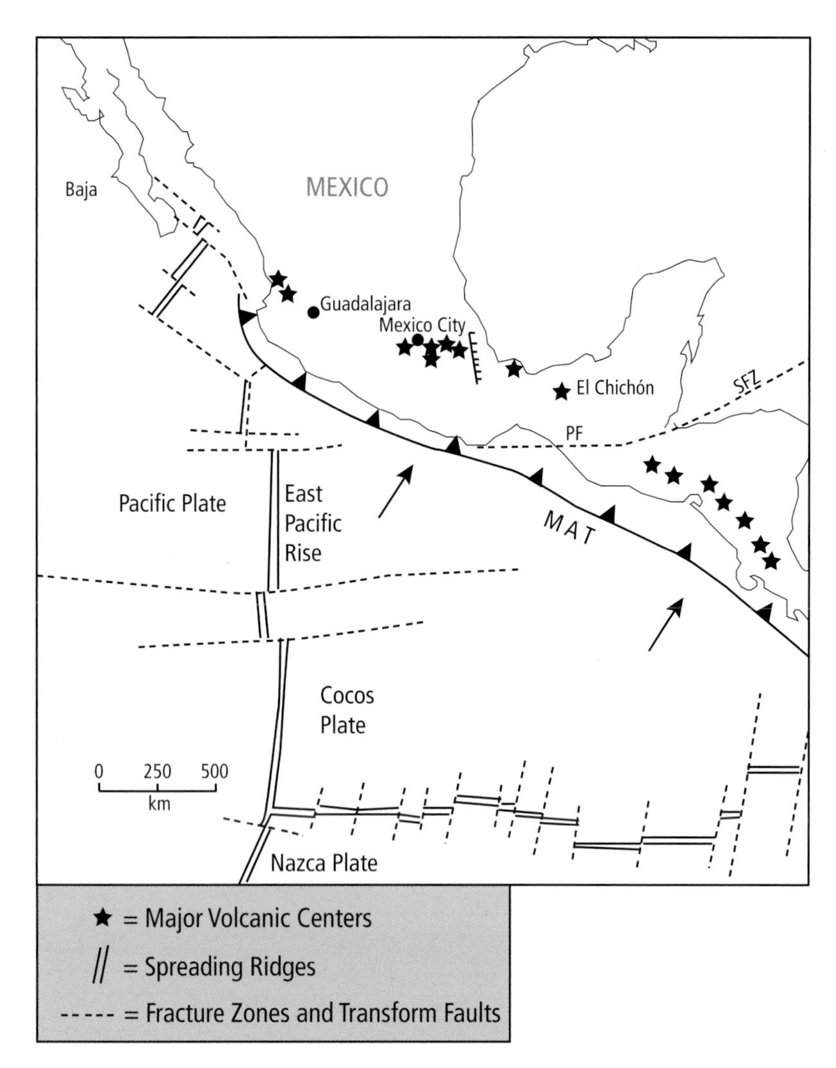

FIGURE 2.9: *Triple junction of the Pacific, Cocos, and Nazca plates in the Pacific Ocean located off of Mexico and Central America. Note the transform faults which offset the East Pacific rise (EPR) and Galapagos ridge axes, subduction of the Cocos plate under Central America and Mexico, and the related Middle America Trench (MAT) and volcanic arc. The EPR becomes a transform fault, which is rifting the Baja Peninsula away from Mexico. SFZ is the Swan Fracture Zone, PF is the Polochic Fault. Major volcanoes are indicated by *. Major cities are Guadalajara, and Mexico City (modified from Luhr et al., 1984).*

Nazca—are simultaneously forming and moving away from each other. The Cocos Plate, in turn, is being subducted under the advancing Central America continental lithosphere. To the north, the Pacific-North American plate boundary becomes a transform fault. (See Figure 2.9.)

Other examples of triple junctions where plates form and diverge can be found along the southern East Pacific Rise (Nazca, Antarctic, and Pacific plates) and in the Atlantic Ocean (North American, African, and South American plates) (see Figure 2.1).

The Central American volcanoes are a product of Cocos Plate subduction. To the north, the spreading center and transform faults are wrenching the Baja Peninsula and southwestern California, now part of the Pacific Plate, away from the North American Plate.

Plate tectonic generation-destruction complexities are found in the western Pacific Ocean in and around Indonesia and the Philippines (see Figure 2.1). Some tectonic settings are hybrids. For example, large basalt-erupting volcanoes—Mount Etna on Sicily and Mount Fuji on the Japanese island of Honshu, for example—have grown in tensional or divergent zones between subducting plates.

The surface of the Earth, then, is a complex mosaic of interacting lithospheric plates of various sizes and shapes, all moving at various speeds and in different directions across the Earth's surface. Some plates, such as the North American, are growing larger. Other plates, such as the Pacific, San Juan, and Nazca, are diminishing in size as they are subducted under continental lithosphere.

In addition, spreading centers such as the Juan de Fuca Ridge will be overridden by advancing continental lithospheric plates. And as far as we know, the Earth is the only one of the inner, small, rocky planets (Mercury, Venus, Earth, and Mars) or satellite moons in our solar system to have active plate tectonics with associated volcanism.

All the inner planets have witnessed volcanism, as has our Moon, but only the Earth and Venus remain volcanically active. The others, because of their small size, have cooled. Volcanism on Mars probably ceased about 500 million years ago, while it stopped on Mercury 2.0 billion years ago and on our similarly sized Moon between 2.5 to 3.0 billion years ago.

Without plate tectonics, no appreciable horizontal crustal movements of the other rocky, inner planets has occurred. Only volcanism and meteoric impacting have brought materials from the deep to the surface, and vice versa. Water erosion long ago, however, exposed Martian subsurface rocks. Venus, enveloped under a dense, hot (850°F), and carbon dioxide- and sulfur-rich acid atmosphere, has suffered intense surface rock weathering of its impact, faulted, and volcano-strewn surface.

Hot Spot Volcanism

Volcanoes typically occur in constructional, destructional, and transform-type fault tectonic plate boundary settings. But volcanism is also found in areas not related to plate tectonics. Examples include the Hawaiian Island volcanic chain, the Yellowstone volcanic system, and scattered volcanic centers in Africa. Each of these examples occurs within the interior of oceanic or continental lithospheric plates, far from plate boundaries.

These volcanic centers result from hot spot activity, a term coined in 1963 by J. Tuzo Wilson. Hot spots have helped us understand plate tectonics. But how can they "tell" us things about plate tectonics if they appear unrelated to the usual tectonic sources of volcanism? The answer: some, if not most, hot spot centers remain essentially in fixed positions within the Earth over long periods of time relative to the lithospheric plates that may or may not move over them. Some hot spots reveal the direction of lithospheric plate motion, changes in plate direction with time, plate velocity, changes in velocity, and whether lithospheric plates even move.

Hot spots appear to result from the partial melting in the upper mantle of rock (peridotite) that produces basalt, which then rises through cooler and therefore denser mantle. (Warmer materials are usually less dense than cool material of the same composition.[8]) Hot spots also occur on or near the axes of constructional plate boundaries. The volcanism in Iceland is one such hot spot, located on the Mid-Atlantic Ridge.

Hot spots are examined in detail in Chapter 5.

Chapter 2 – Notes

1 Lithosphere literally means rigid or rock-like. Earth's 100-km-thick lithosphere consists of rigid continental or oceanic crust and its rigid upper mantle. This slips over the less rigid asthenosphere below.

2 The seafloor is composed of basaltic lava flows that erupt in and near the center of the ridge crests. The magnetic reversal "stripes" are frozen within the basalt upon extrusion. See Paleomagnetism sidebar, page 13.

3 The maximum depth at which mantle melting occurs is believed to be 200 km.

4 The Earth's crust is affected by three main forces: tension, compression, and shear. Tension is the pulling apart of something. When crustal rocks are subjected to tension they may thin, but because they are brittle, they more commonly rupture, resulting in steeply dipping, faults. The faults can be long and parallel, develop perpendicular to the tension force, and result in long linear valleys or grabens. A graben, therefore, is a down-dropped valley formed by tensional faulting.

Crust can also be compressed, or pushed together. Compression folds rocks and can thrust one slab of crustal rock over another at a low angle. This forms thrust faults.

The third type of crustal force is shear. In a shear situation, crustal rocks slip by each other horizontally along strike-slip or transform faults, such as the world-famous San Andreas in California.

5 A movement of four centimeters per year over an 80-year-old person's lifetime would amount to 320 cm or about 10 feet. Over one million years, the displacement is 4 million centimeters or about 25 miles.

6 Lava, because of extrusion under water, commonly forms distinctive structures called pillows. This strikingly distinctive characteristic allows geologists to determine whether basalt, now exposed on land for example in road cuts or cliffs, was extruded under water. (See Figure 3.2D, page 30, for photo). Places on land to see wonderful exposures of underwater extruded pillowed basalt are near Crescent Lake in the Olympic National Park in Washington State, various places in central Washington, and in the Coast Ranges of California.

7 The deepest part of the ocean, where the seafloor is 11,035 m (36,195 feet) below sea level, is the Mariana Trench in the western Pacific Ocean.

8 A major exception to this rule is water/ice. Ice is colder but less dense than liquid water because of the more open structure of the mineral ice.

Chapter 2 – References

Abbott, 2004; Anderson et al., 2002; Bowring, 2014; Engebretson and Richards, 1992; Fisher and Schmincke, 1984; Gilluly et al., 1968; Graham et al., 2001; Hamblin and Christiansen, 2004; Jeanloz, 2000; Kelley et al., 2002; Luhr et al., 1984; McSween,1994; Molnar and Tapponnier, 1980; Patrick and Howe, 1994; Perfit and Davidson, 2000; Poli and Schmidt, 2002; Schmincke, 2004; Shirey et al., 2008; Sigurdsson, 1990; Williams, 1951; Wilson, 1963; Windley, 1977.

CHAPTER THREE

Volcanic Products, Eruption Styles, and Resulting Landforms

Introduction

Most people imagine volcanic eruptions as fiery explosions with lava flows enveloping the countryside. So when Mount St. Helens in Washington State came to life in the spring of 1980, many were "disappointed" by the lack of blazing lava displays. Instead, on the morning of May 18, the volcano unleashed a pyroclastic flow of hot gases and rock covering 600 km^2 (230 square miles), followed by an enormous plume of ash soaring skyward and massive lahars rumbling down the Toutle and Cowlitz Rivers. The mountain's north flank collapsed in the largest debris avalanche known in recorded history.

But Mount St. Helens did in fact produce lava. This occurred during dome-building phases within the new crater after the May 18 eruption. As the dome formed, extremely viscous, slow-moving lava erupted within the 800-m-deep crater. The dome, which looked like a giant, crusted cow pie, continued to grow slowly after blowing apart several times in the early to mid-1980s. Renewed dome growth began in September 2004. Through January 2008, Mount St. Helens occasionally vented steam and ash.

Lava, an Italian word that means sliding rock, is but one of the many materials that volcanoes produce when they erupt. Other products are gases, tephra (particles of various sizes and varying from cold to hot), breccias, pyroclastic flows, lahars (mudflows), landslides, and tsunamis or giant waves. This chapter describes these products and the landforms associated with them.

Gases

Gases play a vital role in volcanic eruptions. Most eruptions are the result of overpressured gases suddenly releasing, and the type of eruption is strongly dependent upon the gas content of the magmas beneath. Generally, the higher the concentration of gas, the more explosive the eruption. Explosivity is also a function of magma chemistry. While magmas initially contain many different gases, they also pick up groundwater and organic compounds as the magma moves through crustal rocks and sediments.

The study of gases has greatly helped volcanologists learn about volcanic processes. For example, the monitoring of gas emissions from craters has helped forecast eruptions (see Chapter 12 for details). The composition of gas also reveals its source(s)—whether the gas originated in the mantle, crust, groundwater, or sedimentary rocks at shallow depth. Gas studies also help determine depths to magma chambers and the nature of the dissolved volatiles contained in that magma.

At depth, magma contains variable amounts of dissolved volatiles, but when it rises through the crust, the gases separate or "exsolve" as pressure drops. The first gas to exsolve is carbon dioxide, then sulfur, chlorine, water (steam), and finally fluorine. Magma composition is important because water is more soluble in silica-rich magmas—dacite and rhyolite—than in more iron- and magnesium-rich, silica-poor basalt. Given their higher water content, silica-rich magmas are generally more explosive because they are more viscous, which allows gas pressures to build up until liberated by explosive eruptions.

While liberated gases vary considerably between even neighboring volcanoes, water (steam) is most abundant with carbon dioxide next. Plentiful amounts of sulfur compounds (hydrogen sulfide (H_2S) and sulfur dioxide (SO_2) also exist. Minor to trace amounts of other gases and elements may also be present, including carbon monoxide, methane, ammonia, chlorine, fluorine, bromine, helium, neon, argon, krypton, xenon, copper, zinc,

Water

Water, vital to life, has amazing properties. Unfortunately we have used and abused water for waste disposal and other industrial purposes. Clean and abundant water will become increasing scarce and more precious during the 21st Century.

But what do we really know about this liquid? Seawater, with a salt content of about 3.5 percent, covers 70 percent of the Earth's surface. It constitutes 97.5 percent of the hydrosphere, or the Earth's watery envelope. Only 1.5 percent of Earth's water is fresh, and most of that is in the form of ice in Antarctica and Greenland. Of the freshwater total, 0.63 percent is within the ground, 0.01 percent is in rivers and lakes, and 0.001 percent is present in the atmosphere. So while the available freshwater portion of Earth's water resource is small, the quantity is still quite large.

Water, fortunately, is unique in many ways. For example, we would not be discussing water if it behaved like most substances for which the solid phase is denser than the liquid phase. If this generality held for water, we would have a very cold world and the oceans would be filled with ice. The key to this reverse phenomenon is the open structure of the mineral ice. Water is about 10 percent denser than ice, reaching a maximum density at 4°C (37°F).

Also, due to its molecular structure, water is capable of dissolving many solids. The molecule has one oxygen (O) atom between two hydrogen (H) atoms in an L-shaped arrangement.[1] Because O has a negative charge, and each H has a positive charge, the molecule's uneven charge distribution can dissolve many compounds. It has, therefore, a dipole structure.

Surface water is an even better solvent than pure water because carbon dioxide (CO_2) in air[2] combines with water, making weak carbonic acid (H_2CO_3). This, however, is a problem for city buildings, especially those made with limestone and marble, which are soluble in carbonic acid.

We also know that pure water freezes to ice at 0°C (32°F) and boils at 100°C (212°F) at sea level.[3] But with the addition of substances such as salt (NaCl), the freezing temperature decreases and the boiling temperature increases. Seawater freezes at about -1.9°C (27°F).[4]

Hot, salty water is also important in the formation of many types of ore deposits because it effectively transports metals, sulfur, and other dissolved compounds in solution. These ore-bearing liquids, called hydrothermal solutions, are responsible for the development (see Chapter 10) of many types volcanic rock-hosted ore deposits. They include the bonanza gold and silver deposits that brought so many prospectors to the Americas.

Water is vital to eruptions: it pressurizes the system and leads to vesiculation and fragmentation of magmas. Furthermore, its presence causes explosive eruptions because the liquid form expands up to 1,600 times when it changes to the gaseous state. This expansion is instantaneous and therefore powerfully explosive.

But water is also very important because it aids lithospheric plate motion, lowers mantle strength and viscosity, and lubricates subduction zones. Furthermore, water, incorporated in weak, hydrous silicate minerals such as micas, decreases rock strength. Water is also important in the formation of basalt because its presence lowers mantle rock melting temperatures. Finally, water liberated during slab descent into the mantle at 1,100°C during subduction causes melting and the formation of basalt.

NOTES

1 The angle is not 90° but slightly greater at 104° 40′.

2 Air has about 400 parts per million carbon dioxide and the amount is rising, primarily due to the ever-growing burning of fossil fuels (coal, oil, and natural gas).

3 When a substance like water boils, it changes phases from the liquid to the gaseous phase. The boiling temperature, however, can be raised by increasing the pressure on the system. Alternatively, the boiling temperature drops as pressure drops. For example, we know water boils at 100°C (212°F) at sea level but at Yellowstone Park, at an elevation of +6,000 feet, or under lower atmospheric pressure, water boils at 93°C (199°F).

4 By increasing the amount of salt, the boiling temperature of water can be increased from 100°C for pure water at sea level to over 700°C.

Killer Gas Eruptions

Before leaving volcanic gases, let's learn about a recently discovered type of killer gas eruption. During the night of August 21, 1986, a massive jet of water and carbon dioxide gas erupted from the 210-m (650-foot) deep, usually quiet waters of Lake Nyos in the uplands of northwestern Cameroon of west-central Africa.

The following morning, at least 1,746 people and 3,000 of their animals around the lakeshore and in Lower Nyos village 3 km away were dead from carbon dioxide asphyxiation. What happened? Lake Nyos sits in a volcanic crater that produced a limnic or lake-water eruption.

Apparently the lake contained large quantities of carbon dioxide derived from a degassing magma chamber beneath the lake bottom. Until the gas eruption occurred, the magmatic gas had been dissolving in the basal lake waters. And this type of carbon dioxide-rich water, being heavier than regular lake water, will remain at depth unless something disturbs it. Something did, which brought the carbon dioxide-rich water closer to the surface. There, under lower pressure at a depth of about 40 m (125 feet), the gas began to come out of solution not unlike a freshly opened 7-Up. The gas and water then rose rapidly to the surface.

As the carbon dioxide was released, the depth of water degassing migrated downward so that by 150 m, 1 liter of water released 11 liters of carbon dioxide gas. In all, 0.17 km^3 of carbon dioxide gas erupted. But the degassing wasn't a gentle phenomenon—a jet of gas and water soared to 100 m (330 feet) above the lake surface. This produced a tsunami that sent waves 25 m (82 feet) high crashing on the southern lake shore.

The resulting heavier-than-air carbon dioxide gas flowed as a 50-m (165-foot) thick cloud away from the lake at about 45 miles per hour. It went at least 25 km (16 mi) from the lake, killing all animals, even including birds and insects, but no plants.

It is not known what caused the eruption. Possible mechanisms that could disrupt bottom water (but not in this instance) include strong winds, violent storms, cool weather, landslides, and earthquakes.

How common is this type of killer gas eruption and what can be done about it? Many crater lakes emit carbon dioxide gas. A classic example is the Laacher See Lake located within a 12,500-year-old caldera in western Germany. But other killer lakes that have erupted (or might) include Lake Kivu on the Zaire-Rwanda border, and Lake Monoun, 95 km from Lake Nyos, which erupted on August 15, 1984, killing 37 people.

Prevention of future eruptions is accomplished by pumping bottom water close to the surface to initiate controlled limnic eruptions or degassing. This is now happening in Lake Nyos.

A similar tragedy, although at a smaller scale, were the deaths of three skiers in 2006 at Mammoth Mountain ski resort in California. While patrolling areas where volcanic heat keeps snow from accumulating on the slopes, two fell into a magma-released carbon dioxide-rich pit. The third skier was also asphyxiated trying to extricate the others.

REFERENCES

Ladbury, 1996.

vanadium, chromium, bismuth, and even gold and silver.

Water is usually, but not always, recycled groundwater—so-called "meteoric" water originally derived from precipitation.[1] Because volcanic rocks are generally fractured and therefore permeable, water easily penetrates to great depths in volcanic systems. Some of this water is incorporated into magma while in near-surface environments. Water rises in temperature before reaching magma, and may be emitted in hot springs and geysers. When water actually contacts magma, the result may be a violent, phreatomagmatic explosion.

Upon approaching an erupting or steaming volcano the smell of sulfurous gases is common. The rotten egg smell is from highly toxic hydrogen sulfide. The biting, gagging acid smell is usually sulfur dioxide gas, also unhealthful. Sulfur dioxide also reacts with water, forming powerful sulfuric acid. Our ability to smell these and other compounds in very low concentrations is superb, although our sensitivity varies with compounds. For example, ethyl mercaptan, a safety warning odorant added to propane, can be detected in concentrations of less than one part per billion (ppb). Dogs, however, can "smell" between 1,000 and 10,000 (possibly up to 100,000) times better than humans and so are used to detect illicit drugs, find missing persons, and so on. Bloodhounds have the most sensitive noses, while Dachshunds are among the least sensitive dog smellers.

Both dormant and active volcanoes emit a wide variety of gases, including organic and inorganic compounds. The content of such emissions varies from one volcano to the next. Minerals[2] precipitate out as gases escape from vents or fumaroles, forming deposits, some vividly colored. Sulfur and iron chloride, for example, are bright yellow, while arsenic is orange. Of the more than 2,800 known minerals on Earth, some are unique to a particular volcano.

Lava

Some of the most beautiful Earthly scenes are of volcanoes erupting at night. From a safe distance, sightseers in Hawaii enjoy brilliant red basaltic lavas[3] flowing as fast as 55 kilometers per hour (35 mph) from Kilauea's vents, along with swirling masses of slightly cooler and darker crusts. Kilauea's flank vents also produce fire fountains of incandescent magma. Similar pyrotechnic displays draw spectators to Pacaya in Guatemala, Arenal in Costa Rica, Mount Etna in Sicily, and, in April 2015, to Calbuco Volcano in southern Chile.

FIGURE 3.1A: *Gases, mostly water and carbon dioxide, emanating from Caliente Crater, Santiaguito domes, Guatemala. The domes arose starting in 1922 in the 1902 blow out crater of Santa Maria Volcano (Lange photo).*

FIGURE 3.1B: *aa (rough textured) lava flow from Pacaya Volcano, Guatemala (Lange photo).*

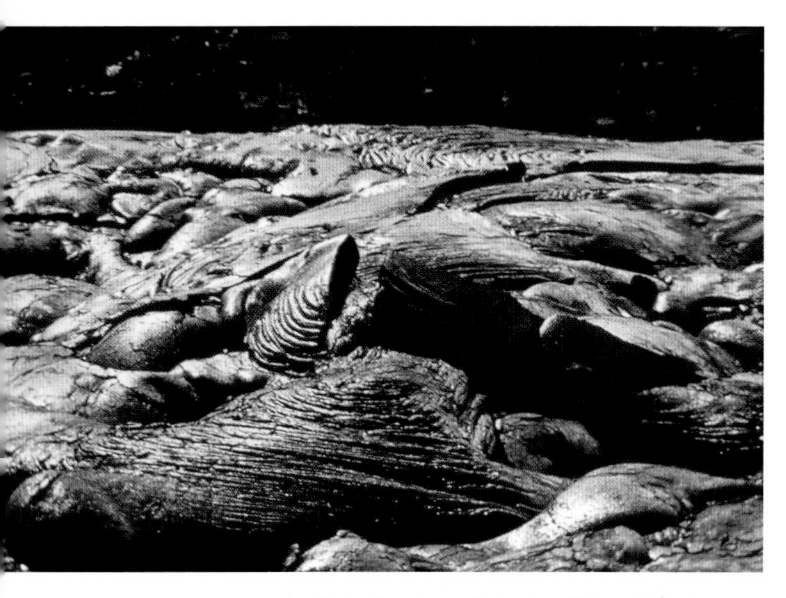

FIGURE 3.1C: *Pahoehoe (smooth textured) lava flow in Hawaii (Lange photo).*

FIGURE 3.1D: *Sakurajima Volcano, Japan, erupting on February 6, 1990. Photograph taken from Sakurajima Volcano Observatory by Tetsuro Takayama. Note the trajectory paths of ejected blocks (permission from Elsevier Limited).*

While glowing, molten lava is fascinating to see. Recently cooled lava is also visually striking. Pahoehoe (smooth, ropy-looking lava characterizing many of the Hawaiian basaltic flows) has its own stark, artistic, almost sensual beauty. Lava that has a very rough surface is called aa (pronounced *ah ah*). Both are, aptly enough, Hawaiian words. Textural gradations between the two exist.

Pahoehoe flows are generally very fluid, and Hawaii flows are one to two meters thick. They veneer gentle slopes, erupt at higher temperatures, and have lower viscosity than aa lavas. Pahoehoe flows contain up to 50 percent vesicles (gas escape holes) and may change into aa lava as the slope angle increases. Some aa flows from Kilauea have changed into pahoehoe flows upon reaching gentler slopes.

Aa flows are generally thicker, varying from 2 to more than 20 m (6 to 65 feet) thick. While their vent flow rate is generally higher than that of pahoehoe eruptions, they advance more slowly than pahoehoe flows.

What causes the different forms of lava? Researchers originally believed that lava formed as pahoehoe or aa based on differences in viscosity or resistance to flow. Today we know that strain[4] within the flow is also important. A higher temperature and lower silica content results in lower magma yield strength. This in turn promotes brittle flow-surface behavior as magma cools, leading

FIGURE 3.2A: *Ship Rock, New Mexico, an eroded, extinct volcano with just the neck or vent feeder, and radially oriented dikes that remain because they are more resistant to erosion than lava flows (USGS photo).*

FIGURE 3.2B: *Columnar basalt, a cooling phenomenon, exposed in the Devils Post Pile National Monument, California (Lange photo).*

FIGURE 3.2C: *Tops of columnar basalt columns exposed at the Giant's Causeway, northern Ireland (Lange photo).*

FIGURE 3.2D: *Ancient basalt "pillows" found on Lopez Island, Washington State. Pillows form when basalt flows into water (Lange photo).*

CHEMISTRY OF THE MANTLE, CRUST AND ARC-TYPE VOLCANIC ROCKS*

Chemistry	Mantle**	Komatite	Basalt	Basaltic Andesite	Andesite	Dacite	Rhyolite Crust	Continental Crust***	Oceanic Crust***
SiO2	45.1	40 to 45	51.7	53.43	59.29	65.13	74.43	57.8	49.34
Al2O3	3.35 to 11.8	5 to 11.8	17.41	17.65	16.89	16.26	13.18	15.2	17.04
Fe2O3			3.34	3.28	2.8	2.24	1.05	2.3	1.99
FeO	8	9 to 11	5.88	6.11	4.35	2.59	1.04	5.5	6.82
MnO	0.1		0.16	0.17	0.13	0.11	0.07	0.18	0.17
MgO	38.1	>18	6.02	4.8	3.18	1.63	0.37	4	7.19
CaO	3.1	2=Ca+NA	9.7	9.26	6.71	4.3	1.34	8.7	11.72
Na2O	0.4		2.85	2.77	3.45	4.23	4.05	3.2	2.73
K2O	0.03	<0.5	0.92	0.75	1.37	2.13	3.66	2.2	0.16
TiO2	0.2		0.98	0.89	0.8	0.66	0.26	0.8	1.49
P2O5			0.22	0.17	0.19	0.19	0.06	0.3	0.16

*=after Hyndman, 1985
**=undepleted or primitive mantle from Francis and Oppenheimer (2004) Table 2.1
***=after Marsh, 1979

TABLE 3.1: *Chemistry of the mantle, basalt, subduction zone-type magmas.*

to breakage and the development of aa lavas.

Lava may erupt from central craters, vents on volcano flanks, and long cracks called rifts or fissures. Hawaiian shield-type volcanoes issue lava from all three. Composite or strato volcanoes, such as those found in the Cascade Range, erupt from craters, whereas the huge basaltic lava flows that flooded central Washington and Oregon 16 to 5 millions years ago issued from fissures (see Chapter 7).

These fissures, after some erosion, form vertical outcrops, or dikes, because feeder fissures are more resistant than flows to erosion. They look like walls "snaking" across the landscape of southeastern Washington.

All magmas contain the same suite of elements but in different amounts. Lava composition affects flow speed, explosivity, color, and the resulting landform. Because oxygen is the most abundant element in magmas, chemical analyses are given in oxide weight percents. Silica (SiO_2—one silicon atom to two oxygen atoms) is the most abundant oxide. It is of major importance in affecting magma viscosity, which in turn controls eruption types and volcano morphology. Table 3.1 lists the major oxides that occur in magmas, along with the rock types and other data.

The silica content of magma affects its characteristics in several ways. The higher the silica content, the more sticky or viscous[5] the lava, hence the slower it flows. Hawaiian basalt, with a silica content of about 50 percent, can reach speeds of 55 km per hour (35 mph), whereas dacite emanating from the dome of Mount St. Helens with 65 percent silica (or 30 percent higher than basalt), barely oozes out of the ground.

Also, the higher the silica content, the lighter the rock color. Yellowstone rhyolite, having up to 80 percent silica, is light gray to white (Table 3.2), whereas Hawaiian lavas rich in magnesium and iron and low in silica are black. The melting temperatures of silica-rich lavas are also much lower than silica-poor lavas (as low as 700°C for rhyolite to as high as 1,250°C for basalt; see Table 3.2).

But how does the silica content control viscosity and flow? When magmas form, and before minerals develop, each silicon atom surrounds itself with and bonds to four oxygen atoms, forming a silica tetrahedron. A single silica tetrahedron, however, is not electrically neutral and so can't exist by itself. Tetrahedrons do not form minerals *unless they bond to other elements or other silica tetrahedrons, thereby forming chains, rings, or other geometric forms.* This covalent or shared bonding linkage of silica molecules to form larger, more complex structures

Magma Type*	Eruption Temperature**	Color	Silica Content (range)
rhyolite (GRANITE)	700-900	white-gray	>72% (72 - 80+)
dacite (GRANODIORITE)	800-1,000	medium gray	~65% (67 - 71)
andesite (DIORITE)	950-1,200	dark gray-black	~58% (57.5 - 67)
basalt (GABBRO)	1,000-1,250	BLACK	~50% (46 - 51)

* intrusive equivalent in capital letters
** in degrees C

TABLE 3.2: *Common Extrusive and (INTRUSIVE) Magma Types.*

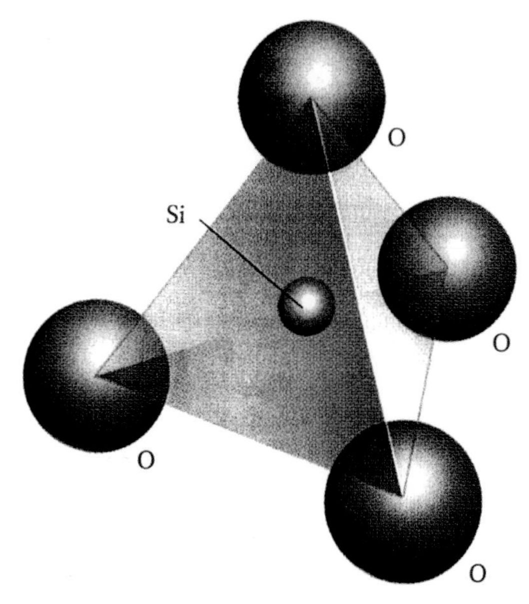

FIGURE 3.3: *The silicon-oxygen tetrahedron basic "building block" of silicate intrusive and extrusive rocks. One small silicon atom is surrounded by four much larger oxygen atoms making a geometrically sound molecule. But, because of the charge imbalance of –4 on the tetrahedron, it must bond with other elements and/or silica tetrahedrons for electrical neutrality (modified from Chernicoff et al., 2002).*

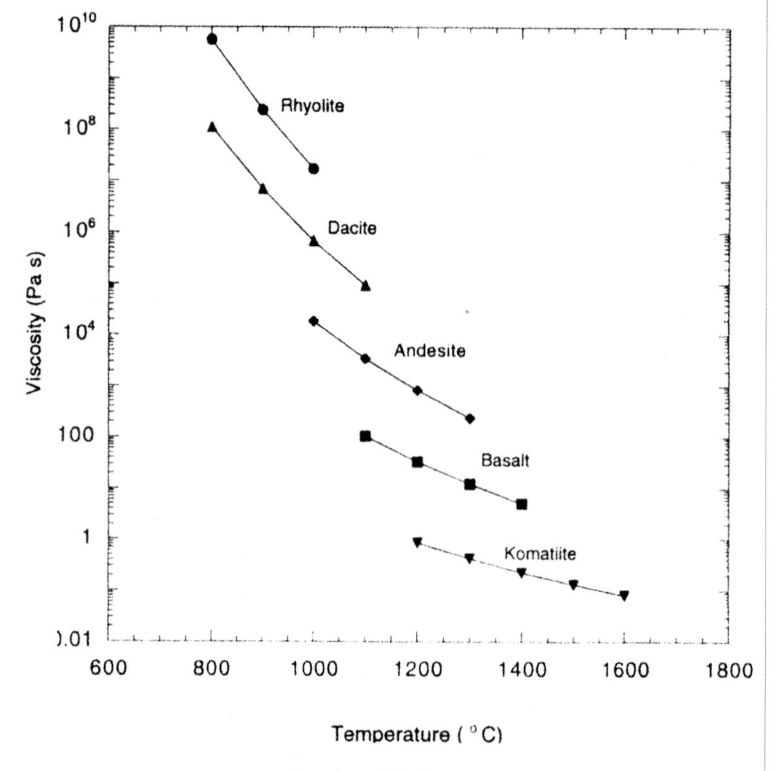

FIGURE 3.4: *Viscosity of different magma types. Note the very high viscosity, or resistance to flow, of rhyolite compared with basalt. Also note how viscosity of each magma type on this exponential scale decreases with increasing temperature (from Spera, 2000).*

in magma is called polymerization. Silica-rich magmas, then, contain more and longer chains and other configurations of silica tetrahedrons prior to the actual development of silicate minerals. It is these silica chains that increase viscosity, or resistance to magma flow.

Viscosity also increases when temperature and pressure decrease and when crystal content and bubble formation are greater. Water dissolved in magmas, however, decreases the viscosity of magmas and lowers melting temperatures because water impedes silica polymerization. Carbon dioxide, on the other hand, increases viscosity because it enhances polymerization.

Back to Basalt

In addition to pahoehoe and aa, basalt may also form other distinctive structures, such as marginal flow levees, similar to the ridges that form naturally along the edges of streams. Stream levees result from the deposition of sediment at the edge of channels where stream velocity markedly decreases. Lava levees develop along the chilled margins of flows and help channel subsequent flows (see photos of levees in Mount Shasta sidebar, page 66).

Another distinctive feature of basaltic and andesitic lavas erupted underwater are pillows. Pillows resemble bed pillows in shape and form due the sudden contraction or implosion of magma contacting water (Figure 3.2D).

Also distinctive are contraction joints that develop columns in basalt and andesite. Magma in volcanic vents may also cool this way: contraction jointing columns are conspicuous in Devils Tower, Wyoming. The joints may be straight and vertical, or curvilinear. The resulting lava columns are generally six sided, although four- and five-sided ones also form (Figures 3.2B and 3.2C).

Columns generally develop downward from the flow surface during cooling because solid rock occupies 5 to 10 percent less volume than magma. Devils Postpile National Monument in California, and the Giant's Causeway in northern Ireland are classic exposures. Columns also occur in Columbia River basalt flows (please see photographs in the Columbia River Flood Basalt sidebar, page 115).

Lava tubes are another interesting phenomenon. They form as lava courses away from vents and a surface crust develops on the flow. The lava beneath, insulated from cooling, can travel miles under the tube roof without losing much heat.

Post-eruption lava tubes remain empty and can be more than 3 meters (10 feet) in diameter. At Kilauea on Hawaii, lava continues to enter the sea

Why Volcanoes Erupt and the Formation of Pyroclastic Deposits

Knowledge of conditions within magma both before and during an eruption helps to understand why and how volcanoes erupt. Great strides have been achieved in these endeavors, especially since the 1980 eruption of Mount St. Helens. Parameters including viscosity, temperature, and gas content can now be approximated by various instrumental micro-examinations of minerals, glass, and fluid-filled inclusions within these materials. Theoretical calculations and remote sensing of eruptions have also added quantitative dimensions to our understanding of eruptions.

Knowledge of fluid dynamics is important to our understanding of eruptions. And fluid mechanics—rheology—is greatly influenced by the viscosity (resistance to flow due to internal friction) of the magma. Viscosity is dependent on magma composition, including contained gases, pressure, and temperature.

Mechanisms that trigger eruptions include (1) magma buoyancy, (2) magma differentiation resulting in a more silicic magma with a concurrent build up of volatile pressure in the upper part of the chamber, (3) injection of hot basaltic magma (low in silica, high in iron and magnesium) into a preexisting magma chamber, (4) water contacting magma, (5) second boiling, and (6) magma chamber decompression due to cone collapse, deglaciation, tidal forces, and earthquakes.

Magmatic buoyancy, which is necessary for upward motion resulting in an eruption, is enhanced by expansion that takes place as magma moves into the lower pressured surface environment. With a continual decrease in pressure, magmas eventually reach volatile saturation. At this stage, a volatile or gas phase separates from the magma, producing an even less dense magma-gas bubble mixture. For example, 1 m^3 of rhyolite magma at 900°C containing 5 percent water by weight expands to a 670 m^3 (2,3661 $feet^3$) magma-water vapor mixture at the surface. This dramatic expansion of the magma-gas is the main driving force of explosive eruptions.

Magmatic differentiation, discussed in detail in Chapter 8, starts with the crystallization of silica-poor minerals. Denser than the magma, these minerals settle in the chamber. The upper part of the chamber, then, becomes enriched in silica and volatiles, with a concurrent buildup of pressure that may eventually exceed the tensile or confining strength of the chamber roof.

Another mechanism—mafic magma injections into preexisting magmas—not only raises the temperature of the intruded system, but also leads to convective mixing and a pressure rise in the top of the chamber. That this mechanism operates can commonly be seen in volcanic rocks exhibiting different magmatic compositions, the result of incomplete magma mixing.

Water coming into contact with magma, or hot rock, is another, very effective eruption triggering mechanism. The result is water flashing (expansion) to steam, which produces a very dynamic, explosive eruption. Surface water, the sea, or subsurface groundwater may be the source. This type of volcanic outburst, termed a phreatomagmatic eruption, achieves maximum energy release when the ratio of water-to-magma is about 30 percent.

Another triggering mechanism, known as second boiling, occurs high in the magmatic system with the formation of late-stage minerals. This event decreases the amount of remaining magma, which in turn increases the gas pressure within the top of the magma chamber.

Decompression of the magma chamber can also trigger an eruption. During the 1980 eruption of Mount St. Helens, people camping within the restricted red zone actually photographed the cone collapse as it happened. The sequence of images shows magma venting during and following a three-stage avalanche of the northern portion of the over-steepened cone into the North Fork of the Touttle River Valley below (see Mount St. Helens sidebar, page 70).

The Formation of Pyroclastic-Producing Eruptions

We know that explosive volcanism fragments magma and rock to form pyroclastic deposits. But how is the magmatic liquid broken into pieces and erupted violently in a very hot gaseous mixture? Knowledge of the characteristics of rising magma in the crust help us understand the mechanism. Very important are magma composition and its dissolved volatile components, mainly water and generally lesser amounts of carbon dioxide.

The solubility of volatiles, or how much fluid can dissolve in magma, is controlled by (1) types and amounts of volatiles, (2) magma composition (3) temperature, and (4) the confining pressure on the magmatic system. Chemical determinations of fresh volcanic glass show that basaltic MORB magmas contain 0.1 to 0.2 percent water, subduction zone basalts 2 to 4 percent, andesitic-dacitic 2 to 5 percent, and rhyolitic magmas contain up to 7 percent water. Magmas are generally undersaturated in volatiles.

Magma composition is a major factor in explosive eruptions due to silicic control of viscosity. High-viscosity magmas allow pressure to build before final vesiculation and explosion of the mixture. With system pressurization, the amount of volatiles that can dissolve in magma increases. Conversely, the greater the volatile content, the greater the depth (pressure) at which separation of the fluid phases (magma from volatiles) takes place.

Confining pressure on magma decreases as it rises. At some point during the upward transit, the volatile vapor pressure of each component will equal the confining pressure on the system. When this happens, gases begin to exsolve or separate from magma, forming gas-filled bubbles.[1] While these bubbles lower magma density and promote magma rising, they also increase magma viscosity and corresponding high internal pressure within the bubbles. Viscosity increases because the magma changes from a one-phase (liquid) to a two-phase mixture (liquid plus gas).

The separation of volatile (gaseous) phases is also induced by the crystallization of volatile-free minerals such as olivine $(Mg,Fe)_2SiO_4$), pyroxene $(Mg,Fe)SiO_3$), and feldspars (K and Ca, Na $Al_2Si_2O_8$) in magma. This both raises the vapor pressure as the amount of magma diminishes and lowers the density of the liquid thereby aiding magma ascent.

With a volatile phase, explosive fragmentation of the magma may commence. This occurs sometime after the pressure within the magma chamber equals or exceeds the strength of the confining chamber wall rocks. The eruption begins when the roof rocks rupture.

A most appropriate analogy is the pressure cooker. These steel pots were popular cooking devices after World War II. Loaded with vegetables or meat and under high water pressure, they would cook foods much more rapidly and with less energy than on a conventional stove. A pressure release valve on the top of the pot controlled the pressure.

Occasionally some unlucky chef used one with a defective valve. The resulting explosion was horrific, with potatoes and steel fragments flying around the kitchen! These happenings did not help sales of the cookers, especially after the introduction of microwave ovens.

With volcanic eruptions, once the magma chamber walls fracture, pressure instantaneously drops as the superheated water and other gases dramatically and explosively expand and vent! One cubic foot of water expands instantaneously up to 1,600 cubic feet of steam. This not only fragments the magma, but drives the material violently upward and out into the atmosphere.

NOTES

1 Volcanic gases are liberated sequentially from magmas because of their varying solubility, with the least soluble leaving first. Carbon dioxide (CO_2), nitrogen (N_2), and noble gases leave the system early and under high pressure. Sulfur compounds (H_2S and SO_2), chlorine (Cl), water (H_2O), and finally fluorine (F) follow. The solubility of most gases in magmas increases with increasing pressure but decreases with increasing temperature.

REFERENCES

Carey and Bursik, 2000; Delmelle and Stix, 2000; Schmincke, 2004; Sparks et al., 1997.

through long tubes. Tubes may be explored south of Mount St. Helens and Mount Adams in southern Washington State, and at Craters of the Moon National Monument in Idaho.

Lava Types and Landforms

Table 3.3 shows the strong relationship between magma chemistry, viscosity (resistance to flow), landform, and eruption type. Basalt forms gently sloped, shield-type volcanoes because of the fluidity of lava. Mauna Kea and Mauna Loa on Hawaii have sub-aerial slope angles of about 4 degrees. However, when volatile- or gas-rich magma vesiculates[6] and erupts tephra (particles), it forms a cinder cone.

The energetics of more than 8,000 eruptions that occurred between 1500 and 1980 were quanti-

Eruption Type Ejecta volume	Explosivity*	Duration of cont. blast in hours	Column height (km)	Land Form	Examples with VEI	Magma Type
Icelandic	Non-Explosive gentle, effusive VEI = 0-1	< 1	0 - 1	fissure eruptions, flood basalts, small shields, cinder cones	Icelandic volcanoes	Basalt
Hawaiian 10(4) - 10(6)m	Non-Explosive VEI = 0-1 gentle, effusive	< 1	<0.1 - 1	fissure eruptions, shields, cinder cones flood basalts	Hawaiian Volcanoes	Basalt
Strombolian 10(4) - 10(7)m	Low Explosivity VEI = 1-3	< 1 to 6	0.1 - 5	cinder cones	Stromboli, Italy	Basalt-Andesite
Vulcanian 10(6) - 10(9)m	Moderate Explosivity VEI = 2-5	< 1 to > 12	1 - 25	Stratovolcanoes, Cinder Cones	Volcano, Italy	Basalt to Rhyolite
Rainierian	Moderate Explosivity			Stratovolcanoes (Intra segment)***	Mt. Rainier, Mt. Baker, Washington	Andesite
Plinian (Pyroclastic flows) (1 - 100 km3)	high Explosivity VEI = 4-8 cataclymal	6 to > 12	10 - >25	Stratovolcanoes, (Segment end)***	Mt. Mazama, Oregon (7), Mt. St. Helens, WA 1980 (5) Mt. Shasta, CA Vesuvius 79 A.D. (6) Tambora 1815 (7) Krakatau, 1883 (6)	Basalt-Andesite- Rhyolite
Ultraplinian (Pyroclastic flows) (100 - >1,000 km3)	High Explosivity VEI = 3-8 cataclymal, colossal	> 6 to > 12	>25	Large Calderas	Yellowstone (8); Long. Valley, California (8) Toba, Indonesia (8)	Rhyolite, +/- Basalt
Pelean (Avalanches) (0.001 - 1 km3)	High Explosivity VEI = 3- ?		0.1 - 5?	Plug Domes, craters	Mt. Pelee, Martinique (4) 1902 Santiaguito, Guatemala	Rhyolite, Obsidian (glass)

(continued)

Eruption Ejecta volume	Magma Viscosity**	Water Dissolved in Magma in %	Magma Temperature	Tectonic Environment
Icelandic	Very low (high Mg)	~0.1 - 1.0	1,100-1,250'C	Hot Spot, rift, spreading center
Hawaiian	Low - melted Ice Cream SiO2 =~50%	~0.1 - 1.0	1,000-1,200'C	Same as above
Strombolian	Low to moderate		800-1,000'C; 1,000-1,200'C	Volcanic arc
Vulcanian	Moderate to high		1,000 - 1,200'C 600-900'C	Volanic arc
Rainierian	SiO2=~58-60	~2 - 3	800-1,000'C	Volcanic arc
Plinian	Moderate to high		800-1,000'C; 600-900'C	Volcanic arc
Ultraplinian	High		600-900'C; 1,000-1,200'C	Hot Spot, Tensional Volcanic arc
Pelean	High-Toothpaste, cold tar SiO2=72%+	~4-6	600-900'C	Volcanic arc

** = Volcanic Explosivity Index number (VEI - 0 to 8 from Newhall and Self, 1982) See explosivity sidebar*
*** = Gases escape easily from basaltic magma; difficult from viscus dacitic to rhyolitic magmas*
**** = Cascade volcanoes, for example, occur in linear segments. Those at the ends are more explosive than those within the segments.*

SOURCES: Newhall and Self, 1982; Sheridan, 1979; Smith, 1979

TABLE 3.3: *Relationships between eruption type, explosivity landform, and magma composition.*

fied by Newhall and Self (1982) in their Volcanic Explosivity Index (VEI) (Table 3.3). Note the striking correlation between the volume of ejecta, the height of the eruptive column, and the explosivity associated with magma composition. Generally, the lower the silica content of the magma, the less explosive the eruption.[7] Basaltic eruptions are the least explosive, whereas rhyolitic magmas form the largest and most explosive eruptions. Basaltic eruptions can be very explosive, however, if water contacts magma.

ANDESITE

Cascade volcanoes such as Mounts Baker and Rainier in Washington, and Hood in Oregon, have slopes of about 30 degrees. Andesite volcanoes have steep

FIGURE 3.5A: *Gentle-sloped basaltic shield volcano Mauna Kea, Hawaii (Lange photo).*

FIGURE 3.5B: *Subduction zone, steep sided arc volcanoes in Guatemala. Left to right Fuego, Acatenango, and Agua. Note how these subduction zone volcanoes line up approximately perpendicular to the direction of the subducting lithospheric plate (see Figure 2.9, page 24, for orientation) (Lange photo).*

FIGURE 3.5C: *Basaltic cinder cone with lava flow that erupted from the base of the cone in the San Franciscan volcanic field in northern Arizona (USGS photo).*

slopes due to the greater viscosity of their lavas that solidify near vents. Andesite flows are commonly blocky, and also may produce columnar jointing, levees, and lava tubes, but not pahoehoe-type flows.

Andesitic volcanoes such as Mount Rainier are known as strato or composite volcanoes due to alternating layers of lava, mudflows, and tephra. This layer phenomenon is well exposed on Mount Rainier, and magnificently within the craters of Mount St. Helens and Mount Mazama (Crater Lake) in Oregon.

DACITE FLOWS AND DOMES

The most viscous and potentially most explosive magmas are dacitic and rhyolitic. Whereas basalt and andesite have 50 and 58 to 60 percent silica respectively. Dacite averages about 65 percent and rhyolite at least 72 percent; Yellowstone rhyolites have as much as 80 percent silica.

Dacite forms lava flows, domes, and flow domes. These lavas move very slowly, have steep fronts, and are commonly accompanied by lots of explosive activity due to volatile pressure buildup within the magma. Dacitic rocks are generally crystal-rich, some more than 50 percent crystalline by volume.

Because of differences in magma viscosities, volcanic dome morphologies vary from the very high-yield strength (mechanically strong) upheaved plug, "Peléean" domes, to the less strong, more rounded-to-flat-topped, "cow flop" domes. Mount Pelée, located on the French island of Martinique in the eastern Caribbean, forms spires hundreds of meters tall. The spires eventually collapse, producing avalanches of cold material or, if still hot, incandescent pyroclastic density flows. Peléean-type domes include the Santiaguito domes, growing since 1922 in the 1902 blow-out crater of Santa Maria Volcano in Guatemala, Mount Lamington in Papua New Guinea, Soufriere Hills Volcano on the British island of Montserrat in the West Indies, and Chaos Crags in Lassen National Park, California.

FIGURE 3.6A: *Flow-banded rhyolite, Panum dome within the Mono craters near Bishop, California (see Figure 3.10A for aerial view of the domes) (Lange photo).*

FIGURE 3.6C: *Nova Erupta Dome, Katmai National Monument, Alaska, which arose following the 1912 caldera-forming Plinian-type eruption of Katmai Volcano. The caldera-forming eruption produced an ash flow which ponded nearby. It has been emitting gases, mostly steam, during cooling and formed the Valley of Ten thousand Smokes (USGS photo).*

FIGURE 3.6B: *Santiaguito domes seen from the slopes of Santa Maria Volcano, Guatemala. The domes rose, starting in 1922, from the 1902 blow out crater of Santa Maria Volcano, and young from left to right. Caliente Crater (Figure 3.1A) is at extreme left (Lange photo).*

FIGURE 3.6D: *Looking north at the summit of Mt. Hood, Oregon. The crater contains a 200 year old dome intrusion. Massive collapse of this dome and that of a 1,500 year old dome triggered pyroclastic flows that raced down southern flank of the volcano creating the smooth landslide surface that now hosts the Timberline Ski Resort (USGS photo - W. E. Scott).*

Flat-topped domes are built of horizontal flows because of lower rock yield strength. The early, post-1980 domes within the crater of Mount St. Helens were classic examples with both endogenous (internal) and exogenous (external) extrusion growth. Short, 200- to 400m-thick flows, or lobes, mostly emanated from the top of the St. Helens dome. During 2006, large spires also rose from the summit area (see photos in Mount St. Helens sidebar, page 70).

The 1910 and 1944 domes associated with Usu Volcano in Japan are examples upheaved plugs. Because of their inherent rock strength, these pushed-up domes don't deform once the rise above the former land surface. In some places, these have "roof rock caps" of all types of rocks.

RHYOLITE

Rhyolite rarely forms lava flows. Instead, domes develop, such as the geologically very young ones (500 to 600 years) at Mono-Inyo Craters of California, and at Valles Caldera in New Mexico. The Yellowstone Caldera in Wyoming, however, does contain rhyolite flows well exposed in the Grand Canyon of the Yellowstone River. Flows may also be seen on South Sister Volcano near Bend, Oregon, and Glass Mountain, associated with Medicine Lake Volcano in northeastern California.

Rhyolitic glass—obsidian—is quite common in regions of rhyolite flows.[8] It has long been used by people for knives and arrow and spear points because it forms very sharp edges. Numerous obsidian sites are scattered around the western United States, including at Yellowstone, Glass Mountain, South Sister Volcano (see SB Figure 2D, Cascades sidebar, page 64), and nearby Newberry Caldera in Oregon.

Spectacular obsidian flow structures may develop, albeit very slowly due to the high viscosity of rhyolite lava (Figure 3.6A).

Eruptive Behavior of Volcanoes

The silica content of magmas also explains the eruptive behavior of volcanoes. Silica-poor, low-viscosity lavas—basalt and andesite—degas more readily than dacite and rhyolite. Therefore, high gas pressures generally don't build up within the upper portions of magma chambers. Certainly, you won't want to get too close to erupting Hawaiian volcanoes because of the intense heat and sulfurous fumes, but basaltic volcanoes rarely erupt explosively, except when water contacts magma. (See Table 3.3 - Relationships between magma type, eruption style and landform).

Tephra (Particles)

Erupted particles or tephra vary from hot to cold and from dust to car-size boulders. **Tephra**, Greek for ash, form from magmatic-, phreatomagmatic-, and phreatic-type explosions. Magmatic explosions, producing vesiculated or swiss cheese-looking tephra, occur in volatile-rich magmas that contain primarily water and carbon dioxide.

Phreatomagmatic explosions result from the interaction of magma and ground, lake, stream, or sea water. The largest explosions occur when about three parts magma encounters one part water (3:1 magma-to-water ratio). **Phreatic eruptions**, also steam-related, develop when water encounters very hot rock. This type of eruption can happen without warning and be deadly. For example, on September 27, 2014, 57 hikers were killed on Mount Ontake, a 3,067-m (10,062-foot) Japanese volcano 200 km (124 mi) west of Tokyo. The blast occurred without warning.

Magmatic explosive eruptions vent both vertically and horizontally. The largest eruptions blast tephra 50 kilometers (31 miles or 163,700 feet) or more into the atmosphere. On May 18, 1980, airline pilots flying between Seattle and San Francisco reported car-sized chunks at altitudes over 18 km (60,000 feet) erupting from Mount St. Helens.

Dust-size ash may remain in the stratosphere for years, producing spectacular red sunsets. Classic ash

Volcanoes and Soil Productivity

A striking correlation exists between the distribution of Central American volcanoes and the most populated countries. The greatest density is in and near volcanic regions because of soil productivity. And soil productivity in these moist, warm countries is due to tephra, or pyroclastic fall particles, and lava flows, which weather rapidly into some of the world's richest soils.

Compare, for example, El Salvador with neighboring Honduras. Honduras is five times larger in land mass but has about the same number of people (6 million) as tiny El Salvador. The population difference, or the density of humans (120/mi^2 in Honduras versus 675/mi^2 in El Salvador), is due to the rich volcanic soils of the latter country that can sustain intense agricultural production of high-protein, caloric foods.

This relationship exists around the world. Note the large populations of the islands of Sumatra and especially Java compared with Borneo and New Guinea. Active or recent volcanism is responsible for the difference in these tropical regions.

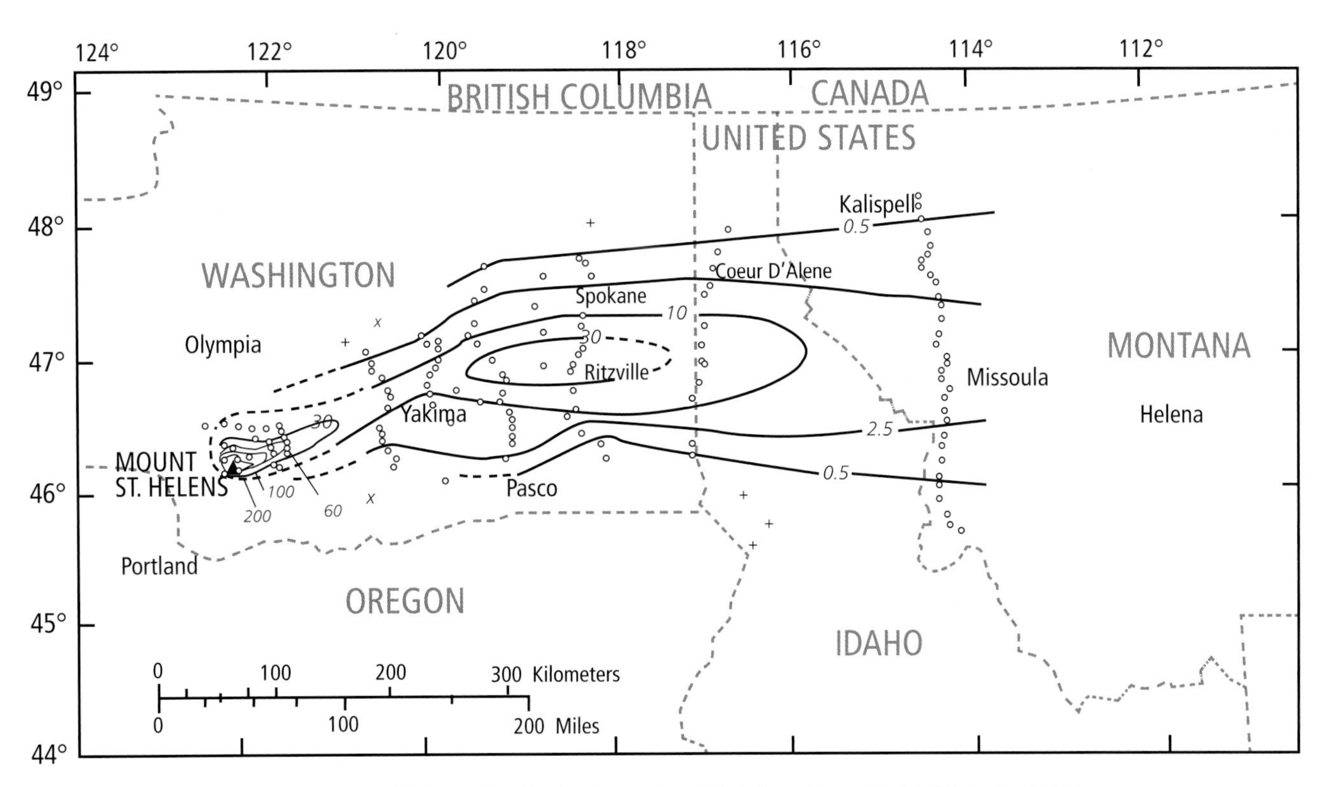

FIGURE 3.7: *Mount St. Helens air fall ejecta distribution in cm (modified from Sarna-Wojciciki et al., 1981).*

sunsets were observed in the Northern Hemisphere after the 1991 eruptions of Mount Pinatubo in the Philippines.

Cold tephra were the first particles erupted from Mount St. Helens in the 1980 eruption, starting in April. These **lithic fragments** were explosively ripped from volcanic vent and summit rocks in phreatic explosions. Subsequent eruptions produced hot, fresh **juvenile particles** from the magma chamber. Some volcanoes have produced juvenile ash so hot that, upon reaching the ground, the particles fused together. These **welded tephra fall deposits** are rare and occur only near vents. One such deposit is located just above today's main boat landing on the Greek island of Thera in the Aegean Sea. (See Figure 3.8D.)

Large tephra pieces don't travel far from the vent, though the largest particles follow ballistic trajectories unaffected by wind. Coarse-grained tephra build **cinder** or **pyroclastic cones** that may exceed 300 m (1,000 feet) in height. They are generally Hawaiian- and Strombolian-type basaltic to basaltic-andesite eruptions.

Cinder cone-building eruptions are usually short-lived, lasting weeks to a few years. Some, however, reactivate after being dormant for many years.

Coarse tephra particles ranging from andesite to rhyolite form blanket-like deposits around erupting vents. These air **fall deposits** rarely exceed 10 m in thickness and are composed of lithics, blocks, and pumice bombs.

Finer grained ash-fall deposits may cover thousands of square kilometers. Such deposits form from ash and Plinian eruptions, and also from phreatic and phreatomagmatic explosions.

But why is tephra erupted instead of lava? Cold tephra forms when water contacts very hot rock. In these phreatic-type explosions, at atmospheric pressure, water expands greatly when it flashes to steam, increasing in volume up to 1,600 times (a water droplet the size of a single grape produces enough steam to fill a large refrigerator). Expansion takes place instantaneously during this phase change and the explosive force can be huge. Expansion into steam decreases with depth within the system.

Juvenile particles both form and look different from lithic fragments. Juvenile particles are composed of glass, are highly vesiculated (contain gas escape holes and tubes), and may have mineral crystals. Pumice is so vesiculated that it floats!

Particles form as magma rises and rock load or **lithostatic** pressure decreases. In the upper crust, volatiles (primarily water and carbon dioxide) dissolved in magma, begin to separate, forming glass-walled bubbles. This decompression or **first boiling** is much like what we see when we take the cap off a bottle of soda: tiny carbon dioxide bubbles immediately form near the bottom of the bottle.

Bubbles grow rapidly in size as their glass walls thin, producing a magmatic foam. Eventually the pressure within the bubbles causes their brittle walls to break as the eruption starts. The formation of these bubbles depends on magma viscosity, a function of parameters such as magma composition, temperature, lithostatic pressure, and the concentration of volatiles.

Theoretically, an explosive eruption begins when the bubble volume reaches at least 65 percent of the magma volume. And the expansion of the growing bubbles increases the ascent velocity of particle-rich gas.

Sustained jets of vesiculating particles lasting hours to days are called Plinian eruptions.[9] They can have incredibly high vent nozzle velocities—greater than 600 hundred meters per second. As bubbles grow, they rise rapidly and continuously from the magma chamber below, no matter what the magma composition. Theoretically, the depth at which vesiculation or magma fragmentation occurs can exceed one kilometer.

The largest erupted particles are room size, but their fallout area is generally restricted to the crater and volcano slopes. At greater distances, tephra deposits become thinner and finer grained until

FIGURE 3.8A: *Cerro Negro Volcano in Nicaragua erupting in 1968. Note two subsidiary cones, one also erupting (U.S. Army photo).*

FIGURE 3.8B: *Tephra from 3,600 B.P. eruption of Santorini Volcano, Greece. Light gray particles are juvenile pumice whereas dark particles are lithic fragments ripped from older flows during the huge Plinian-type eruption (Lange photo).*

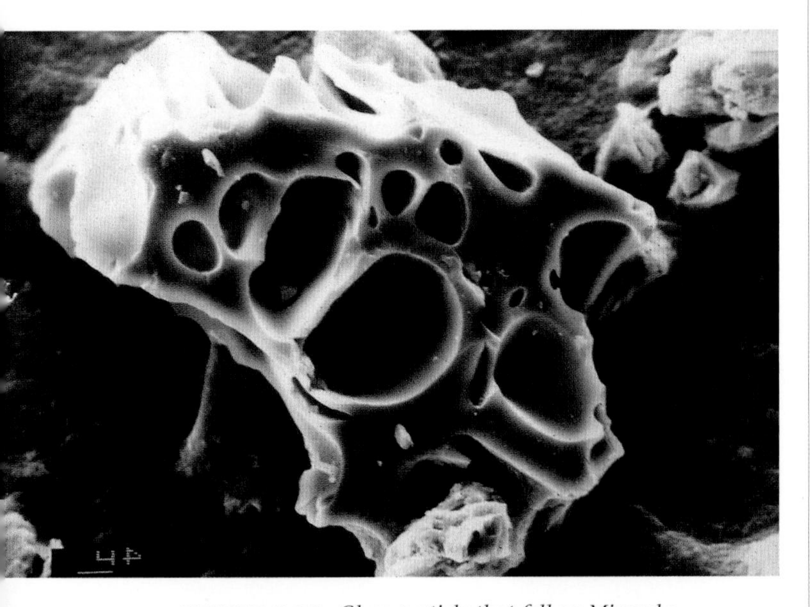

FIGURE 3.8C: *Glass particle that fell on Missoula, Montana, during the May 18, 1980 eruption of Mt. St. Helens, Washington. Note the "frozen in" Swiss cheese-looking gas escape holes. A 4 micrometer scale appears at the photo bottom (SEM photo by Johnnie Moore).*

FIGURE 3.8D: *Welded ash fall from the caldera of Santorini Volcano, Greece. Note the flattening and fusing of particles (Lange photo).*

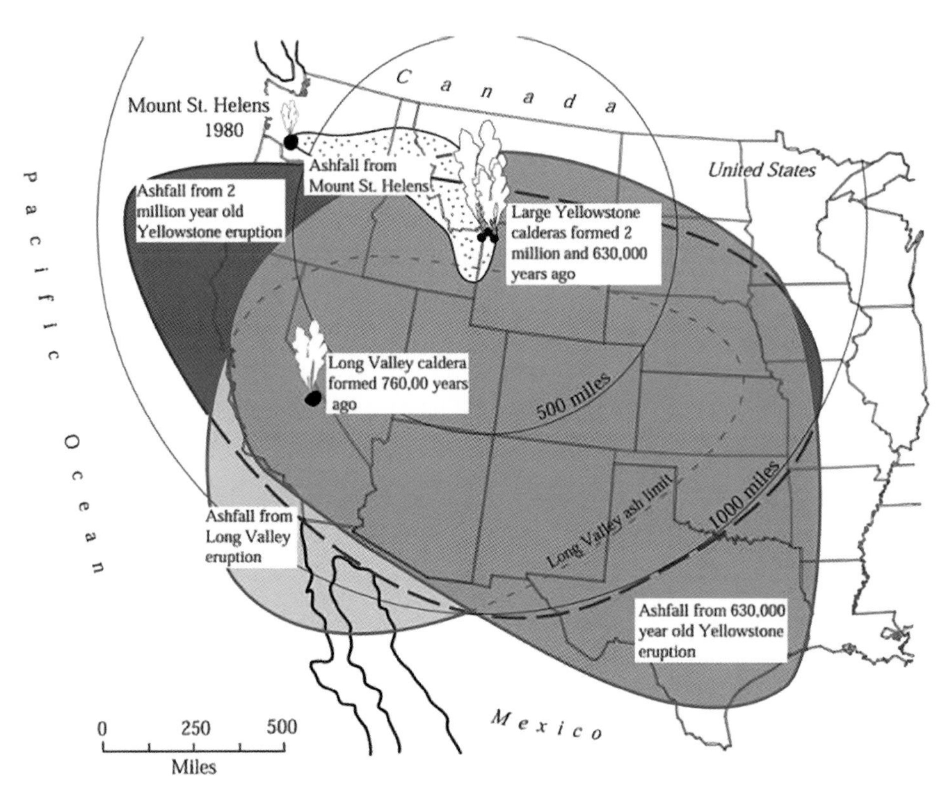

FIGURE 3.9: *Map of western U.S. showing the widespread nature of some tephra deposits erupted during the last two million years. Note the distributions of 2 Yellowstone, Wyoming caldera eruptions (2,020,000 year old Huckleberry Ridge tephra, and the 620,000 year old Lava Creek tephra), the 760,000 year old Long Valley caldera in California, and the 1980 Mt. St. Helens, Washington (USGS diagram).*

only dust-size particles settle. The distribution of fine-grained ash is strongly influenced by wind (see figure 3.7 - 1980 Mount St. Helens tephra).

Ash	< 2.0 mm in diameter
Lapilli	2.0 t 64 mm in diameter
Bombs or Blocks	>64 mm in diameter

TABLE 3.4: *Tephra particle sizes*

Recognizing ash in sedimentary rocks is usually easy because the particles in any particular layer are distinctive and of similar size. However, within any particular deposit, especially near the vent, particle size may vary from top to bottom because of (1) changes in the prevailing wind direction during venting and (2) the dynamics or intensity of the eruption (see Figure Intro 4, page 3).

The coarsest and thickest fraction of the ash naturally falls downwind of and close to the vent. Both particle size and deposit thickness decrease laterally away from this center line. The dynamics of the eruption also control particle size—the more energetic the eruption, the farther larger particles will travel. Many ash deposit exposures, then, show variable grain sizes vertically.

Tephra deposits greatly aid geological mapping because they form good, easily identifiable, and age-definitive geologic units that can be correlated, one to another, over, in some cases, large distances. Deposits can usually be age-dated, and some are widely distributed, both on land and in lake and marine sediments. Furthermore, each tephra unit has its own diagnostic chemical composition and mineralogy. This includes distinctive ratios of one mineral or element to another.

For example, some geologically useful tephra deposits have been found in Mediterranean Sea marine cores. These deposits came from the Santorini Volcano on the island of Thera, and they show how historically active this great volcano has been. In addition to the gigantic 1600 B.C. eruption that destroyed Thera, an even more extensive tephra layer from Santorini erupted 21,000 years ago.

Closer to home is the useful 7,700-year-old-tephra from Mount Mazama, located in southwestern Oregon (today's Crater Lake National Park).

FIGURE 3.10A: *Glass Mountain, CA rhyolite flows (USGS photo).*

Eruption Types and Explosivity

Volcanic eruptions vary from non-explosive, effusive lava extrusions **(Icelandic and Hawaiian)** to explosive, with increasing size and energy seen in **Strombolian, Vulcanian, Plinian,** and ultimately in **Plinian** caldera-forming eruptions. Non-explosive eruptions generally produce basalt, whereas increasing explosivity correlates with higher magma silica content (see Table 3.3).

Icelandic eruptions issue low-viscosity, low volatile-bearing basaltic lavas from fissures. They are commonly accompanied by lava fountaining. The lava flows nearly horizontally, forming large, wide plateaus and small shield volcanoes. Examples include the flood basalt flows found in Washington and Oregon, and in other flood basalt provinces found on continents and the seafloor (see Chapter 7).

Hawaiian-type eruptions are similar but build huge shield-type volcanoes. Mauna Kea, Mauna Loa, Haulalai, Kohala, Mahukona, and Kilauea are classic examples on Hawaii. Lava issues from both fissures and calderas.

Strombolian-type eruptions, named for the volcano in the Aeolian Islands northeast of Sicily, Italy, build scoria or cinder cones that may have accompanying lava flows. These volcanoes are commonly found in regions with basaltic to andesitic volcanism.

In addition to Stromboli, which has erupted almost daily for hundreds of years, numerous cones are located, for example, in the Mohave Desert of California and west of Mexico City. There, near the town of Paricutin, on February 20, 1943, a new volcano was born in a farmer's corn field. The cone grew to over 300 m (1,000 feet) as it destroyed both the towns of Paricutin and San Juan de Parangaricutiro.

Most of these small volcanoes are active for only a few years. Cerro Negro in northern Nicaragua is exceptional. After birth in the 1850s, it erupted several more times and again starting in the late 1960s. Cerro Negro may well be a new strato volcano. Time will tell.

Vulcanian-type eruptions, with magmatic compositions ranging from basalt to rhyolite, produce pyroclastic blasts, viscous lava flows, and mudflows or lahars. Vulcanian eruptions, named for a volcano situated on an island near Stromboli, issue fire and smoke and so reminded Romans of the Vulcan's forge. This Italian volcano has provided the name for all mountains that grow by erupting or pushing up domes.

Table 1. Parameters of the Volcanic Explosivity Index (VEI) of Pyroclastic Ejecta

VEI	**0**	**1**	**2**	**3**	**4**	**5**	**6**	**7**	**8**
Description	**non-explosive**	**small**	**moderate**	**moderate-large**	**large**	**very large------------------------------**			
Volume of ejecta in cubic meters	<10(4)	10(4) to 10(6)	10(6) to 10(7)	10(7) to 10(8)	10(8) to 10(9)	10(9) to 10(10)	10(10) to 10(11)	10(11) to 10(12)	>10(12)
Column Height in Km*	<0.1	0.1 to 1	1 to 5	3 to 15	10 to 25	>25--------------------------------			
Qualitative Description	gentle, effusive---	---------	explosive--------------	---------cataclysmic, paroxysmal, colossal---------------					
Classification	----Hawaiian-----	---Strombolian----		---------Vulcanian--------- ---------Plinian---		--------------Ultraplinian------------------			
Troposphere Injection	negligible	minor	moderate	substantial--					
Stratosphere Injection	none	none	none	possible	definite	significant--			
Average Number per year		100/year	15/year	2 to 3/year	1/2 years	1/10 years	1/40 years	1/200 years	1/50,000 years
Example		Stromboli 1996	Uwzen 1991	Nevado Ruiz 1985	Rabaul 1994	St. Helens 1980	Pinatubo 1991	Tambora 1815	Toba 73,500 B.P.

* = for VEI's 0 to 2 use km above crater; for VEI's 3 to 8 use Km above sea level
Modified after Newhall and Self (1982); Decker and Decker, 2006

SB TABLE 1: The basic assumption of the Volcanic Explosivity Index scale (VEI) is that eruption magnitude (mass of erupted material) and intensity (mass eruption rate) are related. But they may not be. This subsequently led to the formulation of Total Magnitude, Magnitude, and Intensity scales. Whereas Total Magnitude is the amount of eruption ejecta, in the Magnitude (M) scale, M = log^{10} (erupted mass in kilograms) - 7. In the Intensity scale, Intensity (I) = log^{10} (mass eruption rate in kilograms per second) + 3. Note in SB Table 2 how some relatively small magnitude eruptions such as the 1980 Mount St. Helens event at M = 4.8 erupted quite intensely (I = 10.3) (Newhall and Self, 1982).

These commonly very large and symmetrical cone-shaped volcanoes are called strato or composite because they are built of layers of different materials (lavas, tephra, and mudflows). Found in subduction zone-type settings, the Cascade volcanoes as well as those in Central America and the Andes of South America are classic examples.

Plinian-type eruptions issue from strato volcanoes, with the largest originating from large calderas. This eruption type is named for Pliny the Younger who described the 79 A.D. eruption of Mount Vesuvius near Naples, Italy. The eruption killed his uncle, Pliny the Elder, who tried to save people. Plinian-type eruptions form from highly vesiculated andesitic to rhyolitic magmas and are the largest and most devastating Earth-generated phenomenon. (Large meteoritic impacts, such as the one that killed the dinosaurs 65 million years ago, release more energy and thus are more destructive.)

As described in detail in Chapter 6, Plinian eruptions result from highly gas-charged, silica-rich magmas that explosively vent pyroclastic material to very high elevations. Tephra or ash cover vast areas, and extremely hot ash flows traveling at hundreds of kilometers per hour can go up to 100 km (60 mi) from the vent follow.

Famous examples of Plinian eruptions include the 79 A.D. blast of Vesuvius (VEI = 6) (see discussion below) that buried Pompeii, and the 1902 eruption of Mount Pelee (VEI = 4) that killed over 30,000 people in St. Pierre on the French Island of Martinique in the West Indies (see Mt. Pelee sidebar). Others include the Tambora eruption in 1815 in Indonesia, the largest volcanic event since 1500, and the more famous but smaller 1883 Krakatau, Indonesia event with a VEI of 7. Orders of magnitude larger eruptions have issued from the Yellowstone Caldera in the U.S., Toba in Indonesia, and several other large caldera systems.[1]

While these eruption characteristics have long been recognized, the actual size of historic and prehistoric eruptions has recently been estimated. One commonly cited scale is the semiquantitative Volcanic Explosivity Index (VEI) developed by Newhall and Self (1982). After compiling the magnitude of volcanic events that occurred between 1500 A.D. and 1980, a VEI was assigned to over 8,000 historic and prehistoric eruptions.

NOTES

1 See Wikipedia for a list of the largest volcanic eruptions.

REFERENCES

Carey and Sigurdsson, 1989; Decker and Decker, 2006; Newhall and Self, 1982; Pyle, 2000.

ERUPTION	TOTAL MAGNITUDE(1)	MAGNITUDE(1)	INTENSITY(1)	VEI(2)	PLUME HEIGHT (KM)(1)	PEAK ERUPTION IN KILOGRAMS/SEC.(1)
Toba, Indonesia 72,500 B.P.	7 x 10(15)	8.8	?	8	?	?
Tambora, Indonesia 1815	2 x 10(14)	7.3	11.4	7	43	2.8 x 10(8)
Taupo, New Zealand 180 A.D.	8 x 10(13)	6.9	12	6	51	1.1 x 10(9)
Novarupta, (Katmai) Alaska - 1912	3 x 10(13)	6.5	11	6	25	1 x 10(8)
Krakatau, Indonesia 1883	3 x 10(13)	6.5	10.7	6	25	~5 x 10(7)
Santa Maria, Guatemala 1902	2 x 10(13)	6.3	11.2	6	34	1.7 x 10(8)
Pinatubo, Philippines 1991	1.1 x 10(13)	6	11.6	6	35	4 x 10(8)
Vesuvius, Italy 79 A.D.	6 x 10(12)	5.8	11.2	6	32	1.5 x 10(8)
Bezymianny, Russia 1956	10(12)	5.3	11.3	5	36	2.2 x 10(8)
St. Helens, USA 1980	1.3 x 10(12)	4.8	10.3	5	19	2 x 10(7)
Augustine, Alaska 1986	6 x 10(10)	3.8	9.8	4?	12	7 x 10(6)

1. Modified from Table 2. Pyle, 2000
2. The Volcanic Explosivity Index (VEI) assumes that mass of erupted material and mass eruption rate are proportionally related. It has been shown that they are not necessarily related

SB TABLE 2: The relationship between magnitude and intensity and VEI of some historical eruptions. Note that Plinian-type eruptions, as one might suspect, have the greatest M and I values (Newhall and Self, 1982; Pyle, 2000).

FIGURE 3.10B: *Looking south from 35,000 feet at the hay stack-looking rhyolitic Mono-Inyo Craters chain of domes and short flows, some only a few hundred years old, located in east-central California. Note the snow-highlighted obsidian-rhyolitic flow at the bottom of the picture (Lange photo).*

This unit overlies the late Pleistocene or ice-age 11,000-year-old ash sourced from Glacier Peak in the North Cascades of Washington. Together these ash layers have helped geologists and archeologists unravel post ice-age events in the Pacific Northwest. For example, at Indian Creek, south of Helena, Montana (800 km northeast of Mount Mazama), archeologists examined sediment beneath the Mazama ash in order to understand how the inhabitants lived more than 8,000 years ago.

Pyroclastic Flows and Surges

Lava flows and tephra eruptions have undoubtedly caused much human suffering—loss of life and property—but not nearly to the extent of pyroclastic flows or ash flows, surges, and mudflows. While the fastest flowing lava, basalt, reaches speeds of 55 km per hour (35 mph), geological warnings

FIGURE 3.10C: *Tephra layering denoted by different particle sizes and/or compositions erupted 3,600 years ago from Santorini Volcano, Greece. Note the sag structure that developed following the impact of a large lithic block (USGS photo).*

such as earthquakes, steam emissions, warming rock, or other indicators commonly precede eruptions. Pyroclastic flows and surges, however, may issue forth with little advanced warning, and they are often extremely hot and of enormous size. Furthermore, pyroclastic flows can travel at 300 km (180 mi) per hour or more and reach more than 100 km (62 mi) from their source.[10]

The first recorded pyroclastic flow eruption was the 79 A.D. blast from Mount Vesuvius. The eruption killed 4,000 of the 20,000 people who lived in the Roman resort towns of Pompeii and Herculaneum near Naples, Italy. Prior to 79 A.D., Mount Vesuvius was a sleeping giant with a 1,281-m (4,203-foot) summit a mere 10 km (6 mi) northeast of Pompeii. Sheep grazed in the grass-covered crater, its slopes contained vineyards and were logged and farmed, while Roman nobility vacationed in nearby Pompeii. The mountain sprung to life on a lovely summer day, August 24, 79 A.D., shortly after a plume of steam rose above its crater.

The volcano then violently erupted tephra as gas-charged material was blasted high into the stratosphere. The eruption was described by Pliny the Younger, 18-year-old nephew of Roman nobleman Pliny the Elder (who died trying to help others escape). Pliny the Younger described the eruption, now termed the Plinian-type.[9]

"A cloud from which mountain was uncertain at this distance, was ascending, the form of which I cannot give you a more exact description of than by likening it to that of a pine tree [Italian stone pines have an umbrella-like top], for it shot up to a great height in the form of a very tall trunk, which spread itself out at the top into sort of branches; occasioned, I imagine, either by a sudden gust of air that impelled it, the force of which decreased as it advanced upwards, or the cloud itself pressed back again by its own weight, expanded in the manner I have mentioned; it appeared sometimes bright and sometimes dark and spotted, according as it either more or less impregnated with Earth and cinders."

Vesuvius erupted for more than 19 hours, producing 9 km^3 (2.2 miles^3) of pumaceous material that varied from white at the base to gray at the top. There were 6 pyroclastic flows and surges, with the eruptive cloud reaching a height of 36 km (118,000 feet). These devastating events followed several hours of tephra fallout that collapsed roofs. When it was over, Pompeii and Herculaneum[11] lay buried, only to be rediscovered in the 18th century by a farmer digging a water well.

Pyroclastic flows are gas-rich and particle-bearing. They erupt from calderas, composite cones, domes of silica-rich dacite and rhyolite, and even

FIGURE 3.11A: *Plinian-type 1991 eruption from Mt. Pinatubo in the Philippines. Note mushroom-shaped gray, tephra-ladened eruption cloud that reached 35 km (113,000 feet) above the volcano (USGS photo).*

FIGURE 3.11B: *Unwelded Bishop tuff, California, showing the unsorted and unstratified nature of this deposit (Lange photo).*

FIGURE 3.11C: *Densely welded tuff from California. Note the flattening of the once light gray pumice into dark obsidian glass (Lange photo).*

FIGURE 3.11D: *Volcanic mudflow deposit on Mt. Rainier, Washington. Note the unsorted, unstratified nature of this type of deposit that looks similar to unwelded Bishop ash flow deposit seen in Figures 3.11B and Figure Intro 5 (Lange photo).*

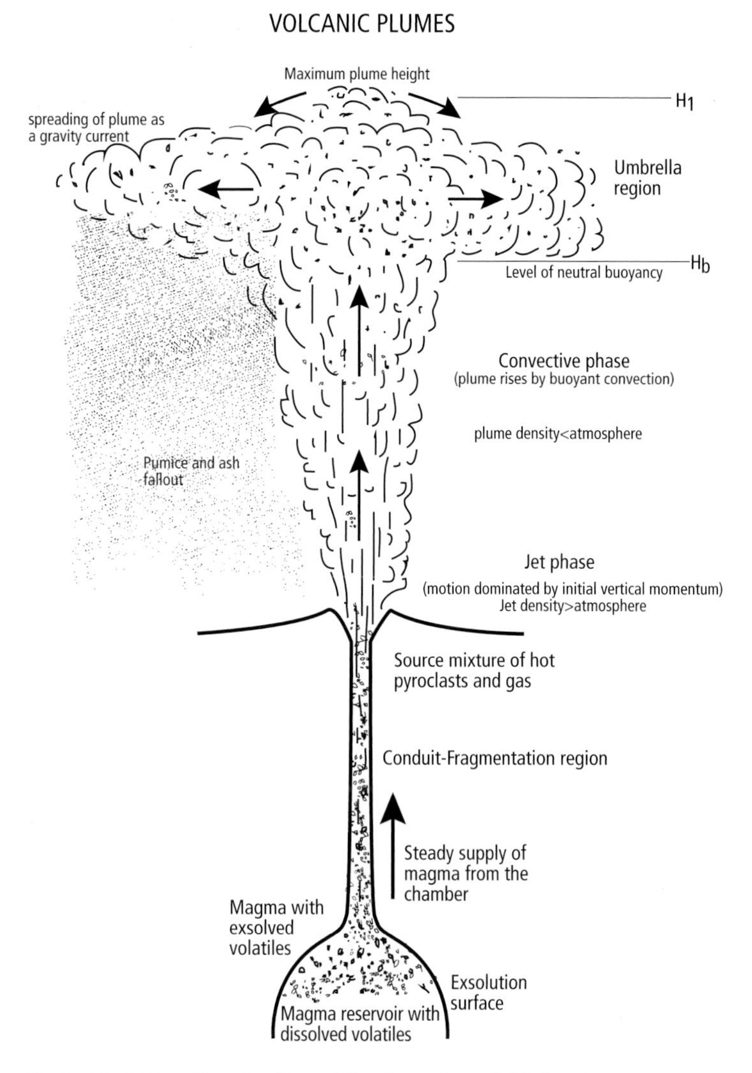

FIGURE 3.12: *Visualization of the formation of Plinian-type eruption. Note the jet or thrust phase below the convecting phase. Tephra fall precedes, and follows eruption column collapse and ash flows which sweep down the sides of the erupting column (modified from Carey and Bursik, 2000).*

from the fronts of slowly moving lava flows.[12] The largest pyroclastic flow deposits, also called ignimbrites, erupt from very large collapse craters or calderas. These giants don't form high constructional features like strato volcanoes, but are fed from huge magma chambers located a few kilometers beneath the surface (see Chapter 6).

After the gas-charged magma erupts, the overlying chamber roof totally to partially collapses into the evacuated chamber, leaving a circular basin or caldera. Some calderas subsequently fill with water, such as has occurred at Crater Lake, Oregon, and Santorini, Greece.

The collapse of high-altitude Plinian eruption columns can cause massive pyroclastic flows, as seen in the great eruptions of Mounts Vesuvius and Mazama. Major pyroclastic flows occurred at Tambora in Indonesia in 1815, Mount Katmai in Alaska in 1912, and El Chichón in 1982 in Mexico.

Smaller, yet equally deadly, are pyroclastic flows generated by collapse of smaller, short-lived tephra eruptions. These include single ash-producing explosions to hours-long events. Altitudes in excess of 10 km (33,000 feet) are achieved by these Soufriere-type eruptions, named for the 1902 eruption of Soufriere Volcano on the British island of St. Vincent in the West Indies.

Mount Mayon in the Philippines, one of the most perfectly shaped composite cones in the world, produces this type of activity. Not surprisingly, large flows have also been observed from Soufriere Hills Volcano on the British Island of Montserrat in the West Indies starting in 1995 and also from Soufriere Volcano on St. Vincent.

Pyroclastic flows may also form from the explosive disruption of growing lava domes. These are called Peléean-type eruptions after Mount Pelée on Martinique, which erupted violently in 1902. The well-documented 1951 eruptions of Mount Lamington on Papua, New Guinea, are other classic examples.

Smaller but locally deadly pyroclastic block flows, also called glowing avalanches, develop during the gravitational collapse of growing lava domes and from the fronts of advancing hot lava flows. These have been named the Merapi-type for the deadly and frequently active Java volcano.

Other famous "Merapi" examples include the 1991 eruption of Mount Unzen, Japan, the 1995 eruptions from Soufriere Volcano on St. Vincent Island, the continuing activity of the Santiaguito domes in Guatemala, and some 1902 eruptions from Mount Pelée that partially filled the valley of Rivière Blanche.

Plinian-type Eruption Development

Plinian-type eruptions commence after volatiles (mainly water and carbon dioxide) build up and pressurize the upper portion of the magma chamber. At some point, (the) pressure exceeds the strength of chamber roof, and the eruption begins through a central volcanic vent and/or through steep fractures that commonly form a circular pattern.

As the magma rises and pressure on it drops, volatiles including water, sulfur dioxide, and carbon dioxide, previously dissolved in the magma at depth, exsolve into a free gas phase encapsulated in glass-walled bubbles.

When the eruption begins, this hot mixture catastrophically vents. Vertically directed Plinian eruptions are composed of a lower gas-thrust phase, and an upper convective thrust phase (Figure 3.12).

The gas thrust phase is dynamically similar to a jet engine, and the thrusted mixture can reach heights of 9 km (30,000 feet). The convective phase rises due to its buoyancy—that is, the very hot gas mixture is lower in density than the surrounding air. This convective phase can reach altitudes of 50 km (164,000 feet) or more.

Either the gas thrust or convective phase may dominate the eruption. The dominant phase depends upon the gas content and composition of the magma, depth of the magma chamber, vent conduit geometry, and degree of fragmentation of the magmatic liquid into particles. Also important is air incorporated into the gas thrust phase. The amount of air can be up to four times the mass of the thrust phase.

What goes up must come down, so when the level of neutral buoyancy is reached, upward convection ceases. The top of the erupting mushroom-shaped column collapses, descending rapidly in pyroclastic flows.

Upon reaching the ground, this very hot gaseous mixture rushes away from the vent at speeds of between 150 and 225 km per hour (90 to 135 mph). The hot gas incinerates the landscape, sometimes also burying it in lithic fragments, pumice, and ash particles.

The flowing pyroclastic mass separates into two phases. One is a ground-hugging mixture of gas, ash, and lithics that rushes downslope under the influence of gravity. This topographically controlled mass fills depressions and valleys. It is a **block and ash flow** if it contains unvesiculated, dense lava clasts. It is a **pumice and ash pyroclastic flow** if pumice-bearing.

The upper, over-riding phase, known as a pyroclastic surge, ash-cloud surge, or nuée ardente,[13] is a less dense, gas-richer, particle-poorer mass. This very hot mixture can sweep over ridges, cross water bodies, and travel more than 100 km (60 mi). Pyroclastic surges hit with hurricane force, and so are capable of unroofing houses and knocking down walls, while incinerating everything in their path, even melting glass. The two pyroclastic phases separate because the larger fragments form a denser conglomeration than the more turbulent, pyroclastic surge mixture flowing above.

The greatest, most recent pyroclastic killer was the May 8, 1902, Mount Pelée eruption on the French island of Martinique in the West Indies. Nuée ardentes engulfed the port city of 30,000 and flowed across the harbor waters, igniting ships and killing sailors. In all, more than 30,000 died, including the governor (see sidebar).

The largest Plinian-type eruptions, such as the 73,500-year-old Toba Caldera eruption in Sumatra, Indonesia, produce pyroclastic deposit volumes of up to 3,000 km^3 (720 mi^3). The Yellowstone system (see Chapter 5) has erupted catastrophically three times. The first produced more than 2,500 km^3 (600 mi^3) of pyroclastic ejecta. The last eruption, over 600,000 years ago, released 1,000 km^3 (240 mi^3) of pyroclastic material. These volumes may be as much as 10 percent of the magma chamber.

The large and destructive pyroclastic eruption of Mount St. Helens in 1980 (see Chapter 4) killed 60 people. Fortunately, Mount St. Helens is far from heavily populated areas. The 1991 Mount Pinatubo eruption was much larger and killed about 300 people. Greater loss of life was prevented thanks to the evacuation of people living on and near the volcano and at Clark Air Base.

How frequently can we expect Plinian-type eruptions? Three to six small ones occur yearly, whereas those with volumes of between 15 and 100 km^3 (3.6 and 24 mi^3) happen perhaps every 100 years. Eruptions of greater than 100 km^3 (24 mi^3) occur every 10,000 to 100,000 years.

Pyroclastic Flow Deposits

Because of the heterogeneous nature of pyroclastic flow mixtures—volcanic ejecta of different sizes and compositions—recognition of these types of deposits is generally easy. However, because mudflow or lahar deposits (discussed next) share some similar characteristics and are found around volcanoes, careful observations are required.

Denser pyroclastic flow-type deposits contain lithic fragments ripped from old flows mixed with freshly erupted volcanic ash and pumice. These

The Tale of St. Pierre and Mount Pelée

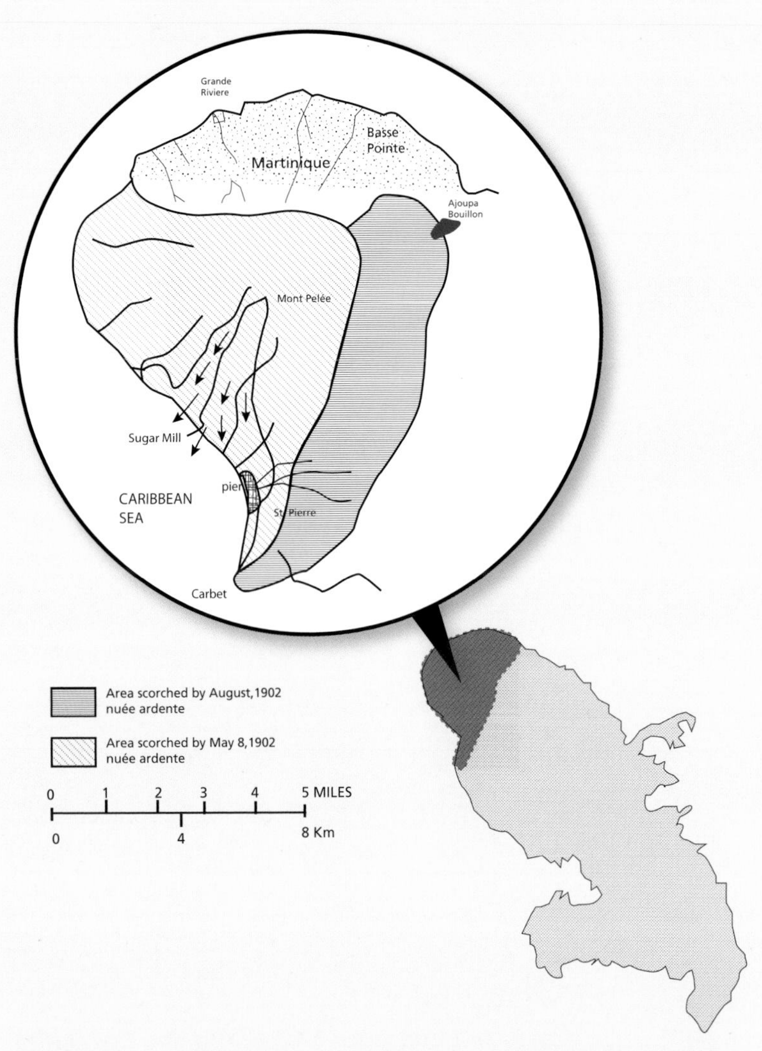

SB FIGURE 1: Map of the island of Martinique showing the locations of Mount Pelée and the city of St. Pierre that was totally destroyed in 1902 (modified from Fisher et al., 1980, Abbott, 2004).

Prior to 1902, St. Pierre was a beautiful city on the northwest coast of the lush tropical French island of Martinique in the West Indies. A sleeping volcano named Pelée sat above St. Pierre. People farmed the slopes, enjoyed the views, and life was good. Prior to 1902, there had been two known eruptions, one in 1792 and the other in 1851, that sprinkled some ash on the city. But life was about to change on paradise island.

The first noticeable activity, steaming fumaroles, began on April 2 and continued until April 23. On April 23, there was a small eruption of ash and sulfurous gases that drifted into St. Pierre. On April 25, explosions and venting of ash commenced and continued until May 5. Meanwhile a governor-appointed commission reported that St. Pierre, located only 10 km (6 mi) from the volcano, was in no danger. However, the frequency of explosions and the roar of venting increased.

Resident doubters thought otherwise, and began closing shops, securing their homes, and leaving. The governor, however, with an important election set for May 10, did not want St. Pierre evacuated. To encourage the citizens that all was well, the mayor and his wife decided to stay in St. Pierre until after the election.

Meanwhile, thick deposits of ash were accumulating on the slopes of Mount Pelée, while sea winds brought heavy rains. The stage was now set for catastrophic events that commenced on May 5.

Between 30 and 150 people were killed when a boiling mudflow (lahar) roared down Riviere Blanche River Valley and destroyed their sugar mill. On May 6, the governor, concerned about voter turn out, ordered the militia to keep citizens from leaving as the ash eruptions, rains, and mudflows from Mount Pelée continued.

May 8, however, dawned clear and sunny, giving hope that the eruption was over. The beautiful early morning, though, was deceiving because magma had blocked the vent during the night of May 7. This allowed the volcano to build pressure that was explosively released at 7:50 that morning.

Witnesses reported four gigantic explosions that produced two great black, tephra-laden clouds. One blew vertically while the other blasted toward St. Pierre. Within two minutes, the city of 30,000 was engulfed in an incandescent, superheated mass of gas and particles. It was subsequently named a nuée ardente, ignimbrite, pyroclastic and ash-cloud surge. Pyroclastic, or block and ash flows, however, did not reach the city.

The ash-cloud surge traveled at more than 160 km per hour (100 mph), unroofing houses, stripping branches and bark from trees, and setting the entire city on fire. All but two ships in the harbor burned as the nuée ardente flowed across the bay. Another nuée ardente struck the city on May 20.

Geologists initially believed that the May 8 and 20 eruptions produced pyroclastic flows and pyroclastic surges due to the collapse of a dome. It is now believed the flows resulted from the collapse of an erupting column back into the crater that then poured southwestward toward St. Pierre through a cleft in the crater rim.

While the resulting pyroclastic deposit was only a few cm thick, the city was destroyed, and all but two of the inhabitants were killed. One survivor was a prisoner in solitary confinement, showing that crime, under the right circumstances, does pay.

Mt. Pelée erupted five more ash flows between June and December of 1902 before activity ceased in 1904.

REFERENCES

Abbott, 2004; Fisher et al., 1980; Young, 1975.

deposits, called ignimbrites or ash flow tuffs, commonly form massive, thick, unsorted, and unstratified conglomeratic sedimentary units. Within a single cooling unit or eruptive phase, there may be one or more pyroclastic flow deposit layers.

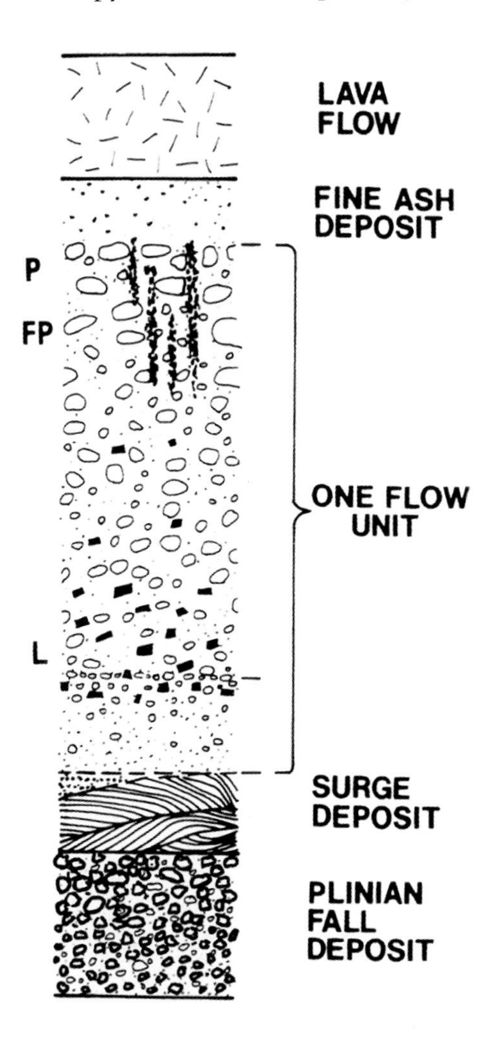

FIGURE 3.13: *Diagram showing the different parts of a complete Plinian eruption depositional sequence that might be found close to a vent. Plinian air fall is overlain by a cross bedded surge deposit followed upward by one ash flow unit, fine air fall ash and possible a younger lava flow. The basal layer of the pyroclastic flow may show inverse grading (coarser particles over finer grained material. Lithic (L) fragments are concentrated near the base whereas pumice (P) concentrate near the top of the unit. FP denotes degassing or fumarolic pipes (modified from Sparks et al., 1973 and Sheridan, 1979).*

Some deposits cover thousands of square kilometers and exceed 100 m (300 feet) in thickness. The thickest deposits occur close to vents and where the flows may pond. Ash flow tuffs occur as single or stacked units.

Some eruptions produce immense pumice rafts[14] that float on the ocean. These can transport plants and animals to new lands. The 1883 Krakatau eruption created such rafts, and some organisms reached Africa's east coast.

Pyroclastic material can be 800°C or more when deposited. Evidence of such heat includes the presence of carbonized wood, a pink coloration due to the formation of iron oxide during emplacement, certain minerals that act as geothermometers, and thermal remnant magnetism.[15]

If emplaced at high temperature, particles may fuse together producing a **welded ash flow tuff** or **welded ignimbrite.** Welded tuff may also form contractional columnar jointing similar to what occurs in basalt and andesite lava flows. (See Figure 3.11C.)

The degree of tuff welding can be estimated by the amount of deformation or flattening of pumice during emplacement. It is determined by measuring the length-to-width ratio of originally equidimensional and porous (vesiculated) pumice clasts. A microscopic examination of pumice reveals that it is essentially all glass, with some containing mineral crystals and/or lithic clasts. Pumice, then is volcanically made glass. As pumice flattens, voids are eliminated, and the pumice reverts to non-vesiculated volcanic glass—obsidian.

Pyroclastic surge deposits are found on top of and laterally away from the denser pyroclastic flows they separated from. These deposits are commonly less than one meter thick, show layering, and may contain unidirectional depositional bed forms, including dune and low-angle cross stratification.

Pyroclastic surges are composed of small particles (crystals, glass or vitric fragments, and lithics) and can erode the ground surface in places without leaving any residue.

Base Surges

Surges or base surges are a type of explosive activity that leaves deposits near vents. The phenomenon was first observed in 1946 following the underwater nuclear explosion that ripped apart Bikini Atoll in the Pacific Ocean. A giant water wave was observed to surge away from the detonation.

The first geologic observations were made in 1965 during a Taal Volcano eruption in the Philippines. Volcanically-derived base surges develop when water and magma (or very hot rock) meet. The ensuing phreatomagmatic and phreatic explosion sends particles rushing away with hurricane force. Deposits from such surges contain vesiculated and and non-vesiculated juvenile fragments, lithics, and crystals. They may be 100 m (330 feet) thick near vents.

Deposits range from massive to bedded in appearance, and may be emplaced either hot or wet and cool. Structures, similar to those found in sedimentary rocks, include laminations, planar bedding, and various sand dune forms (Figure 3.14). Because surge deposits ring their vents, they indicate proximity to ancient vents in old rocks.

Maar–Diatreme Structures

Maar (the German word for lake) are round, dish-shaped, volcanically formed explosion craters hundreds of meters to one kilometer in diameter. These craters generally develop in valleys with high water tables. They result from phreatomagmatic explosions when magma encounters water-saturated ground, and are equivalent to tuff cones that form in water.

Maar craters commonly occur in clusters, and many subsequently fill with water forming circular lakes. A few erupt more than once. Surge and ash deposits ring them, and some have associated cinder cones. Diatremes, or conical bodies of shattered rock, exist beneath maar craters. They may extend up to 2.5 km (1.5 mi) into an irregularly shaped root zone.

Diatremes contain rounded as well as angular fragments of wall rocks, juvenile igneous material, and commonly, wood. The presence of wood demonstrates that material in the diatreme moves vertically, and that material from the surface is swallowed by the diatreme as it develops. Diatremes may also be cut by younger intrusions.

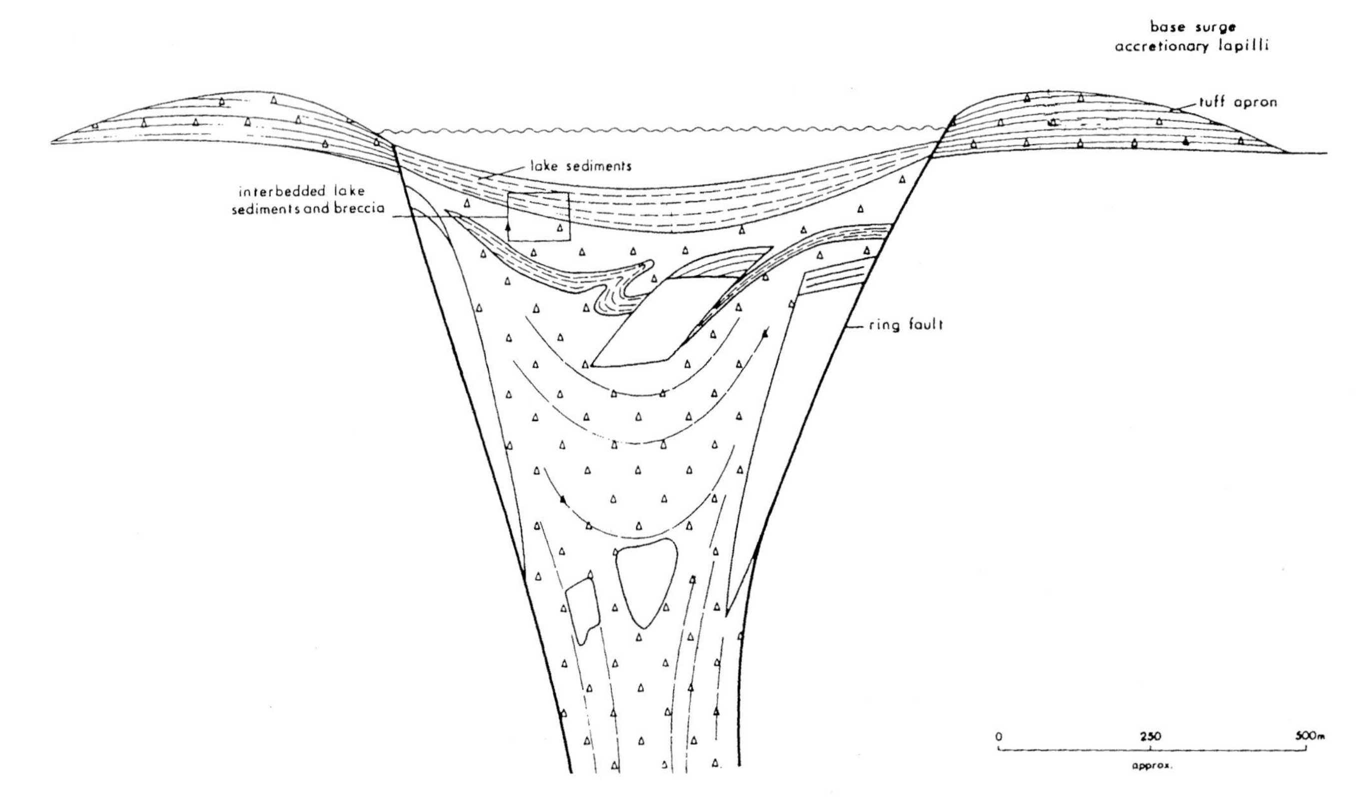

FIGURE 3.14: *Maar/diatreme crater cross section. Note the base surge-tuff "apron" deposits that surround the crater (modified from Baker et al., 1986).*

The envisioned fluidized motion in the pipes abrades and rounds rocks. Maar-diatremes form from all magma types, with basalt being the most common. Furthermore, some diatremes contain diamonds or precious metals such as gold and silver (see Chapter 10).

Maar craters include Diamond Head and Punch Bowl in Honolulu on the island of Oahu, Ubehebe Crater at the north end of Death Valley in California, numerous craters in the Pinacate region of the State of Sonora in northern Mexico, and near Massif Central, France (Figure 3.15).

Lahars (Mudflows)

Lahars, from the Indonesian word for mudflow, are a major volcanic killer. Hot or cold, they issue from vents or develop on volcano slopes. Lahars on slopes form after rapid snowmelt because of volcanic ground heating, heavy rains, or during unusually warm periods. Heavy rains trigger them on snow-free slopes. Lahars move at speeds of up to 50 km (30 mi) per hour. They relentlessly bury everything in their path with a porridge-like material that rapidly sets up like concrete.[16]

In 1985, lahars swept off the slopes of Nevado del Ruiz in Columbia. They killed 22,000 hapless souls in the town of Amero, 45 km (18 mi) from the 5,400-m (17,700-foot) summit. As more people live on mountain slopes and in valleys, the threat to life and property is, unfortunately, increasing.

The Nevado del Ruiz lahars were a mixture or slurry of mud and water that rafted huge boulders, trees, and other debris. The giant, glacier-capped volcano came to life on the stormy night of November 13. Without warning, ash flow eruptions melted ice and snow below the summit, resulting in lahars. The lahars attained speeds of 45 km (28 mi) per hour, an encore to 1845 Nevado Ruiz-generated mudflows that killed over 1,000 people in a much smaller Armero.

An amazing aspect of mudflows is their carrying capacity. Water content varies, but as the amount of mud relative to water increases, so does the carrying capacity. Combine this increased specific gravity[17] with speed and turbulence, and lahars can move cars, houses, railroad engines, and multi-ton boulders.

Geologists have discovered evidence of some really gigantic lahars in recent years. For example,

FIGURE 3.15: *Diamond Head enclosing the Punch Bowl, Honolulu, Oahu, Hawaii (USGS photo).*

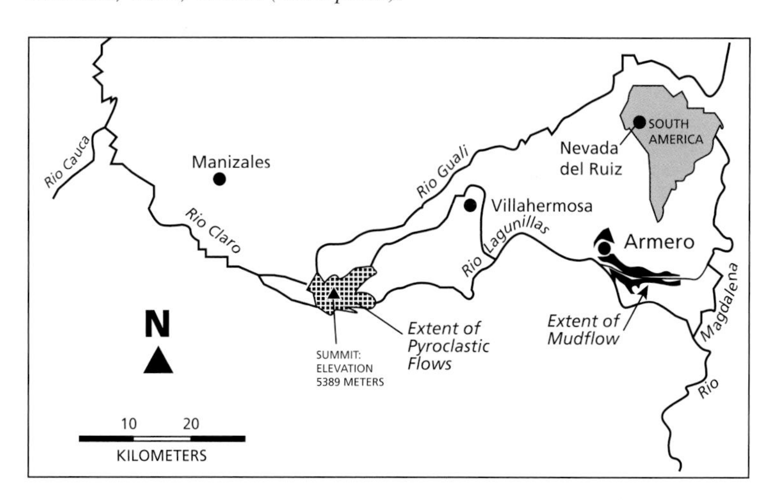

FIGURE 3.16: *Map of the region around Nevado del Ruiz Volcano, Columbia, South America, showing extent of lahars that roared down Rio Lagunilas on November 13, 1985. The city of Armero was partially buried with the loss of 20,000 people (modified from Fisher et al., 1997).*

Mount Rainier in Washington State has produced two monsters. The largest, the 5,700-year-old Osceola flow, swept off the north side of the mountain. Rainier lost about 2,000 feet of its summit in this event. It buried the White River and Green River valleys with up to 30 m (100 feet) of material. The end of the flow reached Kent, 120 km (75 mi) from the summit. Towns including Enumclaw, Buckley, and Auburn are built on this 260-km^2 (100-mi^2) deposit.

The smaller but still catastrophic Electron mudflow roared down the mountain only 500 years ago, burying 48 km (30 mi) of the Puyallup River Valley that drains the west-northwest side of the mountain. Imagine the reoccurrence of such an event today! In addition to the tens of thousands of people living in these valleys, there are now dams

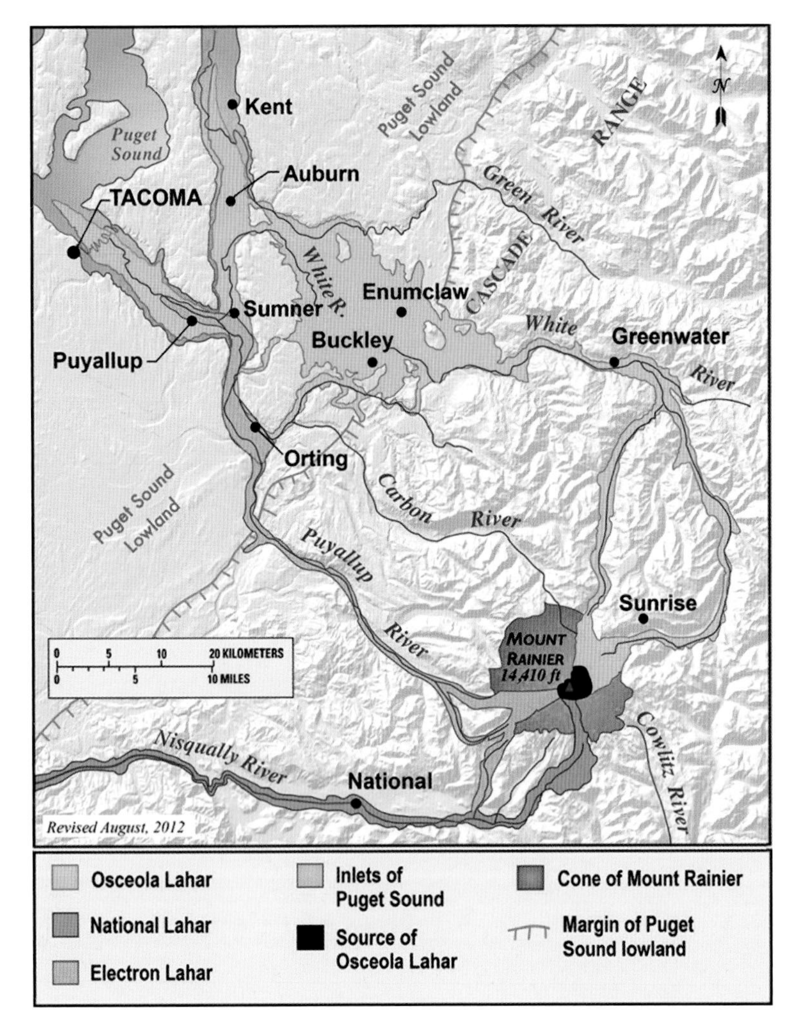

FIGURE 3.17: *Map of the Mt. Rainier region of Washington State showing the extent of the 5,000 year old Osceola and 500 year old Electron mudflows. Those valley regions buried by the mud flows are now inhabited by tens of thousands of people (USGS map).*

and reservoirs. While the dams might not fail, water might wash over them, flooding the unlucky folks living downstream. Unfortunately, mudflows will come careening off Mount Rainier and other Cascade volcanoes again.

Volcano Collapse

A major volcanic hazard, only recently recognized, is partial to total cone collapse. The process is apparently quite common, especially with strato-volcanoes and flow domes. At least 400 volcanoes in different tectonic settings around the world have experienced this fate during the last 10,000 years. Globally, there have been about four per century during the last 500 years.

Central America witnessed at least 40 cone collapses on 24 volcanoes with volumes of at least 0.1 km^3 during the past 10,000 years. The interval between collapses averages 1,000 to 2,000 years, with most known affected volcanoes located in Costa Rica and Guatemala.

The most important reasons for cone collapse are (1) hydrothermal alteration of volcanic rocks, (2) basement rock deformation due to loading by the great weight of the cone, and (3) magma intrusion into the cone.

Hydrothermal (hot water) alteration of volcanic materials decreases rock strength and increases the fluid pore pressure within the volcano. Portions of the cone then "float," leading to slope failure as the cohesiveness of materials diminishes.

The formation of cones also causes loading of, and stress on, basement rocks beneath. The added weight may then exceed the strength of the rocks, causing them to rupture along new or pre-existing faults. This in turn affects the volcanic cone above.

Magma intruding cones, as was clearly documented with photographs during the 1980 eruption of Mount St. Helens, also causes cone failure. Photos clearly show that, before the eruption vented, magma injection caused over-steepening of the northwest side of the mountain, which precipitated a massive landslide. With so much material removed, pressure within the chamber exceeded the strength of the confining walls. The result was an instantaneous, pressure release-type eruption.

One of the largest known terrestrial landslide resulted in the destruction of the ancestral cone of Mount Shasta in California, some 360,000 years ago. The massive volcano, perhaps weakened by hydrothermal waters within the cone, or due to a situation similar to the 1980 Mount St. Helens eruption, collapsed. Forty-five km^3 (10.3 mi^3) of the mountain spread to the northwest, covering 675 km^2 (260 mi^2). This huge debris deposit lies today between the newly rebuilt cone and the town of Yreka, 50 km (30 mi) away. (See Figure 3.18, page 54.)

Much greater land sliding, however, has affected volcanic islands. Classic examples include portions of all the Hawaiian Islands (see Chapter 5), and the Spanish Canary Island Tenerife off the coast of Africa in the Atlantic Ocean. These gigantic slope failures in turn generated massive, destructive tsunamis.

Volcanic Building Materials

Volcanic materials, because of their varied characteristics and availability, have been utilized since the Stone Age. Obsidian—rhyolitic volcanic glass—was used for arrow and spear points, and knives. Beautiful, long Clovis points, for example, may have led to the extinction of great ice-age mammals such as woolly mammoths and mastodons about 11,000 years ago (Lange, 2002).

At the other extreme are cities, housing thousands of people, carved into ash flow tuff[1] or ignimbrite deposits in Asia. The most famous areas include the Cappadocia region of central Turkey where dwellings carved into 20-million-year-old tuff have been inhabited since at least Roman times. Similar habitations also occur in the Caucasus region of Georgia and Armenia.

Tuff is strong, light weight, easy to cut, resistant to weathering, and a good insulator, so people have long used it to build walls and buildings up to several stories high. In fact, 41 large cities, including Rome, Mexico City, and Manila, are partly to totally underlain by tuff,[1] which has been used in their buildings.

The most useful tuffs are those units that are reasonably thick and slightly welded. In these deposits, the glass shards and pumice and rock particles fused together immediately after deposition due to retained heat. (Unfortunately, as the degree of welding increases, so does the difficulty of working the tuff). Alternatively, the particles bonded later due to the precipitation (gluing together) of silica from ground waters that percolated through the deposits.

Addis Ababa, Ethiopia	Medellin, Colombia
Ankara, Turkey	Mexico City, Mexico
Arequipa, Peru	Nagasaki, Japan
Bandung, Indonesia	Naples, Italy
Cuidad de Chihuahua, Mexico	Quito, Ecuador
Durango, Mexico	Rome, Italy
Erzurum, Turkey	Salta, Argentina
Guadalajara, Mexico	San Jose, Costa Rica
Guatemala City, Guatemala	San Luis Potosi, Mexico
Jerevan, Armenia	San Salvador, El Salvador
Kagoshima, Japan	Santiago de Chile
Kumamoto, Japan	Sapporo, Japan
Kumayri, Armenia	Taipei, Taiwan
Managua, Nicaragua	Tashkent, Uzbekistan
Manila, Philippines	Tbilidi, Georgia
Medan, Indonesia	Tegucigalpa, Honduras
	Yogyakarta, Indonesia

(Heiken, 2006)

SB TABLE 1: Cities built on Tuff

Blocks of basalt, andesite, and dacite are less commonly used in building construction but are popular as crushed road and building aggregate. And columnar basalt and andesite have been used for building blocks, and for columns in store and house fronts.

Ash and tuff deposits are also valuable as grit in abrasives, industrial soaps (for example, the workman's delight Lava Soap), and for aggregate in concrete. Coarse ash and volcanic rocks are also finding increased use in gardens and in commercial landscaping.

NOTES

1 Tuff is from the Italian word *tufo*, which means "rock that can be cut with a knife.

REFERENCES

de Rita and Giampaolo, 2006; Funiciello et al., 2006; Heiken, 1979, 2006, Lange, 2002.

SB FIGURE 1: Cappadocia region showing eroded remains of 20-million-year-old ash flow tuff with housing dug into the tuff in the upper left portion of the picture (Lange photo).

SB FIGURE 2: Close up of ancient Turkish housing in the tuff. Some settlements contained up to 5,000 people, plus animals and churches (Lange photo).

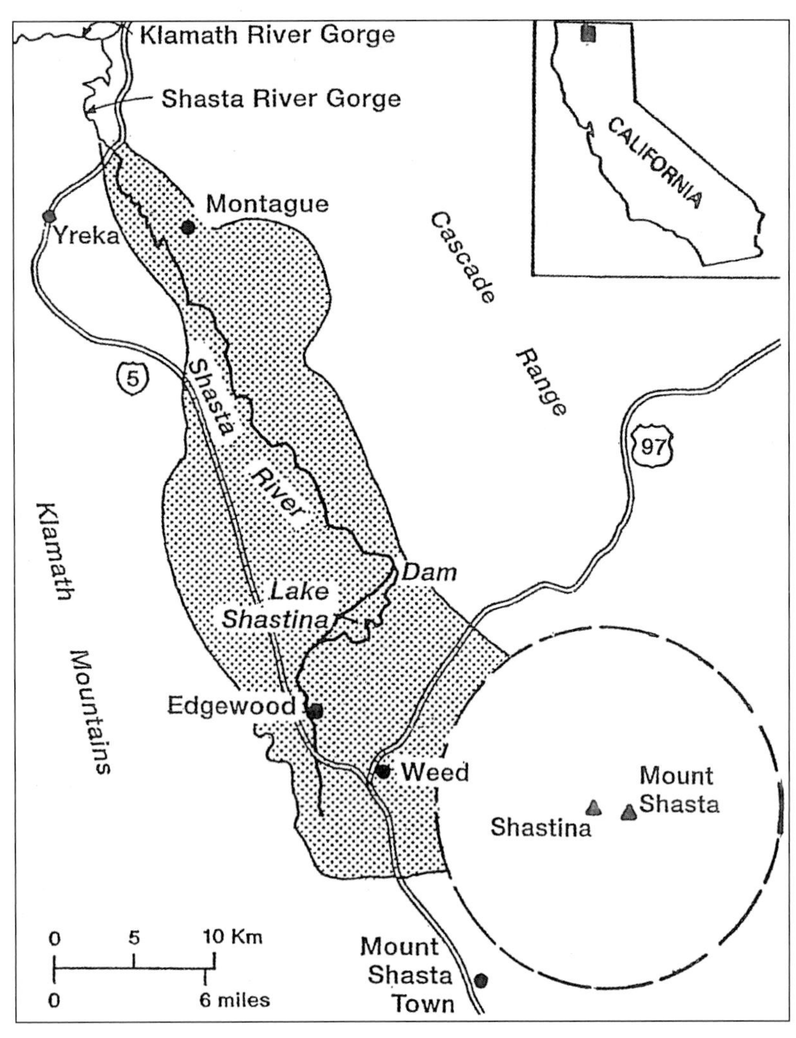

FIGURE 3.18: *Map of 360,000 year old debris avalanche that destroyed ancestral Mt. Shasta volcano in northern California (modified from Crandell et al., 1984).*

Volcano-Generated Tsunamis

Volcano-produced tsunamis, or giant water waves, may be the largest volcano-related structure-destroyer and people killer. People who live along coastal regions near or across oceans from earthquake and volcanic zones live in danger of being swamped by these waves. Tsunamis,[18] which can exceed 100 m (330 feet) in height as they roll onto shore, form by several mechanisms. These include faulting, volcanic explosions, earthquakes, and landslides of sections of volcanoes into the water. Tsunamis in all cases are created by displacements of sea, bay, or lake water.

The most recent killer tsunamis were the December 26, 2004, magnitude 9.2 earthquake-generated waves in the Indian Ocean off of Sumatra, Indonesia, and the March 11, 2011, waves from the magnitude 9.0 quake that devastated the northeast coast of Honshu Island. The Sumatran tsunamis killed at least 250,000 people, and the Japanese tsunamis killed about 25,000 people. These seismically generated waves travel at jet speed but are not a danger to ships at sea. They rise up only as their bottoms begin to drag in shallowing coastal waters. There, the water forms giant waves that may roll inland for more than a mile, destroying everything in their path.

Giant, eruption-generated waves have killed thousands of people and destroyed untold amounts of property throughout history. They are, in fact, responsible for 25 percent of volcano-related deaths. The 1883 eruption of Krakatau in Indonesia, for example, generated several tsunamis during the latter stages of the island-destroying eruptions. One tsunami alone killed 36,000 people as the 30-m (100-foot) wave crashed onto shores along the Sunda Strait that separates Java and Sumatra. Some tsunamis from these eruptions may have formed from ashflows entering the sea.

Active, dormant, and extinct volcanoes can also generate tsunamis when their cones collapse into surrounding waters. The north flank of Tenerife Island in the Canary Islands off the west coast of Africa did this 20,000 years ago. It must have generated huge tsunamis, as 1,000 km^3 (240 mi^3) of the island slid northward. Underwater mapping shows the slide debris covers an area 100 by 80 km (62 by 50 mi). Another slide that large would catastrophically affect Europe, parts of the west African coast, and the east coast of North America.

Landslides of even greater magnitude have affected the Hawaiian Islands. At least 100,000 km^2 (24,000 mi^2) of the Hawaiian ridge are covered by submarine landslides between Kauai and Hawaii (see Hawaii sidebar, page 82). Some debris deposits have lengths of 230 km (145 mi) and volumes in excess of 5,000 km^3 (1,200 mi^3).

Slope failure can occur during all stages of island development, from sea mounts to extinct volcanoes on large islands. Slope failure has resulted in tsunamis reaching 365 m (1,200 feet) above sea level on Lanai and other Hawaiian islands. Since prehistoric times, similar mammoth slope failures have generated huge tsunamis, inundating many coasts surrounding the Pacific Ocean.

Tsunamis also develop from cone collapse into emptying magma chambers, during or following eruptions. This happened 3,600 years ago when Santorini, on Thera in the Aegean Sea, erupted violently and destroyed most of the island and its villages. Tsunamis were generated probably by the explosive eruption, ash flows entering the sea, and by the subsequent collapse of the island into the magma chamber below.

Crete, 115 km (70 mi) to the south and the center of the Minoan civilization, was devastated by Santorini's eruption-generated earthquakes and following tsunamis that racked the northern shoreline. Palaces, ships, and harbors were destroyed, and the Minoan civilization went into decline.

CHAPTER 3 – NOTES

1 Oxygen isotopes are used to determine whether water being emitted from volcanoes is meteoric (from precipitation), or juvenile (derived from deep sources including the mantle, and never having "seen the light of day)." See end note 2 Famous Calderas sidebar, page 104, about O isotopes. Carbon isotopes are similarly used for tracing sources of carbon-bearing compounds.

2 A mineral is inorganic, meaning not now nor ever alive, naturally occurring (not human made), with a crystalline structural arrangement of the atoms of the element(s) composing the mineral. For example, sulfur (S), gold (Au), copper (Cu), and silver (Ag) are each mono-elemental minerals, whereas quartz is composed of the elements silicon (Si) and oxygen (O) in the ratio of one Si to two O, or SiO_2.

3 Basalt, a heavy, black lava, has about 50 percent silica and is richer in magnesium and iron than are granite and rhyolite. It is the most abundant lava on the surface of the Earth (and the Moon).

4 Strain is the deformation of a material, such as magma, under the action of an applied force, in this case shear. But shear force is not necessarily easy to visualize. *Webster's Dictionary* defines shear as "an action or stress resulting from applied forces that cause two contiguous parts of a body [magma in this case] to slide relative to each other in a direction parallel to their contact plane [flow direction in this case]." The San Andreas Fault in California, a transform tectonic plate boundary, also exhibits shear. There, the Pacific plate is sliding northwestward past the North American plate along the contact plane, the vertically dipping and northwest-oriented San Andreas Fault.

5 Viscosity, or a liquid's resistance to flow, is defined as the ratio of shear stress to strain rate, and its value varies by orders of magnitude among liquids (see Figure 3.4). Viscosity was formerly measured in Poises. Now the unit is the Pascal-second (Pa-s), with 10 Poises equaling one Pa-s. For example, compare the flow rates of cold tar, honey, and water. Cold tar is more viscous than honey, which in turn is much more viscous than water, which we know flows very easily. In fact, water is termed a Newtonian fluid because shear stress and strain are proportional. If we compare water with basalt, basalt is about 100,000 times more viscous. And rhyolite is thousands of times more viscous than basalt. Viscosity decreases with increasing temperature, but it increases slightly as pressure decreases.

6 Vesiculation is fragmentation of a magma due to exsolution (separation) of volatiles (fluids, primarily water and carbon dioxide) into a separate phase from a formerly homogeneous, volatile-bearing magma. Exsolution begins when the vapor pressure equals the confining pressure in a magma. The onset of exsolution depends on magma type, confining pressure, and volatile content and composition (see Why Volcanoes Erupt sidebar, page 33).

7 Basaltic eruptions are usually not very explosive unless significant quantities of water contact magma. When that happens, a phreatomagmatic explosion or eruption occurs.

8 Why glass? In order for minerals to form, elements must be able to move through the magmatic liquid and bond to silica structures. This is prevented in liquids of high viscosity, such as in rhyolite, resulting in a frozen and random "collection" of atoms of different elements and silica tetrahedrons. The material is called volcanic glass. Obsidian, even though dark in color, is high-silica glass with a rhyolitic composition.

9 Plinian-type eruptions are named after Pliny the Younger who observed the 79 A.D. eruption of Mount Vesuvius that destroyed the Italian towns of Pompeii and Herculaneum.

10 Temperatures can reach 850°C or more within the erupting cloud.

11 Herculaneum was buried under 20 m (65 feet) of mudflow or laharic sediment several days after Pompeii was destroyed. Fortunately, most residents had already left the city by the time the rains, possibly generated by the Plinian eruption, produced the lahars that swept off the volcano and buried the town.

12 Masaya Caldera near Managua, Nicaragua, is unusual in that it has produced large magnitude Plinian and phreatomagmatic basaltic eruptions.

13 Nuée ardente is a French term that means glowing cloud, as witnessed in the 1902 eruption of Mount Pelée on the French island of Martinique in the West Indies and described by French volcanologist Alfred Lacroix.

14 Pumice rafts from large eruptions, such as Krakatau, are known to have transported both plants and animals great distances, thereby introducing species to different islands and continents.

15 Characteristics of the Earth's magnetic field (direction to the magnetic poles, inclination to the closest pole, and

polarity of the Earth's magnetic field) are frozen into lava as it cools. These characteristics, described in a sidebar in Chapter 2, may allow geologists to determine the age and location of lava at the time it cooled.

16 Lahars (mudflows), like air fall and other sedimentary deposits, have distinctive, recognizable characteristics in outcrops. Because these flows carry material ranging from clay to boulder size, the deposits appear as a mixture of unsorted and unstratified particles (see Figure 3.11D). The greater the ratio of water to mud, the more stratified the resulting deposit appears, and the less capable the flow is of transporting large, heavy objects.

17 Specific gravity is the ratio of the density of one substance to another of the same volume. One cubic centimeter of water weighs one gram, whereas the same volume of granite weighs 2.65 grams. The specific gravity (S.G.), then, of water is 1, and the S.G. of granite is 2.65. If we add mud to water, we can increase the S.G. of this mixture to over 2.0.

18 Prior to about 1960, giant sea waves were called tidal waves even though they are not generated by tidal forces. As the story goes, an American geophysicist at a Japanese geophysical meeting asked his counterparts what they called these giants waves. The Japanese scientists said they call them *tsunamis,* which when translated means "harbor" (tsu) "wave" (nami). "Harbor wave" seems too narrow a term for these monster waves that typically cross miles of open ocean before wreaking havoc on land, but *tsunami* is certainly better than the misnomer *tidal wave,* so the name has stuck.

CHAPTER 3 – REFERENCES

Branney and Kokelaan, 2002: Bullard, 1962; Carey and Bursik, 2000; Carey and Sigurdsson, 1987; Cas and Wright, 1987; Chesner et al., 1991; Crandell, et al., 1984; Decker and Decker, 2006; Delmelle and Stix, 2000; Dragovich et al., 1994; Fisher et al., 1980, 1997; Hon et al., 2003; Francis and Oppenheimer, 2004; Garcia, 2002; Heiken, 1979; Iverson et al., 1998; Lacroix, 1904; Lockridge, 1989; Miller et al., 1982; Moore et al., 1989; Moore et al., 1994; Perez and Freundt, 2006; Peterson and Tilling, 1980; Reid, 2004; Sarna-Wojciciki et al., 1981; Sheridan, 1979; Siebert et al., 2006; Sigurdsson, 2000; Skinner and Porter, 1996; Simkin and Siebert, 2000; Sparks and Walker, 1973; Watts and Masson, 1993; White and Ross, 2011.

CHAPTER FOUR

Convergent or Destructive Plate Boundaries

Introduction to Subduction Zones

An association long known, but not understood until the recognition of plate tectonics in the 1960s, was why volcanoes and earthquakes occur, often together, in long linear or arcuate deformed continental mountain ranges, and in oceanic island arcs. Unknown too was why submarine trenches, the deepest parts of oceans,[1] were located seaward of volcanic arcs, and some within sight of land.

Also puzzling was the significance of the discovery, in Indonesia in the 1930s, that earthquakes tend to occur landward from submarine trenches. Working independently of one another, Japanese seismologist Kiyoo Wadati and American geophysicist Hugo Benioff plotted earthquakes' starting points—their **hypocenters.**[2] The plotted hypocenters followed an incline starting near the surface of the submarine trenches and trending downward beneath neighboring volcanic arcs. They noted that Indonesian volcanoes lie about 100 km (60 mi) above the top of this inclined, earthquake-bearing zone, which became known as a Wadati-Benioff zone.[3] But the full significance of their discoveries would not be recognized for some time.

With the development of plate tectonic theory, we learned how the various "geologic pieces" fit together into a coherent picture. Oceanic lithospheric plates dive or are subducted beneath either continental or oceanic lithosphere forming Andean- and island arc-type systems, respectively.[4] This process results in rock deformation and uplift, earthquakes, and volcanoes. The driving mechanism for these collisions is mantle convection, explained in Chapter 2. Convergent plate boundaries are also called subduction zones or destructive plate margins.

Wadati-Benioff zones were subsequently recognized under other volcano- and earthquake-bearing mountain ranges and arcuate island systems. The descent angle of Wadati-Benioff zones averages about 45°, but it varies from just a few degrees to almost 90°. In the Izu-Bonin-Mariana system in the western Pacific Ocean, the descent angle varies from 50° under Japan to almost 90° in the Mariana Islands to the south.

Furthermore, while the slab descent angle is relatively constant under Honshu (Japan), the Kurile

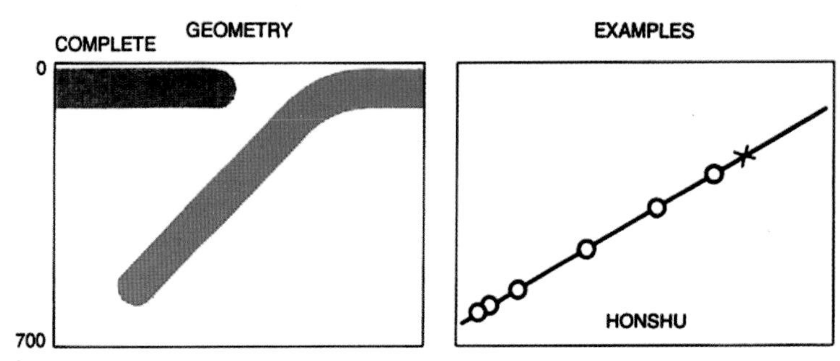

FIGURE 4.1A: *Diagram of a subduction zone showing a slab dipping st a constant angle. This occurs under the Japanese Island of Honshu, the Kurile Islands located north of Japan, and under the Tonga-Kermadec region north of New Zealand (from Toksoz, 1975).*

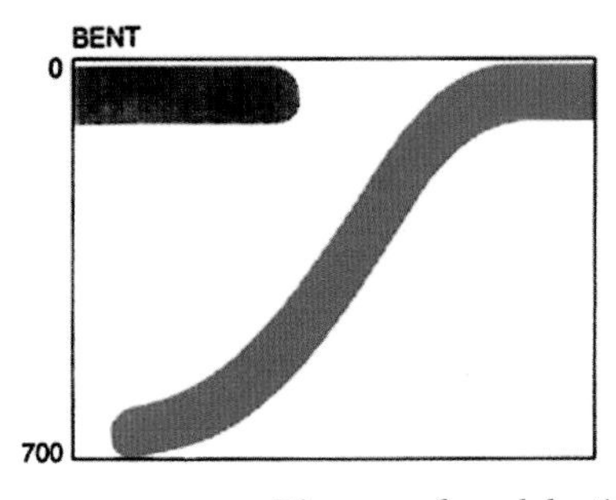

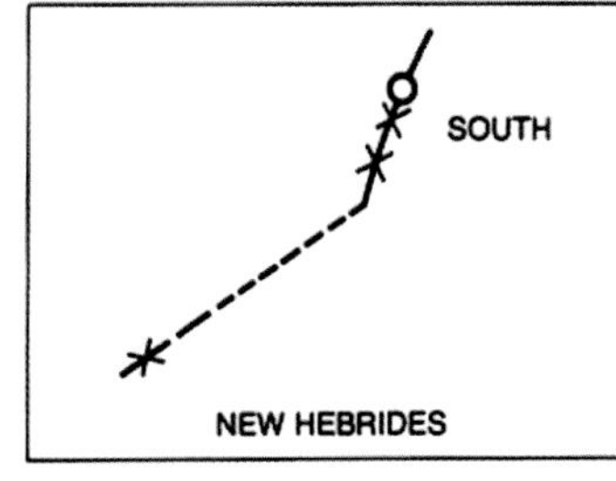

FIGURE 4.1B: *Diagram of a subduction zone with a subducting slab showing a "bent" descending angle. Examples include the New Hebrides Islands in the South Pacific Ocean (from Toksoz, 1975).*

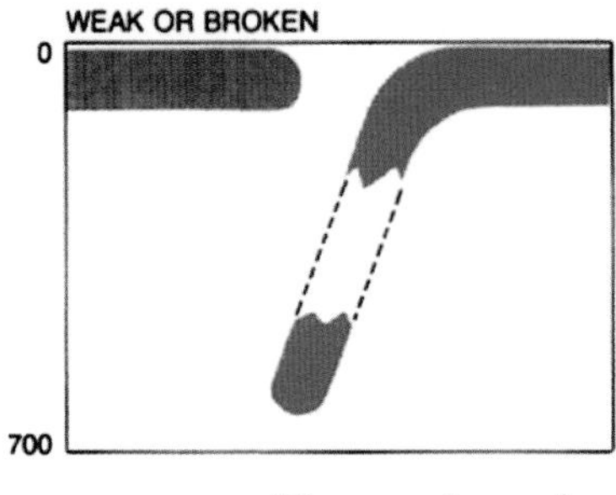

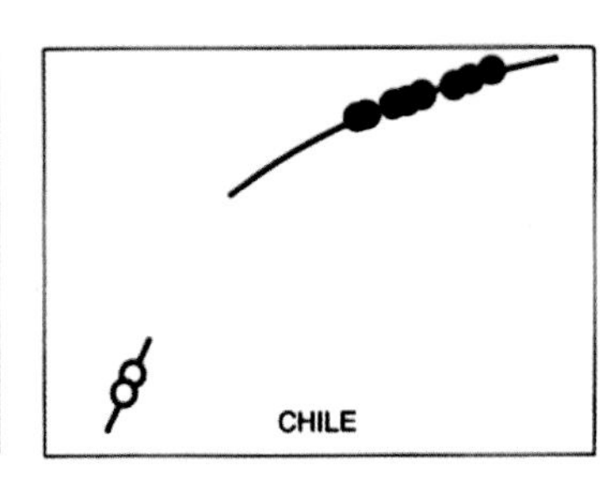

FIGURE 4.1C: *Diagram of a weak or broken descending slab as occurs under parts of Peru and Chile (from Toksoz, 1975).*

(Russian Far East), and the Tonga-Kermadec (north of New Zealand) (see Figure 4.1A), the dip varies with depth (see Figure 4.1B) in other locations such as under the New Hebrides. And the slab descends in pieces under Central America and Chile (see Figure 4.1C).

Descent angles tend to be steeper when subduction rates are 6 cm or greater per year and also when the overriding plate is thickened by compression or by accumulation of material coming off the descending slab (known as underplating). Steep descent angles also develop when oceanic lithospheric plates are more than 15 million years old because, by that age, they have cooled and are therefore denser than younger oceanic lithosphere. This is found in subduction zones far from spreading centers, such as in the western Pacific Ocean. Descent angles approaching 90° occur under the Mariana Islands and in the western Pacific Tonga and Kurile systems. In these "old and cold" convergent systems, the leading edge of the slab may reach depths of 1,000 km.

The descent angle may be low (<30°) if the distance between the spreading ridge and trench is small. In this case, the subducting plate would be more buoyant (less dense) due to its young age and therefore warmth. Part of the Peru-Chile system has this characteristic.

Subduction zone earthquakes are generated by friction between the subducting and overriding, rigid lithospheric plates. The subducting plate warms as it dives into the hot mantle, resulting in earthquakes more centrally located within the subducting, still partially rigid slab. When the slab reaches depths of about 700 km (420 mi), it becomes too warm and plastic for frictionally generated earthquakes.

In the Aleutians, Sumatra-Java, and Central America, deep earthquakes do not occur because subduction is so recent that their slabs have yet to reach great depths. The Mediterranean region also lacks deep earthquakes, but here it's because subduction occurs so slowly. The slab is heated and assimilated into the mantle at shallow depths.

Subduction Zone Magma Generation

The distance between volcanoes and the top of the subducting lithospheric slab generally varies between 100 and 125 km (60 to 75 mi) but can be anywhere from 80 to 270 km. In this zone, water is continually liberated from the descending, water-saturated slab, which in turn lowers the melting temperature of the intruded mantle. The melting mantle forms magma (see Chapter 8 for Magma Generation).

At least five mechanisms heat the descending slab. They include (1) heat flowing into the slab from the surrounding, warmer mantle, (2) heat released during radioactive decay of, primarily, uranium, thorium, and potassium isotopes, and other unstable isotopes of elements within the slab, (3) heat released within the slab as minerals change into more dense and therefore stable mineral phases as pressure increases, (4) frictional heat generated between the down-going slab and surrounding rocks, and (5) increasing pressure on the slab. Slab heating is also enhanced by the more efficient transfer of heat into the slab from the mantle, as the descending slab grows hotter.

Once the slab penetrates 1,100°C rock, the water liberated from the slab causes partial melting (up to 25 percent) of the mantle. Melting occurs because water lowers the melting temperature of this rock—peridotite[5]—by 200°C to 300°C. Basalt magma thus forms. This basalt is less dense than surrounding rocks, so it rises through fractures toward the surface. As magma rises, it encounters and may melt rock that is richer in silica and with a lower melting temperature. This includes granite. Melting of these rocks contaminates the basalt, especially with silica. Andesite is one possible product of this process (see Table 3.2 - Common extrusive and intrusive magma types, page 31).

Magmas: Extrusive and Intrusive Varieties

People enjoy collecting things. We have butterfly, coin, button, fossil, and stamp collections in addition to car, plant, and comic book collections. Some people collect minerals and rocks, distinguished by their different major element chemistries and characteristics. These minerals and rocks are given names: minerals are commonly named for people, igneous rocks for places (See Table 3.1 - Chemistry of the Mantle, page 31).

Island arc volcanoes erupt basalt and andesite. Andesite, named for the Andes Mountains of South

America, contains 58 percent silica in contrast to about 50 percent for basalt. Andesite is most abundant in Andean, subduction-type settings. It also occurs with dacite and silica-rich rhyolite, the fine-grained and/or glassy equivalent of intrusive granite.

When magmatic rocks form deep within the Earth, they cool slowly, allowing time for mineral crystals to form. These are totally crystalline and are called intrusive rocks. Magmatic rocks that crystallize at the Earth's surface are called extrusive. Intrusive rocks form the backbone and bulk of Andean arcs.

Batholiths

Granite, which is totally crystalline (containing only visible mineral crystals and no glass), forms because sufficient time is available at depth for complete mineral formation (crystallization) from magma. When volcanism ceases in Andean systems, erosion dissects the mountains, exposing crystalline rocks. Gabbro is the intrusive equivalents of basalt; diorite is the intrusive equivalent of andesite.

A great place to see subduction-type granites is in

FIGURE 4.2A: *3,000 m (10,000 feet) of exposed Sierra batholith Granite, looking north in Kings Canyon National Park, California (Lange photo).*

FIGURE 4.2B: *Dark brown areas are sedimentary rock roof pendants not yet eroded away by glaciers and water erosion from the top of the Sierra batholith (light gray) which intruded the sedimentary rocks, Mineral King area of California (Lange photo).*

FIGURE 4.2C: *Different phases of granite: coarse- and fine-grained varieties from different localities and with different mineral proportions and chemistries (Lange photo).*

FIGURE 4.2D: *Porphyritic granite found on Mount Whitney in the Sierra batholith, California. Large pink potassium feldspar mineral crystals are phenocrysts, or crystals larger than other mineral crystals that exist within a finer grained groundmass of feldspar, quartz, and biotite (black) minerals. Phenocryst crystals started forming earlier than the smaller minerals surrounding them (Lange photo).*

the Sierra Nevada massif of east-central California. The massif is 650 km (400 mi) long, 100 km (60 mi) wide, and up to 30 km (18 mi) thick. Huge exposures of granite form the rugged landscapes of Yosemite, Kings Canyon, and Sequoia National Parks (Figure 4.2A).

Between about 220 and 75 million years ago, this region of California contained huge magma chambers that fed mostly andesitic volcanoes. These volcanoes were located high above the magma chambers on roofs of sedimentary and volcanic rocks. After subduction stopped off the coast of central and southern California,[6] volcanism ceased, and the magma feeding the volcanoes cooled and crystallized into intrusive rocks of various compositions. With the end of volcanism, roughly 75 to 80 million years ago, erosion by running water and most recently by glaciers has stripped away thousands of feet of sedimentary, volcanic, and granitic rocks, exposing 40,000 km^2 (15,450 mi^2) of this huge intrusive massif.

In places, however, one can still see some of the rocks that roofed the crystallized rocks below. The Mineral King area in the southern Sierras is a classic example (Figure 4.2B). Volcanic pipes cutting vertically through granite that long ago fed the volcanoes high above are also apparent. An examination of the granitic outcrops reveals that (1) the granite varies in mineralogy and composition (Figures 4.2C, 4.2D, 4.3A), and (2) these phases show cross-cutting relationships. In Yosemite Valley, at least 10 visually different intrusive phases, ranging in age from about 225 to 80 million years ago, are exposed.

Another geological aspect of these massifs is that tens of millions of years of intrusive and extrusive igneous activity can be seen. The result is a complex assemblage of rocks that have been intruded, faulted, and in places metamorphosed.[7] And all of this is related to the subduction process.

Because the exposure of Sierra granite exceeds 104 km^2 (40 mi^2), it is called a **batholith**, the Sierra Nevada batholith. Exposures of intrusive rocks of less than 104 km^2 (40 mi^2) are named **stocks.** The Sierra batholith is more accurately termed a composite batholith because it contains many individual intrusions or **plutons** of various sizes, compositions, and ages (Figures 4.2C, 4.2D, and 4.3A).

Root Zones of Batholiths

In many destructive plate localities, especially where subduction ceased long ago, uplift and erosion have exposed the root and deep lateral zones of batholiths. Two distinctive features are found in these once deep crustal regions: swirly masses of dark and light-colored rock termed a migmatite, or mixed rock (Figure 4.3B), and cross-cutting dike swarms (Figure 4.3C).

A good place to see both of these is along the North Cascades Highway (State Route 20) in northwest Washington State. The granitic dikes were derived from intruding batholiths and stocks. Here they cut each other in addition to cutting the highly

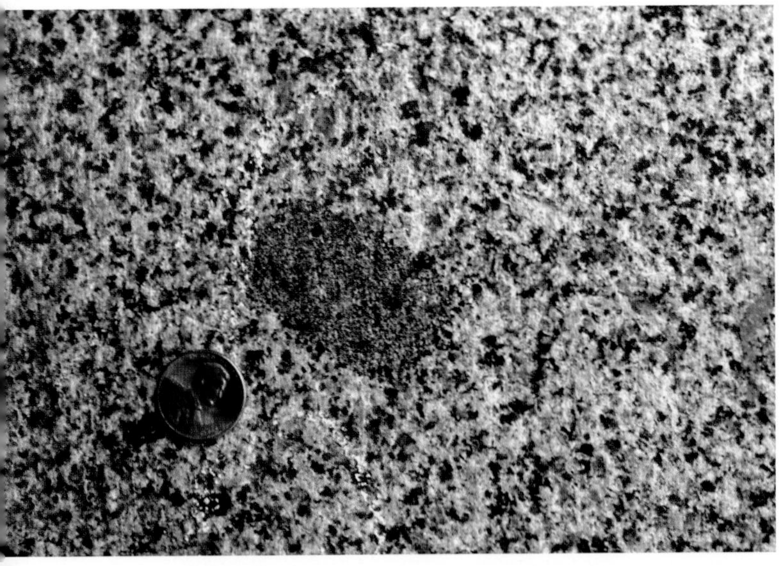

FIGURE 4.3A: *Dark area is an included xenolith or exotic rock fragment which fell into, and was partially digested by the granite magma, and then "frozen" into the intruding granitic magma (Lange photo).*

FIGURE 4.3B: *Migmatite or mixed rock of granitic (light-colored) and more basic (less silica but more magnesium- and iron-bearing igneous rock) exposed on the North Cascades Highway, Washington (Lange photo).*

FIGURE 4.3C: *Light-colored granitic igneous dikes cross cut older, gray-colored batholithic rocks. Note the offsets (small-scale faulting) of some of the dikes. North Cascades Highway, Washington (Lange photo).*

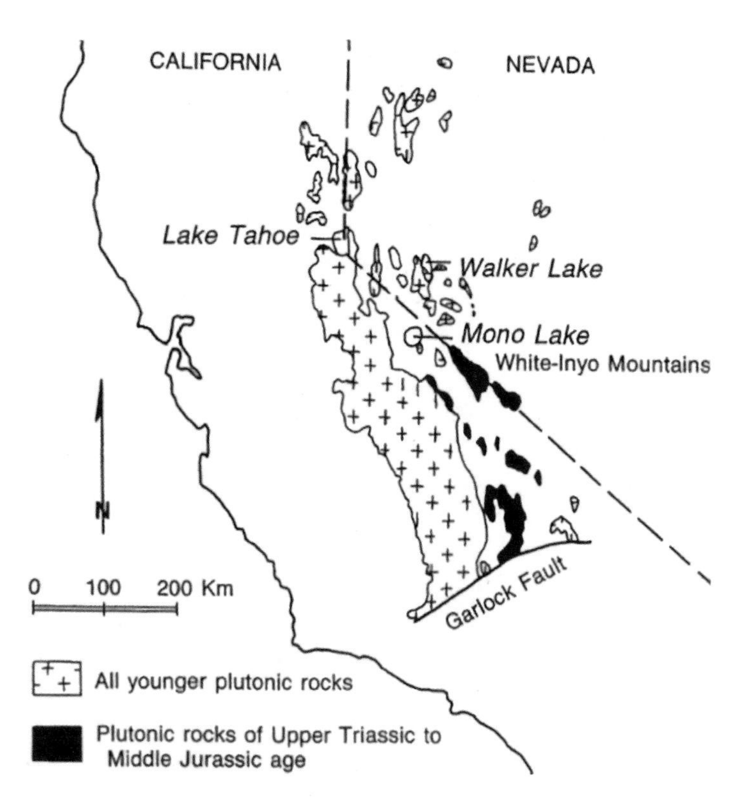

FIGURE 4.3D: *Map of Sierra batholith (> 40 square miles in area) granitic rocks and granitic stocks (< 40 square miles) (modified from Raymond, 2004).*

metamorphosed[7] and intruded older country rock surrounding the dikes. Note that some dikes cross-cut other, older dikes.

In this case, erosion exposed the formerly very deep, high-temperature, high-pressure parts of a batholitic system formed during subduction. Under these conditions, rocks behave plastically, like toothpaste, rather than exhibiting brittle behavior that occurs under lower pressures and temperatures.

While the light-colored rock is granitic, dark rocks are highly metamorphosed and intruded country, or original, rock. These may have originally been sedimentary, metamorphic, or igneous. Note that small displacement faulting (less than one meter) subsequently offset parts of the older migmatite. (Figure 4.3B) Because faulting results from brittle behavior, the offsets seen in Figure 4.3C probably developed later, following uplift when these rocks were closer to the surface.

Outcrops of intrusive rocks are sparse between central California and the North Cascades of Washington State. Granitic stocks and batholiths are prevalent again from northern Washington to southern Alaska. Exposed intrusive rocks are rare to the south because subduction is still occurring off the northwestern U.S. coast (See Cascades sidebar, page 62, for details and a map).

Batholiths no doubt exist under the volcanic cover of the middle Cascades. They will be exposed after subduction and volcanism stop and erosion has removed the overlying rocks. Subduction will cease when the North American plate finally overrides the spreading center located offshore.

The same situation exists in the Andes Mountains of South America where volcanism will continue until the South American lithospheric plate overrides the East Pacific Rise spreading center.

Back to Subduction Zone Volcanism—Volcanic Arcs

Both Andean and island arc-type subduction systems are generally bow-shaped or **arcuate** in map view, with volcanoes aligned roughly parallel to the oceanic trenches. The volcanic arcs are also segmented. In the Aleutian and the Cascade systems, segments contain between 2 and 12 volcanoes. Furthermore, the spacing between volcanic centers is quite consistent. For example, Aleutian volcanoes are spaced between 50 and 70 km (30 and 42 mi) apart, but the spacing between volcanoes that produce unusually large amounts of lava is greater than those that produce less lava. Linear segment ends are commonly offset from each other. Offsets, which are transform faults, may be related to the geometry of the subducting, rather than the overriding, plate. Old basement fractures may also control offsets.

The Cascade Range

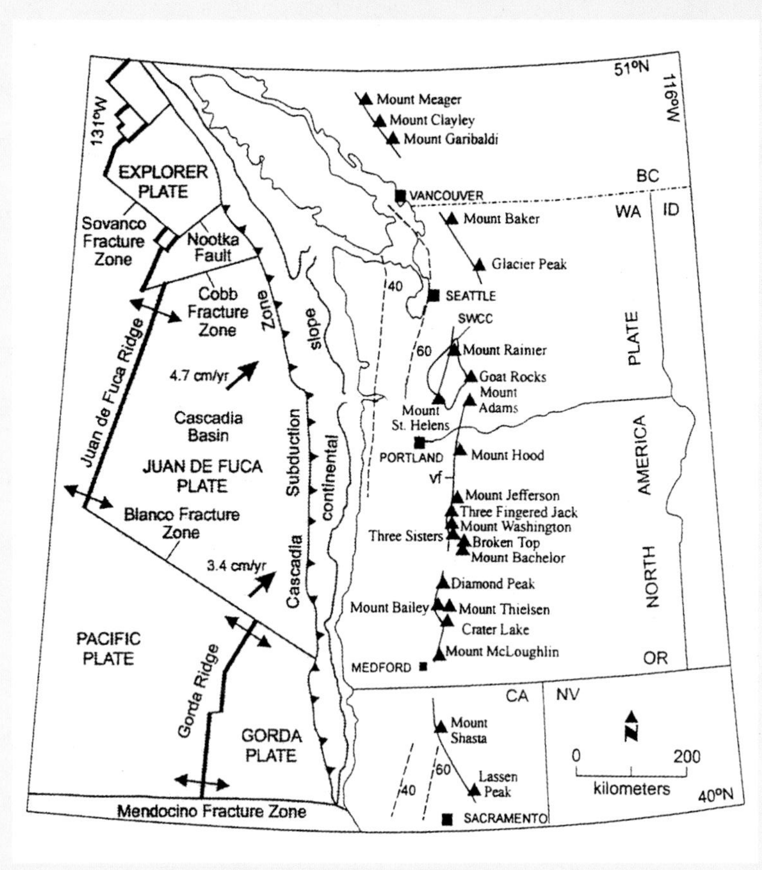

SB FIGURE 1: Simplified map of the Cascade volcanic arc, North American continental margin, and the spreading center responsible for the arc off the Pacific coast. Note the direction of spreading, angle of convergence, and convergence rate with the North American plate that is moving to the southwest. Also note the segmentation of the volcanic arc (modified from Hammond, 1998).

Characteristics

The 1,250-km-(775-mi)-long Cascade Range, extending from the Mount Meager-Garibaldi complexes in southern British Columbia, Canada, south to Lassen Peak in northern California, exhibits most Andean-type subduction zone attributes (SB Figure 1). They include faulting, seismicity, intrusive and extrusive activity, uplift, and metamorphism, but not a prominent seaward trench.

The volcanic arc is the product of at least 40 million years of convergence of the overriding North American plate moving southwestward with a vestige of the East Pacific Rise. Subducting from south to north are the eastward moving Gorda, Juan de Fuca, and Explorer oceanic plates. The western edges of these oceanic plates grow at the segmented spreading centers, located between 250 km and 350 km (155 and 220 mi) off the Pacific coast. Eventually, however, these small plates will be totally consumed, thereby halting subduction, when the North American plate finally overrides the spreading ridge complex. The spreading center is a northern Pacific Ocean vestigial remnant of the East Pacific Rise (see Chapter 2) located to the south and off the west coast of Mexico and Central and South America.

The North American-Pacific plate boundary in this region, following consummation, will become part of the transform fault systems located both to the south (San Andreas fault) and north (Fairweather or Queen Charlotte fault). The North American and Pacific plates will slide by each other along the entire West Coast.

Within the Cascade Range are 20 major volcanic centers, active for tens of thousands to many hundreds of thousands of years. These complexes contain volcanoes that vary from simple basalt and basaltic andesite cinder cones[1] that generally erupt only once, to complex composite cones that witness multiple eruptions, to massive calderas (see SB Figure 1).

Compositions of magmas vary from basalt to rhyolite, with the bulk of the volcanism being of andesitic and dacitic composition. Eruptions vary from non-explosive to cataclysmic, as evidenced by the 1980 eruption of Mount St. Helens, and the far larger Mount Mazama eruption 6,850 years ago that formed the Crater Lake Caldera.

The Cascade Range of snowcapped volcanoes, some active, many dormant, and some extinct, forms a single chain. With a maximum width of 50 km (30 mi), it is one of the Earth's narrowest volcanic arcs. The initially rapid subduction rate of 8.4 cm per year now varies along the front from 3.4 cm to 4.7 cm per year.

The arc (SB Figure 1) is composed of 6 distinct linear segments. They vary in length from 110 to 240 km (70 to 150 mi). Segment boundaries—faults—are parallel to the direction of plate convergence.

Seismic studies show that the top of the subducting plates at both the south (California) and north (southern British Colombia) ends of the arc have dips that shallow to 20° between depths of 45 and 60 km (28 and 37 mi), and maintain that angle until at least 80 km (50 mi) depth.

Geochemical studies of basalt suggest that from west to east near the southern part of the range (from Mount Shasta to Medicine Lake), an increased depth of magma generation takes place. Basalt is generated in the west at 36 km (22 mi), whereas 75 km (47 mi) to the east the depth is 66 km (41 mi).

Cascade volcanoes are classified into three types based upon their magma compositions and eruptive style, and appear related to their segment position. Large volcanoes such as Mounts Rainier, Baker, and Hood, located within the segments, are coherent volcanoes because they produce lavas with only a small compositional range—porphyritic andesite and basaltic andesite. While capable of producing large amounts of lavas and mudflows, these structurally uncomplicated volcanic cones rarely erupt violently.

Glacier Peak, Mount St. Helens, Mount Adams, Mount Mazama (Crater Lake), and Mount Shasta, however, have explosive histories and are called divergent-type volcanoes. Located on or near segment boundaries, they erupt a compositionally wide range of magmas—porphyritic basalt, andesite, dacite, and rhyolite. These volcanoes are more structurally complex than the coherent type, having both central vent and dome eruptions, plus numerous associated basaltic cinder cones.

CASCADE VOLCANO TYPES*

COHERENT** (porphyritic andesite)	DIVERGENT** (basalt to rhyolite) explosive	DIVERGENT BEHIND THE VOLCANIC FRONT** (basalt and rhyolite) explosive
	Mt. Garibaldi, B.C. 8,787′	
Mt. Baker, WA 10,778′		
	Glacier Pk, WA 10,451′	
Mt. Rainier, WA 14,410′		
	Mt. St. Helens, WA 8,000′ (9,677′)	
	Mt. Adams, WA 12,276′	Simcoe Volcanic Field, WA
Mt. Hood, OR 11,245′		
Mt. Jefferson, OR 10,495′		
	Three Sisters Volcanoes, OR 10,358′ to 10,047′	Newberry caldera volcano, OR (shield volcano)
	Mt. Mazama (Crater Lake)	
Mt. McLoughlin, OR 9,493′		
	Mt. Shasta, CA 14,161′	Medicine Lake Highlands, CA (shield volcano)
	Lassen Peak, CA 10,457′	

**= after McBirney, 1968; Hughes et al., 1980*
***= listed north to south*

SB TABLE 1: Cascade volcano types: coherent, divergent, and divergent behind the volcanic front.

The third Cascade volcano type, also called divergent, is located near segment boundaries but behind or east of the main, high volcanic front. These are large, bimodal volcanic systems that erupt primarily basalt and rhyolite. The eruptions issue from calderas with huge magma reservoirs. They do not build high volcanic edifices. From north to south they are the Simcoe Mountain volcanic field in southern Washington, the Newberry Shield Volcano in west-central Oregon (a national monument), and the Medicine Lake shield volcanic field in northern California.

Future Cascade Eruptive Activity

What type, and frequency of future eruptive activity is expected from the Cascade Range? Geological field work and age dating done around the volcanic centers, especially since the 1960s, show the hazardous products to be lava and tephra eruptions and the potentially more dangerous pyroclastic flows, lahars, debris avalanches, and cone collapses.

Since the end of the ice ages about 10,000 years ago, at least 450 lava flows have occurred. Several per century, with some capable of going 10 km (6 mi) or more from their vents, might erupt. Even more tephra eruptions are expected, with some merely producing slight wisps of ash, while others might cover thousands of square kilometers with tephra. With the general prevailing wind direction from west to east and northeast, areas to the east and northeast can expect the brunt of the fall out, as occurred with the 1980 Mount St. Helens eruption.

In the last 15,000 years, there have been at least 325 pyroclastic flows. These dense mixtures of hot gases and rock fragments can travel tens of kilometers down valleys and over ridges, as did the flows from Mount Mazama that went 60 km, destroying or burying everything in their path.

Pyroclastic surges of dilute clouds of rock debris in gases that move at high velocity may also happen. For example, surges from the 1980 Mount St. Helens eruption traveled 28 km (17 mi), destroying almost 600 km^2 (230 mi^2) of vegetation.

Lahars, or water-saturated mixtures of rock debris ranging from hot to cold, are also well represented in the Cascade Range geological record. They constitute one of the gravest threats to property and people living in valleys near the mountains because of (1) their frequency of occurrence, (2) speeds of up to tens of kilometers per hour, and (3) the snow and ice available for rapid melting on the mountain slopes. The remains of at least 445 historical lahars have been discovered. Included is the giant 5,700-year-old Osceola flow from the north side of Mount Rainier that traveled 120 km (75 mi) from the volcano. It reached Puget Sound and buried 260 km^2 (100 mi^2) with up to 30 m (100 feet) of debris.

Also of major importance to life and property are debris avalanches and cone collapses. These sudden, high-speed releases of material result from structural failure of volcanic cones. Set off by earthquakes or eruptions, these gravity-driven masses destroy everything in their path. The most recent example was the 1980 slope failure of Mount St. Helens. But one of the largest known landslides destroyed the ancestral Mount Shasta 360,000 years ago. With a volume of 45 km^3 (10.75 mi^3), it traveled 64 km from the cone and buried 675 km^2 (260 mi^2) north-northwest of the former peak (see Mount Shasta sidebar).

Cascade sidebars of Mounts Rainier and St. Helens are also within this chapter. Mount Mazama (Crater Lake) is discussed in Chapter 6.

NOTES

[1] More than 1,000 cinder cones are found between Lassen Peak and Mount Rainier.

REFERENCES

Hammond, 1998; Harris, 1990; Hildreth et al., 2003; Hughes et al.,1980; McBirney, 1968; Miller, 1990; Sherrod, 1989; Swanson et al., 1989; Tanton et al., 2001.

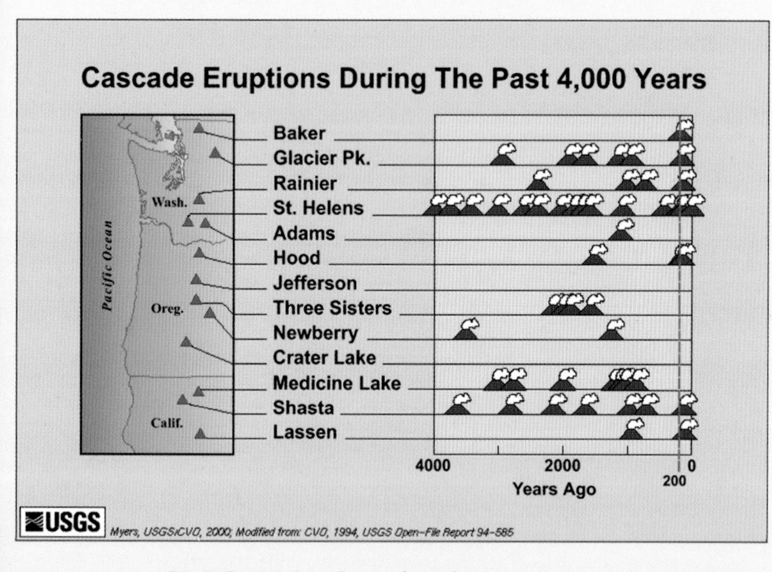

SB FIGURE 2A: Cascade volcano eruptions during the last 4,000 years (USGS diagram).

SB FIGURE 2B: West flank of 3,427-m (11,245-foot) Mount Hood, Oregon (USGS photo).

SB FIGURE 2C: East side 3,185-m (10,451-foot) Glacier Peak, North Cascades of Washington (USGS photo).

SB FIGURE 2D: Looking north at an obsidian flow at the base of 3,157-m (10,358-foot) South Sister near Bend Oregon. Middle and North Sister Peaks are in background (USGS photo).

In addition to Andean (continental margin) and island arc systems, some destructive plate boundaries are composites of the two. The Aleutian volcanic arc is a classic example (Figure 4.4). The inner, peninsular part out to Unimak Island is Andean because the Pacific oceanic lithosphere is subducting under North American lithosphere. Southwest of Unimak Island, the system is an island arc-type because the Pacific plate is diving beneath an oceanic extension of the North American-Asian plate.

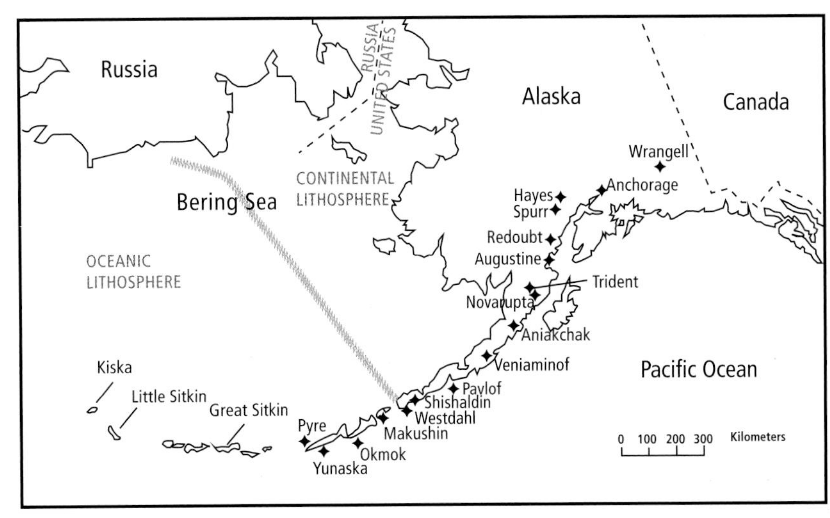

FIGURE 4.4: *The composite Aleutian subduction zone having an Andean component south westward from Anchorage to Westdahl volcano. The subduction zone then becomes an island-arc system further to the southwest (oceanic lithosphere subducts under oceanic lithosphere) (modified from U.S. Geological Survey Circular 1073).*

Other differences between destructive plate margins relate to the speed of overriding plates, and densities of subducting slabs. For example, whereas the collisional zone between plates appears to be compressional, extension may also occur in the overriding plate. This results in basins, and back arc volcanism, where volcanoes are found behind the main arc.

Compressional regimes occur where the upper plate, moving rapidly, encounters young and therefore warm and relatively buoyant oceanic slabs. The collision forms a shallow trench, a shallow subduction descent angle, and an **accretionary wedge** or **prism.** The wedge is composed of seafloor sediments and volcanic rocks scraped off the subducting slab. The assemblage attaches itself to the overriding plate. This situation is found in Chile and in the Coast Ranges wine country of California. There, oceanic sediments, seamounts, some with clinging limestone reefs, and pieces of volcanic seafloor crust and the uppermost mantle are attached to the overriding North American plate. These sections of peridotitic mantle[5] coupled with seafloor crust are called ophiolites. They are found in accretionary prisms elsewhere in the world.

Alternatively, when a more slowly advancing upper plate encounters an older and therefore cooler and less buoyant subducting slab, the result may be a deep oceanic trench, steeper slab decent angle, and thinning of the upper plate. No accretionary wedge forms, but if the overriding plate thins due to extension, back arc volcanism may occur. This situation develops behind, not within, the existing volcanic arc. Volcanically active back arcs are found in the western Pacific, including in the Mariana system.

When does volcanism begin in a new destructive plate system, and are there subduction zones without volcanoes? Volcanoes appear after the leading edge of a subducting plate reaches a depth of about 120 km (75 mi), but apparently volcanism doesn't occur if the angle of subduction is less than about 25°. This situation exists along segments of the Andes Mountains of South America. Also, portions of arcs where transform faulting rather than subduction occurs (such as along the southwestern end of the Aleutian arc), are volcanism-free.

Subduction Zone–Type Volcanoes

Andean-type volcanoes differ in both shape and composition from island arc system volcanoes. Island arc systems erupt mostly basalt, and, with maturity, more silica-bearing magmas, including basaltic andesite. Andean-type volcanoes erupt lavas ranging from basalt to rhyolite. For example, the older, basal flows of Mount Shasta in northern California are basaltic, whereas the youngest activity consists of silicic dacite domes and ash flow tuffs. Overall, Andean-type volcanoes produce mostly andesite.

Andean-type cones are steep-sided (averaging about 30°) due to the viscosity of their silica-enriched lavas. They commonly form perfectly symmetrical cones, such as Mount Fuji in Japan and Mount St. Helens prior to its 1980 eruption.

Mount Shasta

Geologists know that perfectly shaped volcanoes must have erupted recently, and in all likelihood, will again erupt. Had they not erupted recently, erosion from water, glaciers, and slope failure would have carved deep gashes into their flanks. Mount St. Helens, prior to its 1980 cone-destroying eruption, was a classic example. But looks can also be deceiving. For example, northern California's 4,316-m (14,161-foot) high Mount Shasta, when viewed from the south or southeast, shows deep canyons eroded long ago by Pleistocene ice age glaciers.

View the volcano from the west, north, or northeast, however, and an entirely different picture of the mountain emerges—one with two classic, uneroded cones. Eruptive activity that built the two volcanoes, Shasta and Shastina, consisted of lava, mudflows, pyroclastic flows, and plug domes. In fact, Mount Shasta City with more than 10,000 residents and located on the west slope, is situated upon pyroclastic flow deposits that are less than 10,000 years old, erupted out of the main Shasta summit and the pristine satellite cones of Shastina and nearby Black Butte. Also, snow-

SB FIGURE 1A: View of the essentially uneroded northwest side of Mount Shasta-Shastina with the prominent 9,500-year-old Lava Park andesite flow in the foreground (Lange photo).

SB FIGURE 1B: Summit of Shastina looking north. Note the small domes within the crater (Lange photo).

SB FIGURE 1C: Close-up view of the northwest summit areas showing Shastina on the right and Hotlum Cone intruded into the slightly older Misery Hill crater. Snow accentuates a remaining portion of the Misery Hill crater rim (Lange photo).

SB FIGURE 1D: Looking north at Black Buttes and a pyroclastic or ignimbrite deposit that issued from them (Lange photo).

free patches exist on the summit dome, together with a small, sulfur gas-emitting fumarole.

In addition, the north and northeast slopes of the mountain, home to the mountain's largest glaciers, show the least effects of glacial erosion. This can only be due to very recent, post-ice age volcanism. So the largest volcano in the Cascades, and one of the biggest stratovolcanoes in the world with a volume estimated at almost 500 km^3 (120 mi^3), looks like it may be extinct, when viewed from the south and southeast, but dormant from other directions.

This beautiful mountain dominating northern California not only has a complex history, but a life span of at least 590,000 years. In fact, there was an ancestral Shasta cone that collapsed in one of the world's largest terrestrial landslides about 360,000 years ago (see Figure 3.18, page 54). The remains of the former mountain extend 64 km (40 mi) to the northwest as a hummocky deposit, with some hills of debris over 215 m (700 feet) high. Individual fragments of andesite, mud flows, and pyroclastic debris of the former mountain in places measures up to hundreds of meters deep. The entire deposit covers 675 km^2 (260 mi^2), and has a volume of 45 km^3 (10.75 mi^3).

Since the ancestral cone collapse, the present Shasta stratovolcano has grown in its place, incorporating part of the old cone plus materials from four different vent locations. The oldest vent, south of the present summit, is the Sargents Ridge Cone. It is estimated to be less than 250,000 years old. This ice age andesitic cone subsequently suffered extensive erosional damage and became covered in part by younger volcanic rocks.

The youngest part of the main Shasta cone is its highest part, Hotlum Cone. This dome intruded the slightly older Misery Hill cone but left a bit of the older crater (see SB Figure 1C).

The Misery Hill cone is as young as 20,000 to 15,000 years, whereas Hotlum Cone has remained intermittently active since birth about 8,000 years ago. Both summit cones are composed of short, blocky flows of andesite, but the youngest Hotlum Cone erupted lavas are silica-rich dacite.

Growing on the side of Mount Shasta is the 3,750-m (12,300-foot) satellite volcano Shastina. Also built mostly of andesite, it took about 300 years, between 9,700 and 9,400 years ago, to reach its present size.

Shastina's last eruptive phases were dacite domes intruded into its summit crate, and pyroclastic flows that raced down the west, northwest, and southwest

SB FIGURE 2A: Aerial view looking to the southeast (USGS photo).

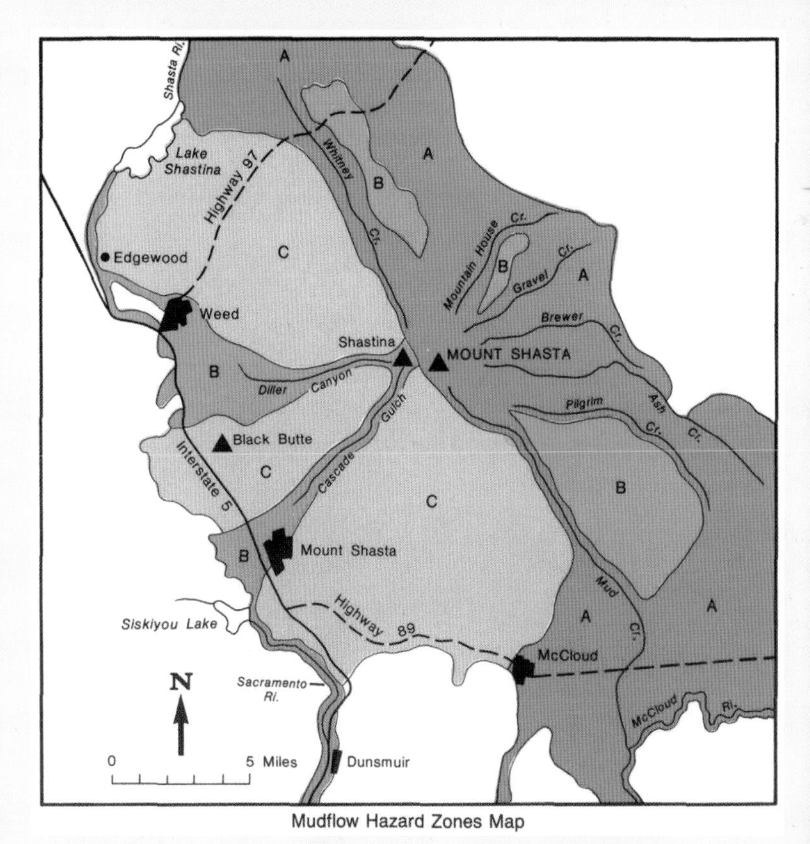

SB FIGURE 2B: Lahar or mudflow hazard potential map (USGS map).

slopes inundating the present-day sites of Weed and Mount Shasta City.

The Lava Park andesite flow, prominently displayed and essentially barren of vegetation, also erupted at this time from the northwest flank of Shastina. The blocky flow, containing levees and a steep terminus 110 m (360 feet) high, is almost 6.5 km (4 mi) long.

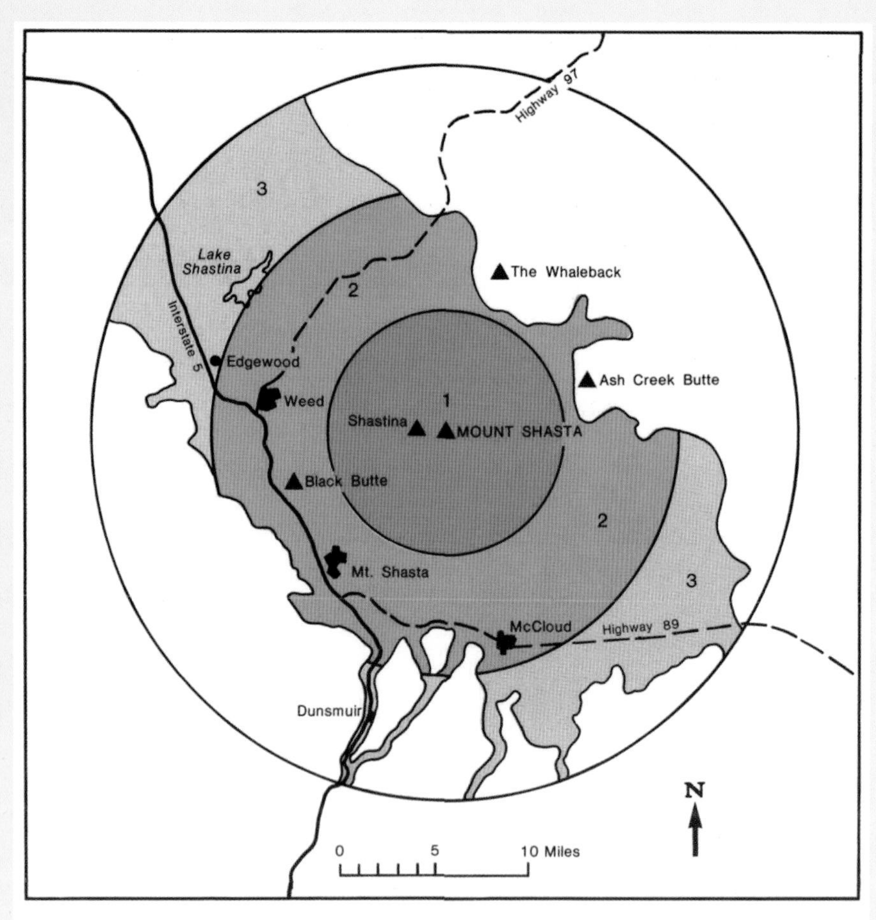

SB FIGURE 2C: Map of lateral blast hazards of Mount Shasta (USGS diagram).

It stopped just short of Highway 97 that connects Weed with Klamath Falls, Oregon.

While Shastina was developing, the cluster of four plug domes 12 km (7 mi) west southwest of Shastina grew to form the 760-m (2,500-feet) high Black Butte complex. These dacite plug domes vented pyroclastic flows that, together with those from Shastina, cover 110 km^2 (43 mi^2) of the region.

So what might be next for the mountain complex? Hotlum Cone has erupted about once every 600 years on average during the last 4,000 years, producing lava, pyroclastic, and mud flows. It may have erupted last in 1786, according to the observation of explorer La Pérouse, who was sailing off the northern California coast. In the future, we can expect flank eruptions producing andesitic cinder cones and lava flows, in addition to the formation of dacitic domes and their accompanying pyroclastic flows.

Ironically, areas most vulnerable to volcanic-caused death and destruction are the places that now have some of the greatest concentration of people: the west and northwest sides of the massif. However, mudflows resulting from non-volcanic causes such as rapid snowmelt or heavy rains can affect all slopes of Mount Shasta, because of its size, type of activity, frequency of eruptions, and location, poses a great danger to life and property. A time bomb, its potential danger is only equal to or exceeded by Mount Rainier to the north. While Mount Rainier is not known for ash- or pyroclastic-type flow eruptions, it poses a major hazard due to its history of huge mudflows. They were generated by either volcanic- or non volcanic-induced melting of the 4 mi^3 of ice and snow that drape the mountain.

Finally, the evolution of magma chemistry from basalt to silica-rich explosive dacite of this divergent-type volcano, and its earlier great collapse, make the possibility of a huge, volcano-shattering eruption likely too. We need only to look north to Mount St. Helens for a possible scenario.

REFERENCES

Christiansen and Miller, 1989; Crandell, 1989; Crandell and Nichols, 1989; Harris, 1988; Miller, 1980.

Many Andean-type volcanoes also attain great elevations, such as Mount Ararat in Turkey at 5,165 m (16,946 feet), Chimborazo at 6,267 m (20,561 feet) in Ecuador, and Mount Shasta at 4,316 m (14,161 feet). Mount Shasta rises more than 3,050 m (10,000 feet) above the non-volcanic mountains nearby. Island arc volcanoes don't achieve these elevations but may form huge calderas, such as Aniakchak in the Aleutian Island chain.

Andean-type volcanoes, also called composite or strato volcanoes, are built of alternating layers of lava, tephra, and mudflows (lahars). These layers are beautifully exposed in the calderas of Mount St. Helens and Mount Mazama (today's Crater Lake National Park) (see Famous Calderas sidebar, page 108, for details and pictures of Crater Lake).

Magmas and Other Things

Island arc volcanoes erupt mostly basaltic and basaltic andesite lavas because basaltic seafloor carrying varying amounts of sediment is subducted under an overriding plate of the same composition. Andean-type systems produce mostly andesite because silica-rich crustal rocks contaminate the mantle-generated basaltic magma as it rises. These same systems produce lesser amounts of other rock types, including basalt, dacite, and rhyolite. Their generation will be discussed in Chapter 8.

In both Andean-type and island arc systems, the "rule of 100 kilometers" holds: volcanoes commonly occur about 100 km (62 mi) above the top of the subducting plate.[8] This relationship is useful because we can use it to determine whether the angle of the subducting plate has changed. For example, one of Ecuador's newest volcanoes, 5,230-m (17,159-foot) Sangay, is well east, or further away from the trench, than Ecuador's highest and older volcano, 6,267-m (20,561-foot) Chimborazo. The rule of 100 tells us that the subducting angle has decreased, resulting in the youngest volcanoes developing farther away from the trench.

The youngest volcanoes in El Salvador in Central America and on the Kamchatka Peninsula of the Russian Far East, however, are closer to the sea and the submarine trench. The subducting angle, therefore, has steepened during the last million years, resulting in volcanism shifting closer to the trench.

Rate of Magma Ascent and Yearly Erupted Magma Volumes

How rapidly does magma rise to the surface, and how much is erupted yearly, worldwide? The answer to the first question varies, and the answer to the second can be stated as an estimate, at best. In Hawaii, Kilauea Volcano provides a well-studied, instructive example for contrast and comparison. We have definitive seismic evidence that Kilauea's magma rises, presumably within fractures, about 1,000 m (1 km) per day. Therefore, about 60 to 70 days after initial seismic activity begins 60 to 70 km (37 to 44 mi) below Kilauea, surface eruptions commence. Seismologists also track the upward movement of Kilauea's magma and can determine the location of the impending eruption.

While Kilauea is well studied, it is only one volcano and represents only one type of volcanic activity. The eruptive behavior of subduction-type volcanoes is more complex and less well known because of (1) the more complex rock sequences magmas move through, (2) the greater viscosity of most arc magmas, (3) the greater vertical distance magmas travel to the surface (60 to 70 km in Hawaii versus 100 km or more for arc volcanoes) and, (4) the long dormancy between eruptions.

Compared with Kilauea's "layer cake" of basaltic seafloor, the folded, faulted, and intruded assemblage of sedimentary, metamorphic, and igneous rocks in a subduction zone presents a greater challenge to rising magma. Also, the much greater viscosity of these more silica-rich magmas impedes upward movement.

Finally, our knowledge of eruption mechanics of arc-type volcanoes is limited because of the low eruption frequency of these volcanoes. Historically, Kilauea erupted on average about every three years, and then continuously since 1982. Contrast that with Mount St. Helens, one of the most active Cascade volcanoes, which has erupted, on average, every 100 to 120 years over the last several millenia.

How fast does magma rise in subduction zone settings? We know that, to prevent solidification during its journey, magma must rise at minimum rate of between 10 to 100 m (33 to 330 feet) per year. The total trip, then, between mantle generation and the surface for these silica-rich magmas takes from 1,000 to 10,000 years. If this is correct, the magma that erupted in 1980 from Mount St. Helens might have begun its upward trip as recently as 1000 A.D., or about the time the Vikings visited North America and 500 years before Columbus's trip to the New World.

Seismic studies show that magma movement, prior to an eruption, takes place in at least two stages. First, magma forms in the mantle and migrates upward, pooling in a magma chamber. Then, when pressure in the chamber grows too great for magma containment, the volcano erupts. Mount Rainier's eruptions 2,200 years ago originated from two magma chambers. One chamber was 3 to 5 km (2 to 3 mi) below the surface, and the other was 8 km (5 mi) deep. In 1980, Mount St. Helens derived its magma from chambers between 8 and 11 km (5 and 7 mi) deep.

Calculations show that, during Mount St. Helens' May 18 eruption, magma rose at velocities of 2 to 3 meters per second (2 to 3 m/s), whereas magma rose at only 0.004 m/s to 0.015 m/s during the post-eruption dome building phases. During the eruption, then, magma would have taken only about 70 minutes to rise 8.5 km (at 2 m/s).

Mount St. Helens

Once upon a time, not long ago, but prior to May 18, 1980, Mount St. Helens was one of the most beautiful volcanic peaks in the world. It was named by Captain George Vancouver in 1792 for his friend Alleyne Fitzherbert who had been recently knighted Lord St. Helens by King George III of England.

Referred to as the Mount Fujiyama of North America because of its perfect, uneroded conical shape, the 2,950-m (9,677-foot), glacier-capped sleeping beauty towered over Spirit Lake to the north and the huge Douglas-fir forests at its base. But geologists and northwest Indians understood that this dormant volcano was potentially very dangerous.

Having erupted at least 20 times during the last 4,500 years, and averaging 225 years between eruptions, St. Helens was known as Loowit, or Lady of Fire, to the Indians. Geologists looking at eruptive deposits, and using radiometric dating and tree ring analyses, knew that this was a very young and recently active volcano that would no doubt erupt again, most likely soon. In fact, in 1978 U.S. Geological Survey geologists Dwight Crandell and Donal Mullineaux predicted that St. Helens would "erupt violently and intermittently just as it has done in the recent geologic past . . . an eruption is more likely to occur within the next 100 years and perhaps even before

SB FIGURE 1A: Aerial view of Mount St. Helens from the northwest before the 1980 eruption (Lange photo).

SB FIGURE 1B: Mount St. Helens from the northeast in 1975. Note the low tree line due to recent volcanic activity (Lange photo).

SB FIGURE 1C: Mount St. Helens caldera taken in September 1980 from just west of Spirit Lake. Note how the blast splintered the tree stumps and blew away the trunk as would occur in a nuclear blast (Lange photo).

SB FIGURE 1D: North Fork of Toutle River drainage after May 18, 1980, showing former mountain top now forming a debris avalanche deposit that clogs the river drainage (Lange photo).

the end of this century." Little did they know that only two years later. . . .

The oldest volcanic rocks in the area are about 200,000 years old. The oldest Mount St. Helens rocks have ages of about 37,000 years. The last eruptions prior to 1980 occurred between 1830 and 1857. In fact, artist Paul Kane on March 26, 1847, painted an eruption emanating from Dogs Head dome. However, the venting was apparently from Goat Rocks dome instead. The painting hangs in the Royal Ontario Museum in Toronto, Canada. The mountain had also been quite active between about 470 and 370 years ago and also about 1,170 years before present.

Mount St. Helens, a divergent-type volcano, is located at the south end of a linear volcanic segment that includes Mount Rainier; Glacier Peak is at the north end. Its eruptive products show it to be not only the most active volcano in the Cascade Range, but together with Mount Shasta in California, potentially one of the most explosive. The mountain and the immediate surrounding region contain olivine-bearing basalts and pyroxene-bearing andesite lavas in addition to ash (tephra), lahars, and pyroclastic flow deposits and recent domes of silica-rich dacite.

SB FIGURE 2A: Mount St. Helens caldera from Spirit Lake in the mid-1980s (USGS photo).

SB FIGURE 2B: Dome spires within the St. Helens Caldera 2006 (USGS photo).

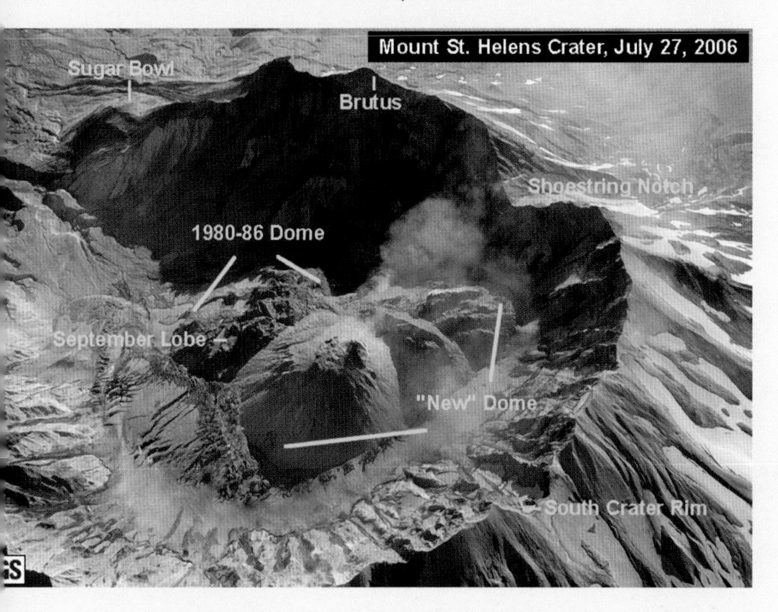

SB FIGURE 2C: Domes growing inside the St. Helens Caldera July 27, 2006 (USGS photo).

SB FIGURE 2D: View from the St. Helens caldera north-northeastward toward Spirit Lake. Note the lack of vegetation in an area that was heavily forested prior to the eruption, trees floating in the lake, and the explosion craters in the avalanche material on the south shore of the lake. The craters resulted from glacier ice coming into contact with hot volcanic material during the eruption. Photograph taken in September 1980 (Lange photo).

1980 Eruption

The 1980 eruption of St. Helens was preceded by a magnitude 4.2 earthquake located under the mountain at 3:47 P.M. on March 20. Swarms of quakes followed, reaching an intensity peak on March 25 of 47 quakes of magnitude 3 or more within a 12-hour period.

The first summit venting was in the form of small phreatic or explosive steam eruptions on March 27 at 12:36 P.M. These phreatomagmatic explosions[1] fractured and expelled summit rocks as ash or tephra lithic particles. These early explosions formed a crater 70 m across on the glacier-covered summit. By March 29, a second summit crater began to form as the volcano self-digested and erupted part of its summit.

On April 1, seismographs recorded the first magma-induced quakes or continuous ground vibration rather than sharp, fault-type earthquakes. Both seismic activity (more than 10,000 tremors prior to May 18) and steam and ash eruptions continued, resulting in hundreds of explosions blasting steam, ash, and gases (carbon dioxide, sulfur dioxide, hydrogen chloride, and hydrogen sulfide) up to 3 km (1.8 mi) above the summit. During the following weeks, as many as 50 earthquakes per day of magnitude 3 or larger occurred as the two summit craters coalesced into an oval basin 500 by 300 m and 200 m deep.

But on March 27, observers noted that as the northern part of the summit had started to subside into a crater, the northwest flank of the cone started to bulge. By April 12, the bulge, almost 2 km (1.2 mi) in diameter, had swelled outward almost 100 m (330 feet). In effect, the mountain was growing a magmatic tumor below the summit.

By late April, the bulge was expanding outward at about 1.5 m per day, thereby over steepening the northern side of the volcano. Volcanic activity might cease or subside, as commonly happens, or result in an avalanche, an eruption of variable size, or some combination or the last two phenomena. On May 7, after about two weeks of little vent activity, small summit steam explosions resumed. And at 8:32 A.M. on May 18, geologists in a light plane above the mountain, campers northeast of the mountain, and climbers on Mount Adams 55 km (33 mi) to the east recorded on film the spectacular eruption.

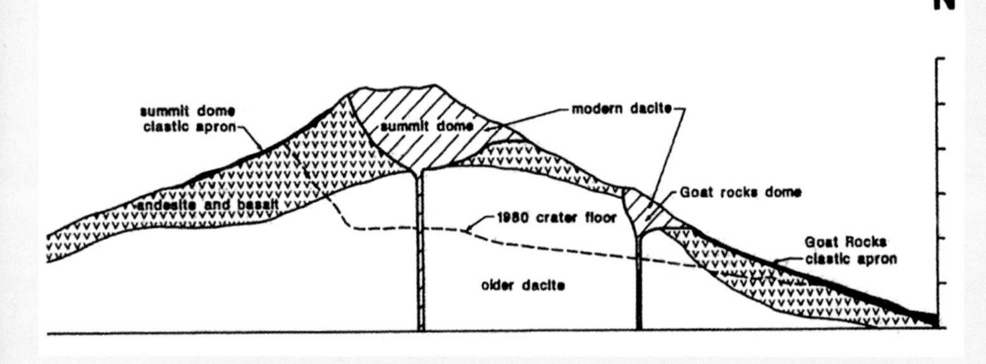

SB FIGURE 3A: Top - North-south cross section through Mount St. Helens before eruption with post-eruption 1980 crater floor (dashed line) superimposed (from Glicken, 1990).

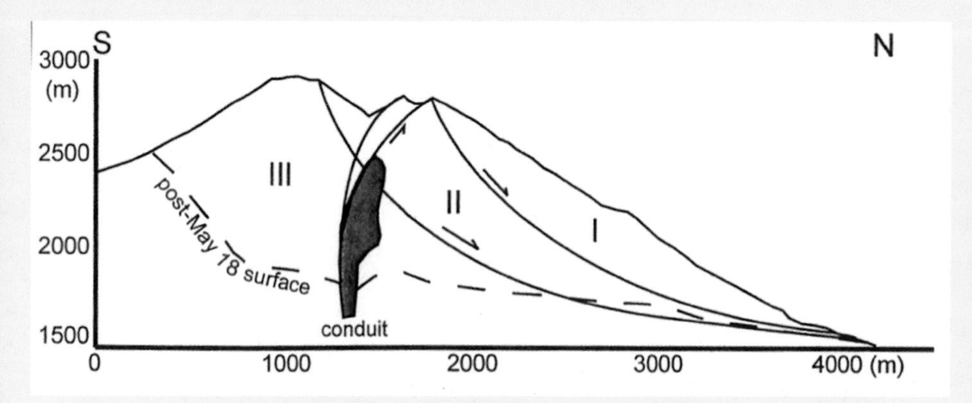

SB FIGURE 3B: Diagram shows the intrusive dome and the three parts of the May 18, 1980, landslide which started moving (stage 1) before venting was observed. The slide relieved pressure on the intruding magma, which then explosively vented (from Glicken, 1990).

The eruption started with an earthquake of magnitude 5.1 under the north flank of the volcano. This unleashed at least three successive massive rockslide-debris avalanches composed of ice, snow, and rock. The 2.5 km^3 mass, fluidized by the incorporation of air and steam, reached speeds of at least 250 to 270 km per hour (155 to 165 mph) as it raced north toward Spirit Lake and the North Fork of the Toutle River.

Blocks of glacier ice flashed to steam when encountering hot rock and magma in the avalanche near Spirit Lake. In the North Fork Toutle River Valley where the bulk of the debris came to rest, the same process created explosion craters up to 100 m (330 feet) across. The former mountain top and north side now formed a steaming hummocky deposit 21 km (13 mi) long, 1 to 2 km (0.6 to 1.2 mi) wide, and up to 150 m (490 feet) thick that temporally blocked the North Fork of the Toutle River drainage.

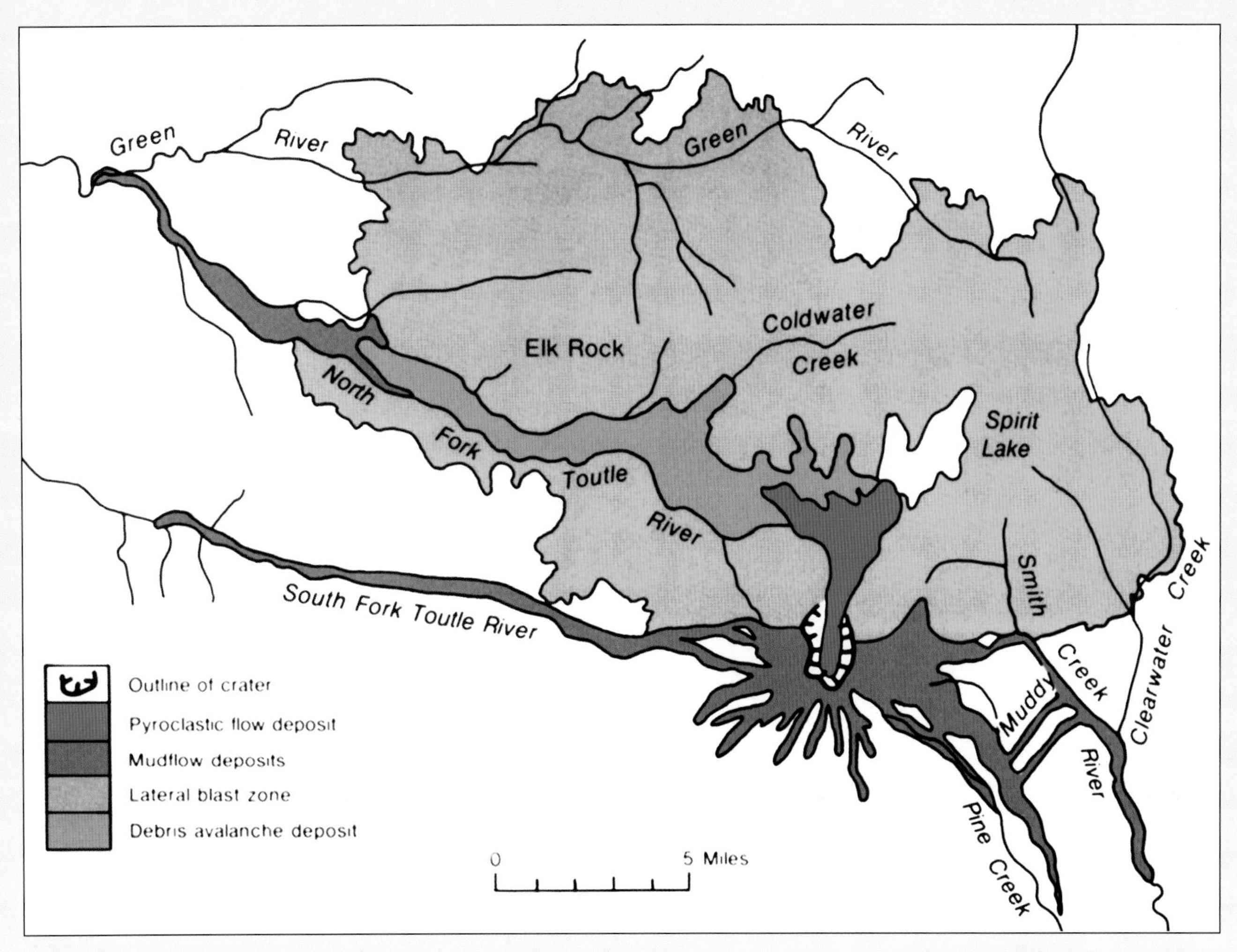

SB FIGURE 4: Map of the blow-down devastated areas and sedimentary deposits in the valleys below resulting from the cataclysmic 1980 eruption and destruction of 10 percent of the Mount St. Helens cone (modified from Kieffer, 1981; Tilling, 2000).

The effect of the landslides was to dramatically decrease pressure on the magma chamber below. This allowed groundwater to flash explosively to steam and the gas-rich magma to explode. What ensued moments after the landslide commenced was vertical- and north-directed blasts of debris-entrained steam and gases up to 300°C that traveled at speeds of 320 to as much as 1,000 km per hour (195 to 600 mph). The debris-laden blast actually overtook the avalanche, arriving ahead of the avalanche in places along the Toutle River.

The blast devastated 550 km^2 (100 mi^2) of forested terrain northeast, north, and northwest of the mountain. Within a few kilometers of the venting, trees meters in diameter were sheared off or up rooted and blown away. Further away was a zone 10 to 15 km (6 to 9 mi) wide with blown down and sheared off trees; the outer perimeter of tree damage consisted of needle-bare, heat-singed trees.

Ash released during the 9-hour eruption totaled 520 million tons and blew northeastward. It spread across about 57,000 km^2 (22,000 mi^2) of the northwestern United States. Blasted debris reached an altitude of 20 km (65,000 feet). Ash deposition disrupted life in central Washington, Idaho, and western Montana for days until, fortuitously, heavy rains occurred. Ritzville, Washington, 330 km (205 mi) east of the volcano, received 7 cm of ash, while Spokane, 430 km (270 mi) east, accumulated 5 mm. Even Denver, Colorado, received a dusting of ash.

Other major eruption features were floods and mudflows that flowed westward down the South and especially the North Forks of the Toutle River. They closed north-bound lanes of Interstate 5 at Castle Rock, Washington, 55 km (35 mi) from the volcano, and clogged the Cowlitz River and portions of the Columbia River to the southeast. Subsequent dredging reopened these rivers to shipping.

The energy released by the eruption equaled a 400-megaton nuclear explosion.[2] Fifty-seven lives were lost, and it cost Washington State at least $970 million.

Mount St. Helens witnessed 21 distinctive eruptions following the May 18 eruption through 1986. Each event was preceded by seismicity and ground deformation. Early eruptions included those of May 25, June 12, July 22, August 7, and October 16-18, 1980. While the small dacite domes that had developed in June and August were blown apart, the dome that surfaced on October 18 grew and by 1986 measured 1,060 m by 860 m (3,475 by 2,820 feet) and was 267 m (875 feet) above the crater floor.

An 18-year hiatus of eruptive activity ended on October 14, 2004, when a new, vigorous, and continuing phase of dome building began in the crater. The 750°C pyroxene-hornblende-bearing dacite, containing 65 percent silica, issued from a vent about 30 m (100 feet) in diameter. Seismicity suggests this lava rose from a 850°C magma reservoir 7 to 12 km (4.4 to 7.5 mi) below.

What's Next?

Based upon past behavior, and that the cone prior to the 1980 eruption was believed to be little more than 2,000 years old, Mount St. Helens might be restored to it's former glory in as little as 100 years. This estimate is based upon the present rate of magma extrusion of a little less than 1 m^3 of dacite per second.

A Russian Parallel

Following the 1980 eruption of Mount St. Helens, geologists realized that essentially the same happened to a subduction zone +9,000 volcano named Bezymianny in The Kamchatka Peninsula of the Russia Far East. This mountain erupted in 1956 following a landslide on its north side, resulting in the development of a north-facing caldera. The valley below filled with the former mountain top.

Shortly after the caldera-forming event, a dome started to rise in the caldera. This dome continues to fill the caldera and is rebuilding the cone.

NOTES

1 When water is heated to boiling (turns to steam), it expands instantaneously about 1,600 times its original volume under atmospheric pressure. In other words, one cubic centimeter of water expands to 1,600 cubic centimeters of steam. The liquid to gas phase change is explosive.

2 Geologists Robert and Barbara Decker also estimated the energy yield to be about a Hiroshima-sized atomic bomb per second, for a total of more than 27,000 bombs.

REFERENCES

Brantly, 1990; Decker and Decker, 1981; Glicken, 1990; Francis and Oppenheimer, 2004; Harris, 1988; Hopson and Melson, 1990; Hughes et al., 1980; Kieffer, 1981; Malone, 1990; Malone et al., 1981; Mullineaux et al., 1981; Palister et al., 1992; Swanson, 1990; Yamaguchi et al.,1995.

TABLE 4.1 – YEARLY WORLD MAGMA PRODUCTION

	Mid-Ocean Ridges	Intra Oceanic Plates	Subduction Zone	Intra Continental Plates
Extrusive component	3	0.4	0.65	0.1
Intrusive component	18	2	8	1.5
Total with % of total	21 or 62%	2.4 or 7%	9 or 26%	1.6 or 5%

* values are in cubic kilometers with % of total
The yearly total is estimated to be approximately 34km3
Values from Schmincke, 2004

During dome building, magma rising at an average rate of 0.08 m/s would have traveled the same distance in about 30 hours.

Furthermore, as magma rises, decreasing pressure allows gases (mostly water with lesser amounts of carbon dioxide, hydrogen sulfide, sulfur dioxide, and chlorine) dissolved in the magma to exsolve into a gaseous state. But if magma viscosity is very high, or the vent plugs, then eruptions may be explosive and tephra-producing, rather than effusive lava flow eruptions.

With regard to mantle-generated magmas, ocean (constructional) ridges are estimated to produce 18 km^3 to 21 km^3 (4 mi^3 to 5 mi^3) of magma per year, or 62 percent of the yearly world total of about 34 km^3 (8 mi^3). Volcanic (destructional) arcs produce about 26 percent of the total, or about 8 km^3 (2.16 mi^3). Of this arc total, 8.4 km^3 (2 mi^3) is intrusive and 0.65 km^3 (0.16 mi^3) is extrusive. The Hawaiian hotspot, in comparison, while steadily increasing its magma generation during the last 35 million years, erupts about 0.18 km^3 (0.043 mi^3) per year.

Pacific Ring of Fire

Let's now view volcanism in the Pacific Ocean Basin. Long called the Ring of Fire, a large part of this ocean is rimmed by subduction zone volcanoes. Within the basin are also hot spot-type volcanoes such as on Hawaii (see Chapter 5).

Note on Figure 4.5 how easy it is to discern subduction-type volcanoes: they not only form linear and/or arcuate distribution patterns, but they generally have spatially associated prominent submarine trenches.[9] As our eyes travel southwestward from the Aleutian Island arc of Alaska, we encounter the Kamchatka Peninsula of Russia, then the Japanese volcanic arc, and then south to the Tonga and New Zealand systems in the south Pacific Ocean. In the Western Hemisphere starting in North America, we see the Cascade Range, and then the volcanoes of Mexico that trend southeastward through Central America to and including Costa Rica. Finally, the Andes of South America from Colombia to Chile and Argentina contain both active and extinct subduction-type volcanoes.

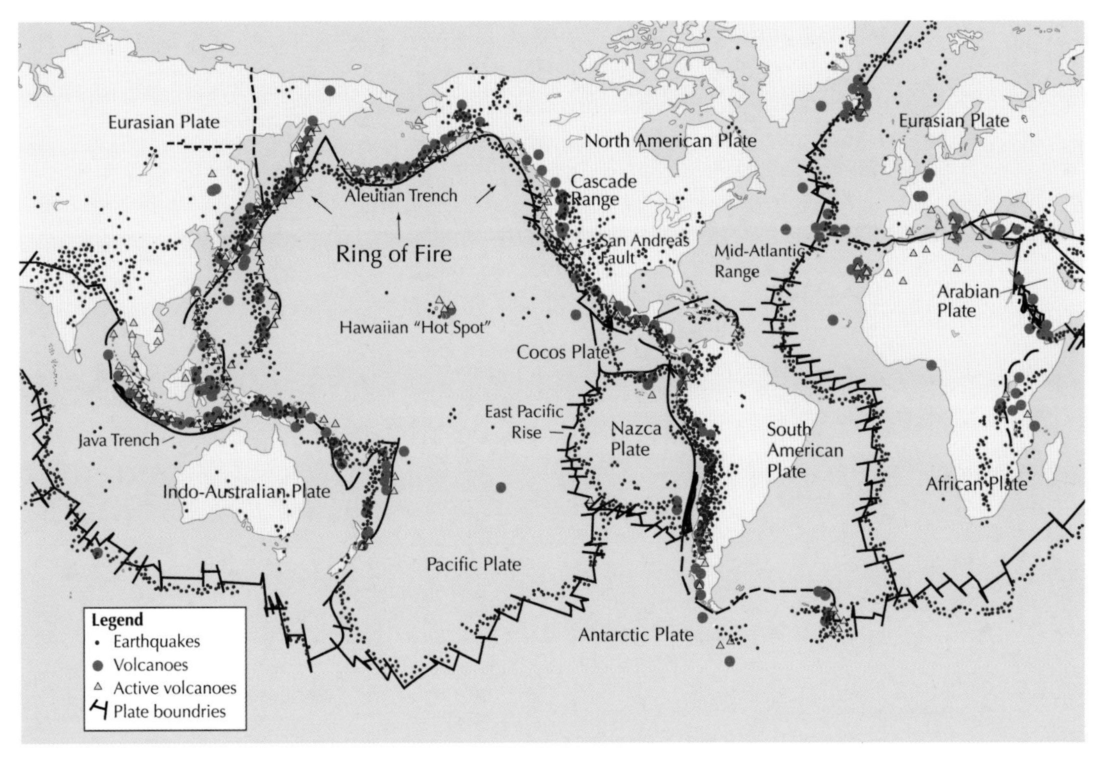

FIGURE 4.5: *"Pacific Rim of Fire" showing major tectonic plates, spreading centers (constructional plate boundaries) in black with transform fault offsets, major volcanoes in red, and the North American-Pacific plate transform San Andreas fault boundary (USGS diagram).*

What can be generalized about the Ring of Fire volcanic arcs? Most volcanic arcs are 50 km to 275 km (31 to 170 mi) wide and 125 km to 250 km (77 to 155 mi) landward of associated trenches. In the eastern Pacific Ocean, the trenches lie 1,300 km to 13,500 km (810 to 8,380 mi) from the East Pacific Rise spreading center. Subduction rates vary from 3.4 to 10.8 cm per year, and the longest volcanic arc is the Peru-Chile at 6,700 km (4,160 mi).

Volcanoes closer to the trench are generally larger, more closely spaced (30 km to 60 km (18 to 37 mi) apart), and are bigger lava-producers compared with more landward volcanoes. Furthermore, volcanoes are aligned parallel to the trend or strike of the subducting plate.

CHAPTER 4 – NOTES

1 The average ocean depth is about 3,810 m (12,500 feet). The deepest locality in the ocean is 10,994 m, 6.831 miles, or (36,070 feet) deep in the Mariana Trench in the western Pacific Ocean.

2 A hypocenter is the exact position within the Earth where an earthquake occurs. Hypocenters may occur at the surface and down to maximum depths of 700 km (420 miles). Earthquake epicenters are the geographical locations of earthquakes.

3 The Wadati-Benioff zone is named for Kiyoo Wadati of the Japan Meteorological Society and Hugo Benioff of the California Institute of Technology, who independently and essentially simultaneously in the 1930s discovered the zones of earthquakes dipping under volcanic arcs.

4 A type of volcanic arc with characteristics intermediate between Island and Andean types exists. This intermediate type starts out as Island arc system, but with time it accumulates a large mass of sediments and both extrusive and intrusive igneous rocks, all of which become metamorphosed. This results in a massive amount of young arc crust. Examples include the Cascade Range of the northwestern U.S., the Japanese volcanic arc, and the Kamchatka Peninsula of the Russian Far East.

5 Peridotite is the common rock type in at least the upper mantle. It is magnesium-iron-rich and composed predominantly of the minerals olivine and pyroxene.

6 Subduction along what is now southern and central California stopped because the North American lithospheric plate overrode the East Pacific Rise located off the coast. The North American-Pacific plate boundary then became the San Andreas, plate-bounding, transform fault. Eventually the spreading center presently located off the coast of northern California northward to southern British Columbia will be overridden by the North American plate too, resulting in the linkage of the San Andreas and the Queen Charlotte (Fairweather) fault

to the north into one, very long transform fault system (see Figure 2.8, page 23).

[7] Metamorphism, literally meaning change (meta) in shape (morphism), involves changing textures and/or mineralogy of rocks with increasing temperature, pressure, stress, time, and the addition and/or subtraction of components, including water and carbon dioxide, from the original starting material. Minerals change and coarsen, and rocks may assume a more layered look due to the formation of planar crystals.

[8] The distance ranges between 100 and 150 km (62 and 93 mi).

[9] The Cascade system, located between northern California and southern British Columbia, Canada, is exceptional in that it does not have a prominent trench. This may be due to a slow rate of subduction and a high rate of sedimentation.

CHAPTER 4 – REFERENCES

Bailey et al., 1986; Hammond, 1998; Hughes et al., 1980; Lay, 1994; Marsh, 1979; Perfit and Davidson, 2000; Poli and Schmidt, 2002; Raymond, 2002; Rutherford and Gardner, 2000; Schmincke, 2004; Sigurdsson, 2000A, 2000B; Toksoz, 1975.

Mount Rainier

Few Seattle visitors will forget the view while flying into the city on a clear day. As the plane descends into Sea-Tac International Airport, the enormous bulk of Mount Rainier and the incredible amount of snow and ice coating it is breathtaking! Named for Rear Admiral Peter Rainier of the British Navy in 1792 by Captain George Vancouver, it is also called Tahoma by northwest Native Americans.

Not only does the volcano have more glacier ice on it than all the other mountains in the lower 48 states combined, but less than 10,000 years ago Mount Rainier was possibly 4,865 m (16,000 feet) tall, not it's present 4,392 m (14,410 feet). This mountain now has three distinct summits. Liberty Cap and Point Success are remnants of an older, higher cone, and the true summit, Columbia Crest, is part of the rim of a recently formed lava cone.

The volcano, which probably reached its maximum size 75,000 years ago, is at least 500,000 years old. But it rests on an ancestral Mount Rainier composed of 1.4-million year-old lava, mud flows, and volcaniclastic rocks of andesitic and basaltic composition. Also, outcropping around and beneath the volcano are granodorite and quartz monzonite intrusions of the 17.5- to 14.1-million-year-old Tatoosh intrusion.

In terms of size and age, presently dormant Mount Rainier and Mount Shasta in northern California at 4,316 m (14,161 feet) are the two highest and among the oldest Cascade volcanoes. Mount Shasta has the greatest volume, followed by Mount Adams south of Mount Rainier, with Rainier in third place. Whereas Mount Shasta sits on the end of a linear segment and erupts the gamut of lava types (basalt to rhyolite), Mount Rainier, a coherent-type volcano, resides within a volcanic linear segment. More than 90 percent of Mount Rainier's 270 km^3 (64.5 mi^3) of flows are andesite.

Major cone-building phases of Mount Rainier occurred between 500,000 to 420,000 and 280,000 to 180,000 years ago. The last 120,000 years have witnessed the eruption of cone-building lavas, but at a much reduced eruption rate. One exception is a 90,000-year-old flow. Estimates place the effusive eruption rate over the last 40,000 years at about 0.1 km^3 (0.02 mi^3) per 1,000 years.

The first Mount Rainier flows were voluminous pyroxene-bearing andesite lavas that, because of the large, valley-filling glaciers, flowed down ridges along glacial margins like spokes radiating from the center of a wheel. The oldest of 23 huge flows, some up to 610 m (2,000 feet) thick and 24 km (15 mi) long, is the massive 365-m (1,200-feet) thick and 9.6-km (6-mi) long Burroughs Mountain flow.

The upper part of the present cone is composed of hundreds of andesitic lava flows, mostly 15 m (50 feet) or less in thickness. These are interbedded with mudflows or lahars and thin tephra deposits, with approximately 90 percent of the cone composed of flows.

But because the volcano was born during the ice ages, it has undergone serious glacial erosion and has had to rebuild to survive the glacial onslaught. Both glacial erosion and land sliding have removed perhaps one-third of its original volume. At least 60 lahars have swept off the mountain since about 10,000 years ago, with four notable ones in the last 7,000 years. Starting at 6,600 years ago, the Paradise and then the Greenwater lahars removed about one-fifth of a cubic mile of mountain.

The biggest mudflow, carbon dated at 5,040 years, however, is the Osceola. This monster removed 3 km^3 (more than 0.7 mi^3) and 610 m (2,000 feet) of the summit of Rainier with the collapse of a 1.75-km (1.1-mi) wide section of the northern portion of the summit. Part of the debris swept down the White River Valley, while another lobe traveled down the West Fork of the White River. The two lobes then coalesced, flowing another 100 km (60 mi) to the west, eventually reaching Puget Sound. Approximately 325 km^2 (125 mi^2) of the Puget lowlands were buried, in some places with 30 m (100 feet) of debris. See Figure 3.17, page 52.

The most recent major mudflow was the 530-year-old Electron. It flowed off the southwest flank and inundated the Puyallup River Valley to beyond the town of Sumner.

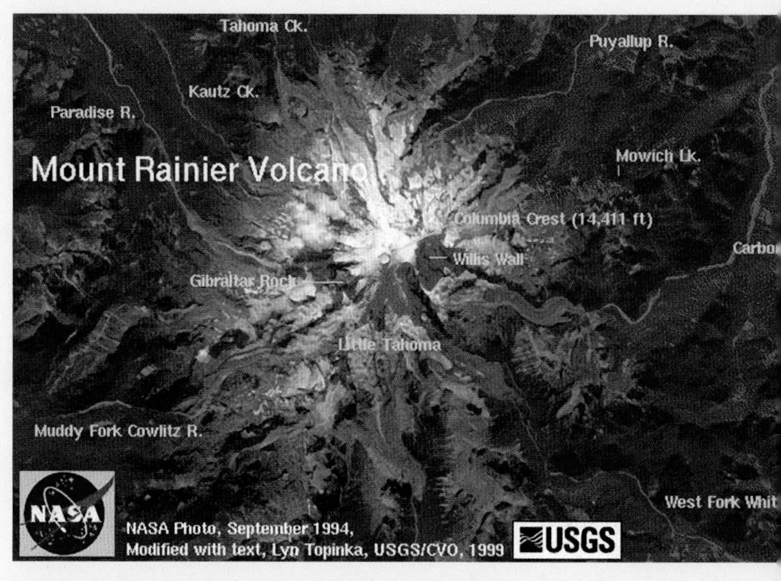

SB FIGURE 1A: NASA aerial view of Mount Rainier (north is to the right) (NASA/USGS photo).

Post Ice Age Eruptions

Mount Rainier has erupted at very irregular intervals during the last 10,000 years. It was especially active between about 6,500 and 4,000 years ago. Not only did this interval include the Osceola lahar, but 8 of 11 post-glacial tephra eruptions.

A hiatus of activity occurred between about 4,000 and 2,600 years ago, but then volcanic activity resumed between 2,600 and 1,600 years before present during the Summerland phase. The most voluminous tephra deposit of the 11 was then erupted as was partial filling of the summit depression caused by the Osceola summit collapse some 2,500 years before. Two overlapping cones composed of glassy, blocky lava and tephra grew to form the new high point, Columbia Crest, of the volcano.

SB FIGURE 1B: Northeast side of mountain showing Little Tahoma Peak on the left, an erosional remnant of the former larger (possibly 16,000-foot high) mountain on the left and the large, Emmons Glacier and to its right the Winthrop Glacier (USGS photo).

What does the future hold in store for Mount Rainier? Certainly the battle between the destructive forces of land sliding, glaciation, and water erosion and volcanic reconstruction will continue. That the mountain is dormant is supported by activity that occurred in 1820(?), 1841-1843(?), 1854(?) (unconfirmed but suspected eruption dates), 1879, and 1882. Presently there are seven significant thermal anomalies on the flanks above 3,350 m (11,000 feet), including the steam caves along the rims of the new, eastern summit craters.

It is reasonable to expect that within the next few hundred years, Mount Rainier will witness floods and mudflows (some volcanically related), tephra eruptions, and possibly lava flows.

SB FIGURE 1C: Northwest side of Mount Rainier showing glacial (erosional) damage (Lange photo).

REFERENCES

Fiske et al., 1963; Harris, 1988, 2005; Lange and Avent, 1973; Mullineaux, 1974.

CHAPTER FIVE

Hot Spot Volcanism

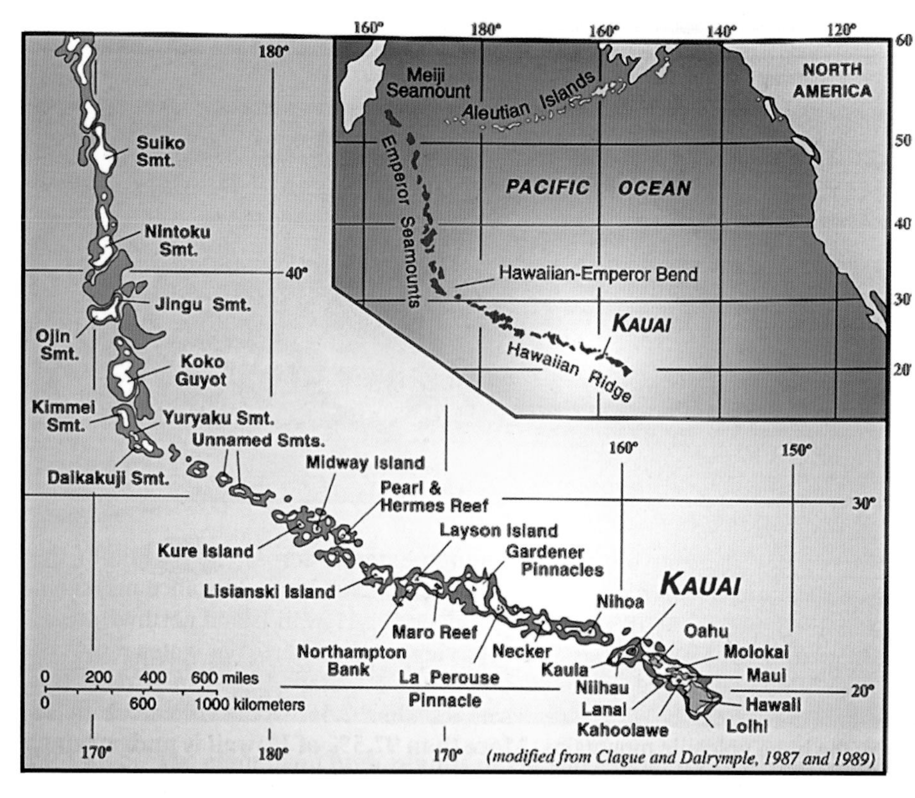

FIGURE 5.1: *Hawaiian (northwest trending)- Emperor (north-trending) hot spot track in the North Pacific Ocean together with the newest, southeast portion containing the Hawaiian Islands. The age of the islands is show from southeast to north in millions of years. Midway is 27.2 million years old, Daikakuji (at the bend in the hot spot trend) is 42.4 million years old, and the oldest seamounts are about 75 million years old (from Blay and Siemers, 2004; Clague and Dalrymple, 1987, 1989; Sharp and Clague, 2002).*

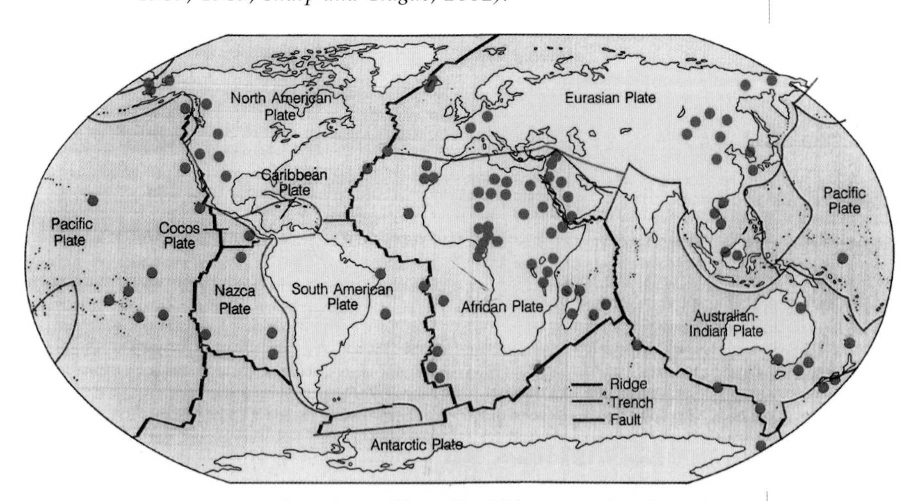

FIGURE 5.2: *Locations of long-lived hot spot volcanism. Not shown are 11 on and around Antarctica, and one in the Arctic Ocean (modified form Skinner and Porter, 1995; Burke and Wilson, 1976).*

Introduction

Previous chapters explored volcanism-producing plate tectonic environments: constructive and destructive plate boundaries and transform faults. The plate tectonic model, however, does not appear to explain volcanic centers such as the Hawaiian Islands, Yellowstone, Iceland, and the many volcanic centers in Africa, eastern Australia, and elsewhere. These volcanic centers, commonly isolated from other areas of volcanism, are called hot spots. As mentioned in Chapter 2, hot spots, while geologically unusual, have helped geoscientists understand how plate tectonics operate.

In 1963, J. Tuzo Wilson of the University of Toronto recognized hot spots and speculated about their origin. He observed that the Hawaiian Islands are both older and more eroded to the northwest, away from the Big Island, where volcanoes are active. Wilson envisioned the northwestern motion of a crustal slab over "a jet stream of lava" to account for the islands' age progression. Imagine a piece of paper (representing the Pacific plate) passing over a lit candle—the flame would leave a track of toasty spots parallel to the paper's line of movement.

Globally, at least 122 hot spots have been active during the last 10 million years; 40 are currently active. Sixty-nine of the 122 are located on continental lithosphere, and 53 lie on oceanic lithosphere. Fifteen of the 53 lie on spreading or constructional plate boundaries—9 near ridge crests and 29 in ocean basins.

Hot spots are of different sizes, have varying discharge rates over time, and may operate for as long as 100 million years. While the position of some hot spots appears fixed within the Earth, many have changed locations. For example, the Hawaiian hot spot moved a short distance south to its present position about 40 million years ago.

Distinctive Characteristics of Hot Spots

How does hot spot volcanism differ from common, plate tectonic-related volcanism? The most important aspect is that hot spots appear unrelated to plate tectonic environments. They tend to occur far from subduction zones and other plate boundaries. What about Iceland, then? The Icelandic hot spot is located near the Mid-Atlantic Ridge, 240 km (149 mi) east of the spreading center's axis. We know, however, that the Icelandic hot spot "migrated" across Greenland, so its present location close to a ridge is apparently coincidental (see Figure 5.3).

Other characteristics also distinguish the Icelandic hot spot from typical plate tectonic volcanism. It erupts lavas at a much greater rate than any part of the nearby spreading center. It has left a track of progressively older volcanic rocks across Greenland and in Baffin Bay, west of Greenland.

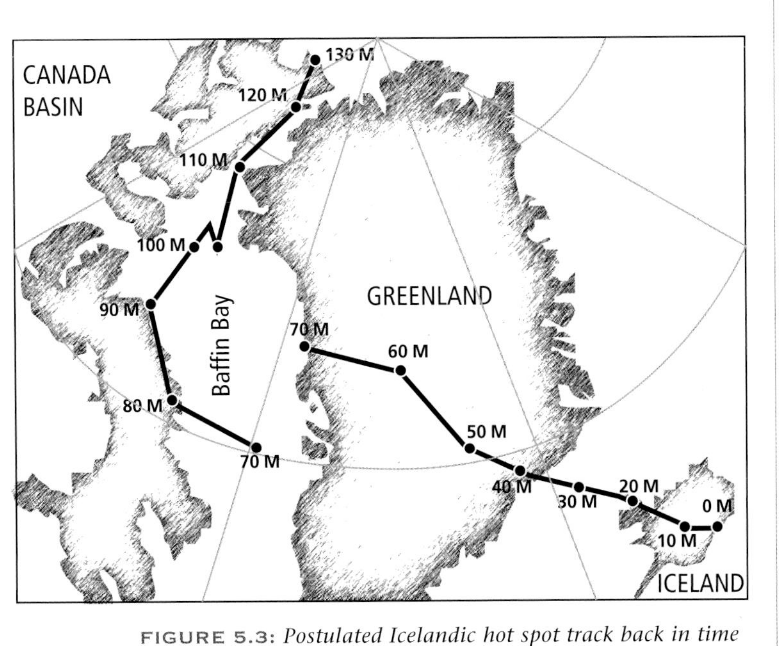

FIGURE 5.3: *Postulated Icelandic hot spot track back in time (millions of years) across Greenland and north into the Arctic Ocean. Note how the opening of Baffin Bay, starting 70 million years ago, has separated the hot spot trend (see text for explanation). Baffin Island is west of Baffin Bay and Axel Heiberg Island with the oldest hot spot ages is the large island to the north (modified from Lawver and Muller, 1994).*

While this situation is analogous to the Hawaiian hot spot track, the Icelandic track is more irregular. More accurately, the path of the North American plate has been much more irregular than that of the Pacific plate unless the position of the hot spot has changed.

Hot spots such as those in Iceland, Yellowstone, and Africa are regions of broad crustal uplift. With average diameters of 1,200 km, hot spot uplifts cover an impressive 10 percent of Earth's surface.

The Yellowstone Caldera, for example, is centered on a domed region at least 500 km across. At 2,100 m (7,000 feet) above sea level, the caldera sits well above the Snake River Plain to the west. The Yellowstone Plateau, then, is a huge thermal bulge or welt in the northwest corner of Wyoming.

Lavas that erupt from these hot spots tend to have different chemistries than mid-oceanic ridge basalt spreading center (MORB) basalt. Hot spot lavas may be unusually rich in elements such as lithium, sodium, and potassium, and they have high neodymium and tritium/helium-4 isotopic ratios (among others) when compared with MORB basalts. These data suggest that hot spot basalt originates at various levels in the mantle.

These uplifted regions also tend to exhibit "positive gravity anomalies," meaning that there is more than the normal amount of Earthly mass beneath them. When hot spot activity stops, however, these regions sink, and Earth re-achieves lithospheric **isostasy** or **flotational equilibrium**[1] on the plastic mantle below.

Why Hot Spots?

Earth scientists have pondered the origin of hot spots since they were first recognized. The mantle plume hypothesis was proposed in the early 1970s and was soon accepted by most geologists (although recent research has reopened the debate). A mantle plume, as originally defined, is a hot, active upwelling of mantle material originating at the mantle-core boundary. Such plumes are huge, have bulbous heads and trailing narrow tails, and are at high temperatures, allowing them to reach the Earth's surface.

Proponents of the mantle plume model postulate that parts of the lowermost mantle are hotter than average due to anomalous high-heat releases from the core below. Because these mantle regions

TABLE 5.1. HOT SPOTS BELIEVED TO HAVE FORMED FROM SHALLOW MANTLE TECTONIC MECHANISMS*

Hot Spot	Lithosphere	Location
Many on African plate	Continental	African plate
Bouvet	Oceanic	South Atlantic Ocean
Cape Verde	Oceanic	Atlantic off NW Africa
Cook-Australs (MacDonald)	Oceanic	Equatorical Pacific Ocean
Galapagos	Oceanic	Equatorical Eastern Pacific Ocean
Kerguelen	Oceanic	Southern Indian Ocean
Marquesas	Oceanic	Southwestern Pacific Ocean
Pitcairn	Oceanic	Equatorial Pacific Ocean
Reunion**	Oceanic	Western Indian Ocean
Samoa	Oceanic	Western Souh Pacific Ocean
Society Islands (Tahiti)	Oceanic	Western South Pacific Ocean
Tristan**	Oceanic	South Atlantic Ocean
Yellowstone	Continental	Western United States

* Most started on spreading center (ridge), fracture zone, or both
** Debate continues as to whether these are deep mantle plume-derived
(after Anderson, 2005)

are unusually warm, they expand and so are less dense than surrounding mantle. The hot material rises. This hot mantle, shaped like an upside down rain drop, or **diapir**, is a few hundred kilometers in diameter. It slowly convects upward at speeds of 5 to 20 centimeters (2 to 8 inches) per year, as the plume rises through the surrounding plastic but cooler mantle. Hot spot magma is at least 225°C hotter than the average mid-ocean ridge basalt.

Proponents cite a number of hot spot characteristics or criteria to support the theory that hot spots form from deep mantle plumes. These include: (1) a relatively fixed location on the Earth's surface, (2) volcanic chains showing age progression on moving lithospheric plates, (3) parallelism of island volcanic chains on these moving plates, (4) large igneous provinces, (5) high magmatic eruption temperature, (6) elevated heat flow on and around hotspots, (7) lithospheric plate thinning, (8) uplift of these large igneous provinces, (9) continental uplift, (10) evidence of lithospheric plate rejuvenation, (11) detectable low seismic velocity within the plume compared with the adjacent mantle rocks, and (12) chemical and isotopic signatures, including primordial mantle neodymium-isotope ratios.

Recent studies, however, show that many of these criteria are no longer considered unequivocal evidence for a mantle plume origin. The same criteria may also accompany other, more shallow, formational mechanisms. Now, combinations of some of the criteria, and not necessarily the same criteria, are used to support mantle plume origins. In fact, today researchers generally agree that most hot spots are not the result of mantle plumes. Instead most have shallow, plate tectonic explanations. Alternative hypotheses include leaky transform faults, incipient lithospheric plate boundaries, tectonically reactivated plate sutures (plate attachment boundaries), fracture zones, cracks in extensional or pull-apart terrains, small scale convection associated with plate motion, and heterogeneity (variations in the composition, volatile content, and melting) in the asthenosphere or upper mantle close to its melting point.

The original list of hot spots believed to be deep mantle plume-related has shrunk to just a few, and some researchers have their doubts about the origin of these. The list presently includes Iceland, Hawaii, Easter (Island), Louisville (in the southwestern Pacific Ocean), Afar (in the horn of Africa), Reunion (in the Indian Ocean), and Tristan (in the South Atlantic Ocean).

Each hot spot has its own "origin" problems. For example, one tenet of plumes is a low seismic velocity region beneath a hot spot. This includes a pancake-shaped zone lying just beneath the 650-km-deep mantle phase boundary. This characteristic is not present in the topmost 1,000 km (620 mi) beneath the Iceland, Easter, Hawaii, Tristan, and Afar hot spots.

Whether or not they are age progressive, volcanic chains can be explained by a variety of mechanisms, including incipient or new plate boundaries, propagation of cracks, leaky transform faults, abandoned ridge, spreading center or ridge propagation, and small-scale convection. In fact, the absence of a crustal swell favors crack control of volcanic chains. In contrast, the presence of a large igneous province where initial eruptions began favors the mantle plume hypothesis.

Hot Spot Tracks, Large Igneous Provinces

HAWAII

The Hawaiian Islands lie in the middle of the Pacific Ocean on the Pacific lithospheric plate (Figure 5.1). The Hawaiian chain is just part of a linear, northwest-trending group of islands, reefs, and atolls that includes Midway Island, which dates from 27 million years ago. This linear group stretches 3,450 km

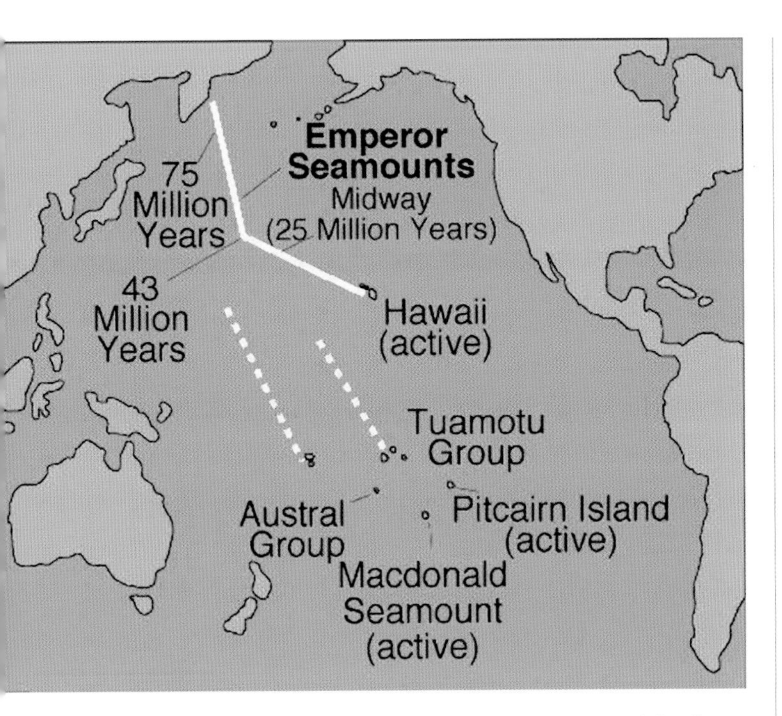

FIGURE 5.4: *Note how the different Pacific Ocean island hotspot tracks mimic each other, supporting the theory that the Pacific lithospheric plate moved northward from at least 70 million years ago until about 43 million years ago. This was followed by a directional change to the northwest (from Burke and Wilson, 1976).*

(2,070 mi) from the big island of Hawaii to Yuyaku Seamount, a submerged extinct volcano. From there, seamounts extend another 2,300 km (1,380 mi) almost due north to Meiji Seamount.[2] In all, there are at least 107 individual volcanoes in the 5,750-km (3,450-mi) chain.

The north-oriented linear array of submerged and extinct volcanoes of the Emperor Seamount chain is part of the line of volcanoes that were once each situated above the Hawaiian hot spot. The Pacific plate at first moved due north from at least 75 million years ago to 43 million years ago. Then, from 43 million years ago until the present, the plate moved northwestward over the hot spot.

What developed was a progressively older, north-oriented string of islands connected to a northwest-oriented chain, with the Big Island of Hawaii sitting over the stationary hot spot.[3] And the Big Island is presently moving off the hot spot as another submarine volcano, Loihi, grows on the south, submarine flank of Mauna Loa Volcano. Its summit, one kilometer below the sea surface, contains fresh lava and thermal springs.

By knowing the age of seamounts/volcanoes, their orientation, and the length of the chains, we can determine the rate of (Pacific) plate movement and direction. The Hawaiian hot spot has remained almost stationary within the Earth during the last 40 million years. The average rate of plate movement during Emperor chain formation was 7.2 +/- 0.3 cm per year. During development of the Hawaiian chain, the plate moved faster at 9.2 +/- 1.1 cm per year. Most recently, the plate moved 13 cm per year between Maui and the west shore of the Big Island.

That the Hawaiian chain reflects the motion of the Pacific plate during the last 43 million years and before is reinforced by the Tuamotu and Austral island hot spot trends. These trends end with Pitcairn Island and Macdonald Seamount, respectively (Figure 5.4).

However, the location of the Hawaiian-Emperor hot spot has not remained fixed in the Earth since its birth. A change in location could be due to either change in lithospheric plate stress or changes in plate motion. Therefore, volcanic chain orientation cannot, necessarily, be viewed as an indicator of absolute plate motion (Anderson, 2005).

HOT SPOTS AND CONTINENTAL BREAKUP

Hot spots have helped us understand how plate tectonics operate and may be responsible, totally or in part, for the break-up of continental lithosphere. Some 300 million years ago, all the continents were joined in a single, super landmass we call Pangaea. Pangaea, meaning "all land," was first coined by Danish meteorologist Alfred Wegener.[4] Plate tectonic motion brought all the pieces together, and for 100 million years, Pangaea existed as a huge, north-south-oriented continent extending from the Arctic to the South Pole.

FIGURE 5.5: *The Earth's single landmass, Pangaea, 250 million years ago, showing the southern portion, Gondwana, and the northern landmass, Laurasia (Lange, 2002).*

Hawaii

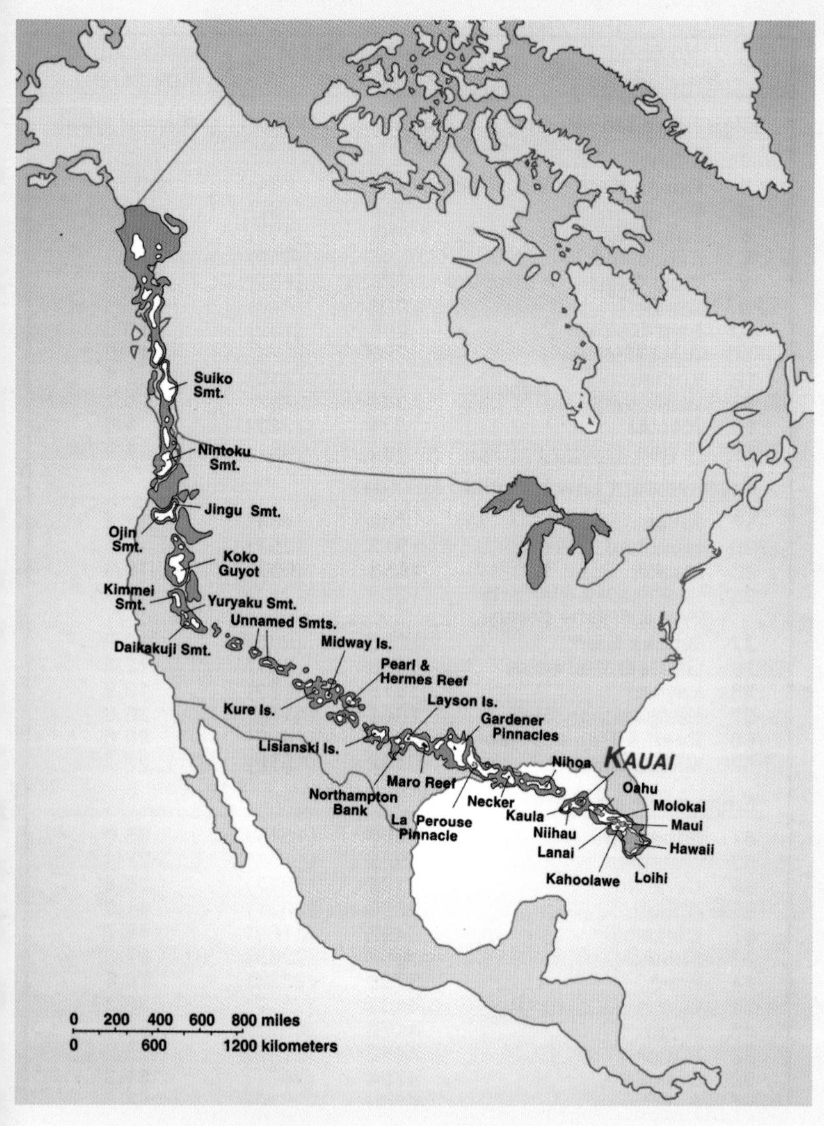

SB FIGURE 1

Volcano Development

Age in ka	Magma Type	Notes	
0	early alkalic stage		shield building
100	tholeiitic stage		stage
200			
300	– – – – –	breaches S.L.	
400			
500			
600	late alkalic stage		volcano max. size
700		volcano shrinks in size	

SB FIGURE 2

Let's return to the Hawaiian Islands for a more in-depth examination. SB Figure 1, compliments of Blay and Siemers (2004), depicts the extent of the Hawaiian-Emperor hot spot track when superimposed on North America–a truly impressive visual! SB Figure 2 shows that each island volcano grows in stages as one of two distinctive types of basalts is erupted. Initial eruptions take place on the ocean floor in about 20,000 feet of water. Then, using the Big Island as an example, some 300,000 years after "birth" the volcano breaches the sea surface and an island is born.

But, as we have seen elsewhere,[1] life is tough for the emerging volcanic island: ocean waves batter the small pile of erupted materials. The risk of a catastrophic phreatomagmatic explosion is high until the volcanic vent surrounds itself with enough material to prevent seawater from entering the upper level magma chamber.

Studies reveal that these islands develop in stages. Hawaii consists of six coalescing volcanoes with a seventh, Loihi, now adding to its mass. The youngest volcanoes are Mauna Loa, Kilauea, and the submarine volcano Loihi along the southeastern margin.

The first 500,000 years are the shield-building stage. Vigorous island growth takes place when at least 95 percent of the volcanism occurs. The next 300,000 years are the post-shield-building phase. The amount of volcanism declines, while, due to isostasy, the island starts to subside into the mantle under its own weight.

Presently, Mauna Kea is in the post-shield-building stage, whereas Mauna Loa and Kilauea are still vigorously shield-building. Table 1 lists the volcanoes that constitute the Big Island and their dates of terminal shield building. Estimates place the growth rate of the big island at about 0.02 km^3 per year over the past 600,000 years.

Note the difference in slope angle between the underwater, shield-building phase and subaerial growth. Very fluid basaltic lavas travel far from the vent on land, producing gentle slopes of only 4 degrees. If erupted underwater, however, lava cools rapidly, resulting underwater slopes of about 13 degrees.

The type of basalt erupted correlates with the phase of volcano development. Alkali olivine basalt forms during the first 100,000 years of growth. This is followed by huge eruptions of tholeiitic, or olivine mineral-free basalt, for the next 400,000 years. The final, post-shield-building phase of 300,000 years is characterized by a return to alkali olivine basalt.

Why the different types of basalt? How can we tell them apart? Basalt is the result of partial melting of magnesium-rich mantle rock (peridotite) about 60 to

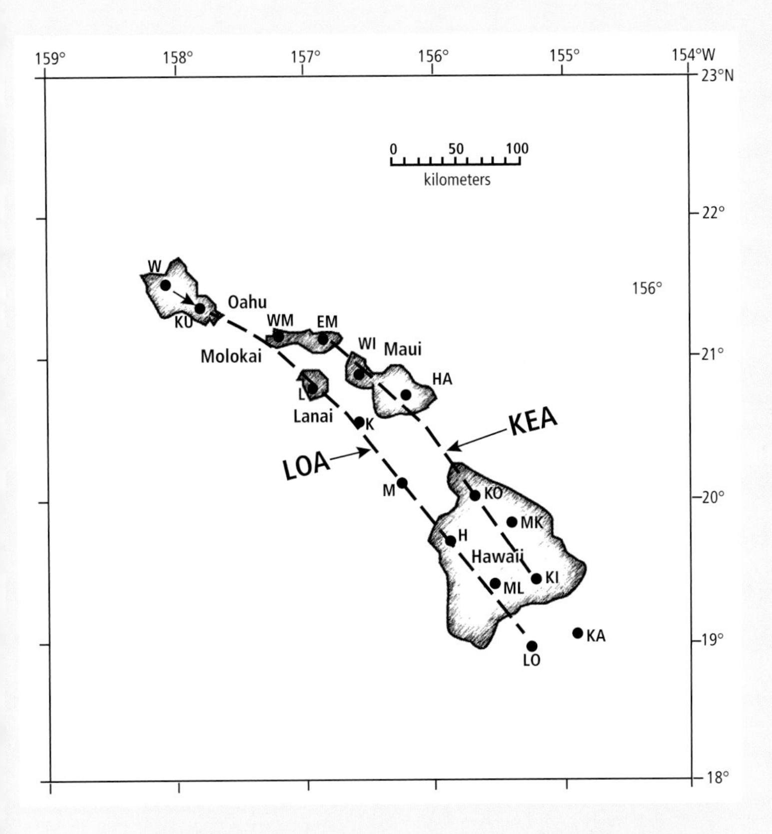

SB FIGURE 3: Shows that from Molokai to Hawaii there are two parallel volcanic tracks, the Loa and Kea. After Oahu formed, the Hawaiian hot spot split into two parallel tracks, forming Lanai at about the same time the western end of Maui developed. This dual trend of volcanic growth appears to be continuing: Loihi is growing southeast of Mauna Loa, with a theorized Keikikea (KA) in the waters southeast of Kilauea. Each volcanic center (island) is between 60 km and 80 km apart (modified from Moore and Clague, 1992; Moore et al., 1989).

TABLE 1. HAWAIIAN ISLAND SHIELD BUILDING

Volcano	End of Shield Building
Loihi	+500,000 years ago
Kilauea	
Mauna Loa	
Hualalai	130,000
Mauna Kea	130,000
Kohala	245,000
Mahukona	465,000

Moore and Clague, 1992

70 km beneath the surface (see Chapter 8 – Magma Generation). The initial, limited mantle melting produces alkalic basalt, which is lower in silicon dioxide but richer in potassium and sodium than the later-erupted tholeiitic basalt (see Table 2.1, page 15). It also contains the pretty, green, glassy-looking mineral olivine.

After intensive melting of mantle occurs during the 100,000-to-500,000-year interval of vigorous shield growth, a greater percent (up to 25 percent) of the melting mantle produces tholeiitic basalt. The presence of olivine mineral crystals only in the alkali olivine basalt easily distinguishes the two basalt types.

The rate of magma ascent through the upper mantle and oceanic crust in Hawaii is known because of seismic, earthquake-generated "noise" produced by rising basaltic liquid. Within 60 to 70 days of the first, deep tremors, lava erupts from the Kilauea system. Magma, then, rises about 1 km per day through cracks or fissures in the upper mantle.

Furthermore, seismologists are able to tell where lava will erupt by carefully monitoring the hypocenter (location within the crust) and epicenters (geographic location) of the small, magma-induced earthquakes. Lava erupts within the Kilauea crater and/or along the northeast rift system. Kilauea has been erupting since 1982, making it one of the longest eruptions in known history.

The lovely green sand beaches of the Big Island are derived from the post-shield-building phase. (Beautiful olivine jewelry is available to buy in Hawaii).

The Fate of Volcanic Islands

If the Pacific Plate is inexorably moving to the northwest, what fate awaits these oceanic islands? Once their volcanoes become extinct, the islands' size

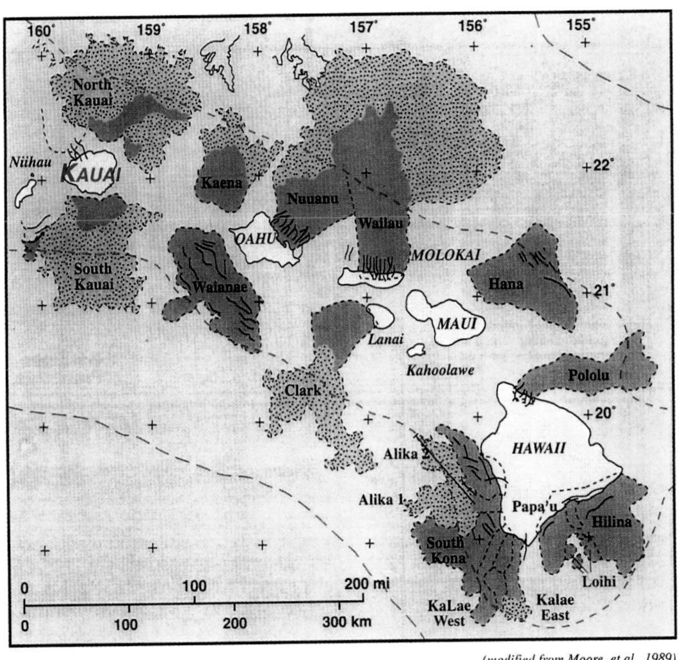

SB FIGURE 4: Not only are the islands sinking under their own weight and being eroded by waves and running water, but huge portions are sliding into the deep water around them as seen in this figure. These are among the largest known landslides on Earth. The Hawaiian Islands are shown in yellow and most landslides are labeled. Green portions indicate chaotic landslide deposits; gray areas indicate hummocky debris flow deposits. The red dashed interior line delineates the axis of Hawaiian Deep; the outer red dashed line shows the crest of the Hawaiian Arch (from Moore et al., 1989; Blay and Siemers, 2004).

decreases for four reasons. First, because of the huge mass of lavas piled on the ocean floor and the elastic nature of the crust, all the islands, atolls, and seamounts continuously sink into the oceanic lithosphere beneath them due to isostasy. This is known because some seamount summits, thousands of feet beneath sea level, are ringed with dead coral reefs that grew only near sea level. (See end chapter note 1)

Second, in addition to isostatic sinking (which also creates moats around islands), the island shores are continuously being battered by waves, forming steep cliffs. Ultimately this develops flat, planed surfaces at sea level. Third, rainwater erodes the slopes into gullies and canyons while moving material out to sea. Waimea Canyon on Kauai is a great example of the erosive power of water. Fourth and finally, huge, periodic landslides have destroyed large portions of all the Hawaiian Islands. These deposits were mapped on the seafloor using side-scan sonar (Figure 5.4, page 81).

The ultimate, inglorious fate of these islands, after wave-planing and sinking, is subduction into the mantle. Note that Meiji Seamount is close to the west end of the Aleutian Trench and located just east of the Kamchatka Peninsula of the Russian Far East. If the Pacific Plate continues on its trajectory, as expected, there is little hope for seamount survival. Therefore, the age of the Hawaiian hot spot is unknown: seamounts older than 80-million-year-old Meiji, if they existed, have already been subducted.

Some island remnants or seamounts escape being subducted only to be obducted, or scraped onto the overriding plate. This happened, for example, to seafloor that contained seamounts and reefs now located in the Coast Ranges of California.

NOTES

1 Surtsey Volcano, near Iceland, erupted in 1963, The nascent island struggled to survive as waves battered it and seawater met magma, causing phreatomagmatic explosions.

REFERENCES

Blay and Siemers, 2004; Ernst and Buchan, 2001; Foulger et al., 2005; Hyndman, 1985; Moore et al., 1989, 1995; Moore and Clague, 1992; Sherrod et al., 2003.

FIGURE 5.6: *Iguazu Falls, shown during the dry season, spills over 132-million-year-old horizontal basalt flows of the Paraná magmatic province of Brazil. The lavas issued from the Tristan da Cunha hot spot. The eastern part of the volcanic field, the Etendeka province, is located in Angola, southwestern Africa. Iguazu Falls, 60 to 82 m (197 to 269 feet) high, are situated between Brazil and Argentina, next to Paraguay. Niagara Falls, in comparison, drops 50 m (165 feet) (Lange photo).*

The southern part of Pangaea, Gondwanaland, included present-day Africa, South America, Antarctica, Australia, Madagascar, and India. The northern part, Laurasia, contained North America, Europe, and Asia (see Figure 5.5).

South America started to separate from Africa about 130 million years ago with the initiation of igneous activity along what was to become the southern Mid-Atlantic Ridge. Hot spot Tristan da Cunha, now a British-controlled island, was born and, as some hot spots do, created an incredibly large continental flood basalt province. (The oldest basalts flows are 128 to 138 million years old).

The province subsequently rifted apart with the establishment of the Mid-Atlantic Ridge spreading center and the opening of the South Atlantic Ocean. Spreading followed in what would become the north-central Atlantic Ocean. Remnants of the South Atlantic province include the 132-million-year-old Paraná region of Brazil and the 132-million-year-old Etendeka region of Angola, southwestern Africa, over 6,000 km (3,660 miles) to the east.

But how do hot spots break apart continents? Hot spots that cause or contribute to continental breakup commonly develop three arms or rifts that radiate from the domed volcanic center. Tristan da Cunha, for example, developed three rifts: one oriented eastward, one to the south and the third to the north. The north arm connected to the rift "arm" of the hot spot to the north where the island of St. Helena is now located.[5]

TABLE 5.2. FLOOD BASALT PROVINCES, HOT SPOTS, AND CONTINENTAL BREAK UP

LOCATION	Age 1	Volume 2	Area 3	Rifting Age 4	Related Hotspot 5	
Columbia Basin	16 +/- 1	~250,000	~160,000		Yellowstone?	yes
Ethiopia-Arabia - 4	31 +/- 1	~1,000,000		30?	Afar, Africa	yes
North Atlantic - 4 Europe-Greenland	57 +/- 1	>1,000,000		53	Iceland	yes
India - Seychelles Deccan Traps	66 +/- 1	>2,000,000	~500,000	60	Reunion Island, Indian Ocean	yes
New Z.-Antarct./Austr. -4	~84 - 100	extensive		~84	Balleny	no
Madagascar-India - 4	~88 +/- 1	extensive			Marion	no
Antarctica/Austr.-India 4 Kerguelen	~116 +/- 1	?	~1,600,000	115	Kerguelen	no
Ontong Java, Indian Ocean	120	~36,000,000				
Parana, Brazil- Etendeka, SW Africa - 4	132 +/- 1	>1,000,000	~1,500,000	127	Tristan da Cunha, S. Atlantic Ocean	yes
Ferrar, Antarctica - Karroo, SE Africa - 4	183 +/- 1	>2,000,000	~140,000	170	Bouvet/Crozet	no
Central Atlantic - 4 NW Africa-E. U.S.	201 +/- 1	>1,000,000	>2,500,000	175		no
Siberian	249 +/- 1	>2,000,000	~1,500,000		Jan Mayen?	no
Emeishan (SW China)	258	?	~250,000	?		no

1 = age of peak igneous activity in millions of years
2 = Volume in cubic kilometers of basalt
3 = Area in square kilometers of basalt
4 = rifting - continental breakup - denoted by basaltic seafloor where known
5 = Presently active hot spot

REFERENCES: Courtillot, et al., 1999; Hooper, 2000; Huffman1990; MacDougall, 1988; Rampino and Self, 2000; Rampino and Stothers, 1988; Storey, 1995

Commonly, a third rift arm of a hot spot stops before it is fully developed, but the two active arms may link up to other hot spot rift arms and then initiate seafloor spreading.[6] (The "failed" arm is called an aulacogen). In this way, active hot spots and two of their linked rifts (arms) may have led to the breakup of Gondwanaland, starting 130 million years ago, and the birth of the South Atlantic Ocean.

While the timing is close between hot spot volcanism and the South America-Africa rifting, it is difficult to establish true causes and effects. The plume (or not)-driven Tristan hot spot may have initiated continental breakup, or provided extra force that drove rifting, or simply been a consequence of seafloor spreading.

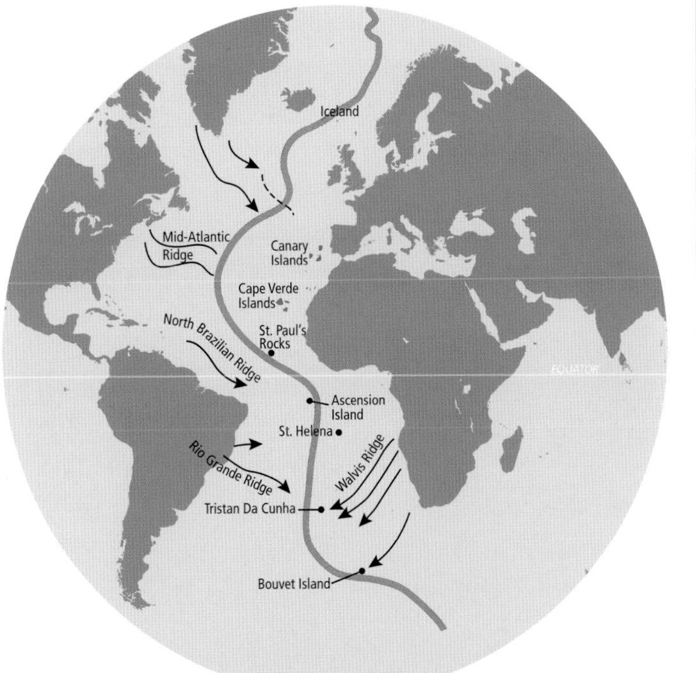

FIGURE 5.7: *Hot spots and the formation of the Atlantic Ocean. Note the various hot spots—the island of Tristan da Cunha in the South Atlantic Ocean, and the Rio Grande and Walvis volcanic Ridges—that mark the paths of South America and Africa, respectively, away from each other after the initial break up starting about 128 million years ago. Note that both Tristan da Cunha and St. Helena (oldest flows date to 132 million years ago) are east of the present spreading center axis. The Mid-Atlantic spreading center "jumped" westward to its present location about 30 million years ago when the African plate essentially stopped moving eastward (modified from Burke and Wilson, 1976).*

Atlantic Ocean seafloor spreading and continental lithospheric breakup proceeded northward, resulting in Africa separating from South America and then North America, eventually forming the North Atlantic Ocean. About 57 million years ago, the North American lithospheric plate finally separated from the Eurasian plate along what is now the east coast of Greenland and west coast of Norway. Vast outpourings of basalt accompanied this final split, a sample of which may be seen in Northern Ireland at the Giant's Causeway.

Seafloor spreading continues in the Atlantic Ocean but at different rates along the ridge. The spreading in the North Atlantic is about 4 centimeters per year, or about the growth rate of human fingernails. If the Atlantic Ocean is widening, it must be at the expense of the Pacific Ocean. In fact, the Pacific Ocean is narrowing and will eventually disappear as the west coasts of North and South America attach themselves to the east coasts of Asia and Australia. The Atlantic Ocean will be huge, and cover everything in between the coasts of the "new Pangaea."

Is continental breakup still occurring? Yes. The best example is the rift system of east Africa. There, lithosphere on the east side of the rift is separating from the bulk of the continent to the west. Interestingly,

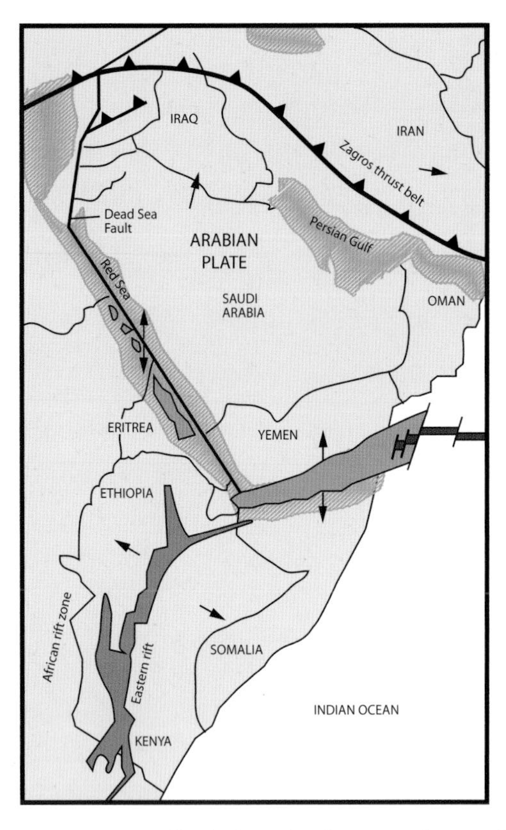

FIGURE 5.8: *East Africa/Afar Triangle - Note the complexity of rifting in east-central Africa. In places, two rift zones—the East African Rift Valleys located south of the map area, exist. Rifting is widening the Red Sea and the Gulf of Eden. The Afar triple junction is where the Red Sea, Gulf of Aden, and Djibouti come together. Arrows denote directions of extension. The African lithospheric plate itself is slowly moving northward, impinging on Europe, and closing the Mediterranean Sea as the eventual suturing of northern Africa to southern Europe occurs (modified from Chernicoff et al., 2002).*

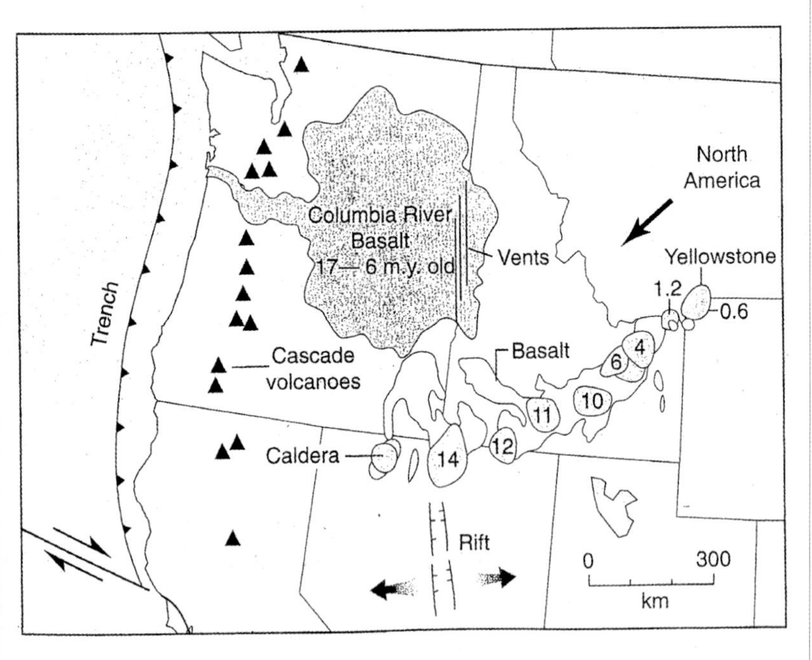

FIGURE 5.9: *Western U.S. with Cascade subduction zone-type volcanoes, the Yellowstone hot spot track, and the Columbia River Flood Basalt province. Note the age progression of hot spot volcanism from southwest to northeast across southern Idaho, in millions of years, due to the motion of the North American lithospheric plate to the southwest (note arrow). Presently North America is moving at about 4 centimeters per year (modified from Murphy et al., 1999; Hamblin and Christiansen, 2004).*

long before plate tectonics was recognized, people referred to the two long, narrow, north-trending valleys in East Africa as the East African Rift. The mechanism causing this phenomenon was unknown at the time. Geologists, however, understood that valleys containing long lakes (Albert and Tanganyika) and volcanoes (Mounts Kenya and Kilimanjaro) were in a vast region undergoing extension or rifting[7] that started 30 million years ago.

The Afar Triangle containing Ethiopia, Somalia, and Eritrea lies over a hot spot. Here there are three active arms: (1) the south end of the Red Sea, (2) the north end of the East African Rift, and (3) the Gulf of Aden. Time will tell which rift arm fails (or if failure occurs), and so becomes an aulacogen or long, sediment-filled continental valley.[6]

YELLOWSTONE HOT SPOT

◆ Regional Overview

Rather than examining 120 hot spots, we'll look at the Yellowstone and Icelandic centers.

The reason for Yellowstone volcanism was long considered an enigma because of its nature and distance, over 1,000 km (620 mi) from the north-trending Cascade Range of subduction-type volcanoes. Also puzzling was the series of calderas, each successively older, in the Snake River Plain. Starting from the 14 to 16-million-year-old caldera at McDermitt, Nevada, that is 800 km (480 mi) southwest of Yellowstone, they are younger as you travel to the northeast. Discovered in part by drilling, the calderas and the bulk of their very silica-rich (rhyolite) out-pourings, were veneered by younger basalt flows. Some of these flows are only a few thousand years old and include those at the Craters of the Moon National Monument.

The northeast-aligned volcanic systems, like Yellowstone, erupted mostly rhyolite tephra and ash flows or ignimbrites. These ejecta outcrop along both edges of the Snake River Plain.

How long has this hot spot system been operating? What will happen next? The "Yellowstone hot spot" may be 60 million years old. Presently, North America is moving to the southwest over the hot spot at 4 centimeters per year. Therefore Billings, Montana, 200 km (125 mi) to the northeast, will lie above the hot spot in 5 million years. Billings will sit atop the thermal welt, as Yellowstone now does. The Yellowstone plateau, however, will subside as

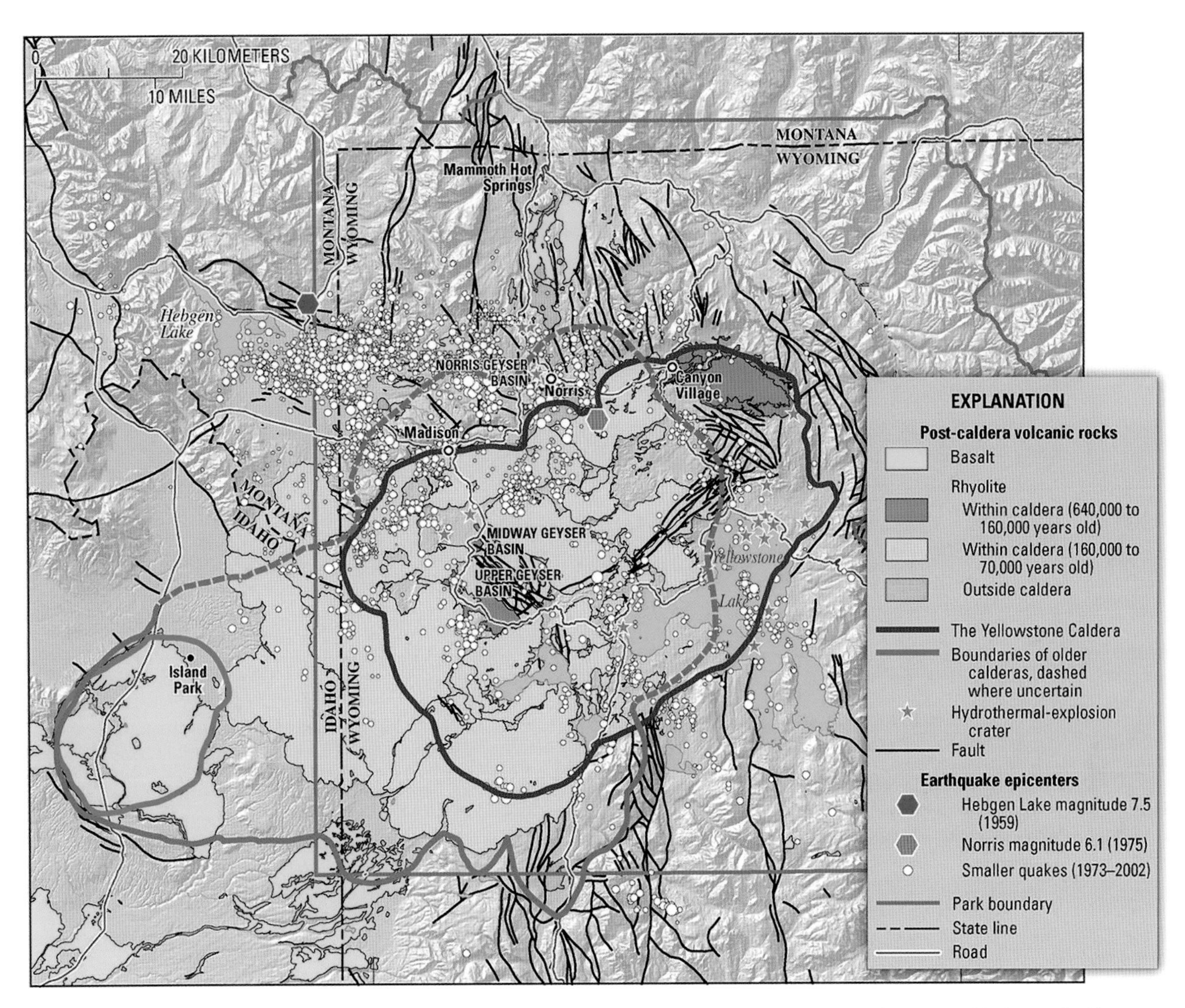

FIGURE 5.10: *Yellowstone Calderas - Note the approximate boundaries of the 3 calderas: 2.2-million-year-old Big Bend Ridge (I), 1.3-million-year-old Island Park (II), and 630,000-year-old Yellowstone (III). Two resurgent domes presently are within the Yellowstone Caldera. Irregular-shaped areas have geothermal activity. Regional faults and geothermal areas are shown as black, irregular patches (USGS, from Christiansen, 1984).*

it cools and joins the topographically lower Snake River Plain to the southwest.

But did Yellowstone form from a deep mantle plume? Because a deep low velocity zone does not exist under the hot spot, the answer may be no. It appears that hot spot melting is within the upper mantle and probably no deeper than about 200 km (124 mi).

◆ Yellowstone Caldera Activity

Volcanic activity started in Yellowstone National Park approximately 2.1 million years ago with huge, Plinian eruptions. About 2,500 km^3 (600 mi^3) of rhyolitic ash flow ejecta erupted, resulting in the development of the Big Bend Ridge Caldera. One of the world's largest, the caldera measures 30 km by 60 km (19 mi by 37 mi). In comparison, the 1980 Mount St. Helens eruption produced about 1 km^3 of ejecta and a caldera only 1 km (0.6 mi) across.

The 2.1-million-year-old ash flow deposit is called the Huckleberry Ridge Tuff. While much of this tuff has eroded away, outcrops still may be seen in and around the national park.

The second major Yellowstone caldera-forming eruption took place 1.3 million years ago. This smaller Plinian eruption created the Island Park Caldera west of the first caldera. It produced the 280 km^3 (70 mi^3) Mesa Falls Tuff.

The last caldera-forming eruption, about 640,000 years ago, was within the earlier Big Bend Ridge Caldera. This huge eruption released about 1,000 km^3 (240 mi^3) of ejecta,[8] resulting in the Lava Creek Tuff deposits and the Yellowstone Caldera. (The distribution of ash deposits from these three huge eruptions covered vast areas of the United States, as shown in Figure 3.9, page 41).

Since the last Plinian eruption, numerous intr-acaldera outpourings of rhyolite have occurred.

How Geysers Work

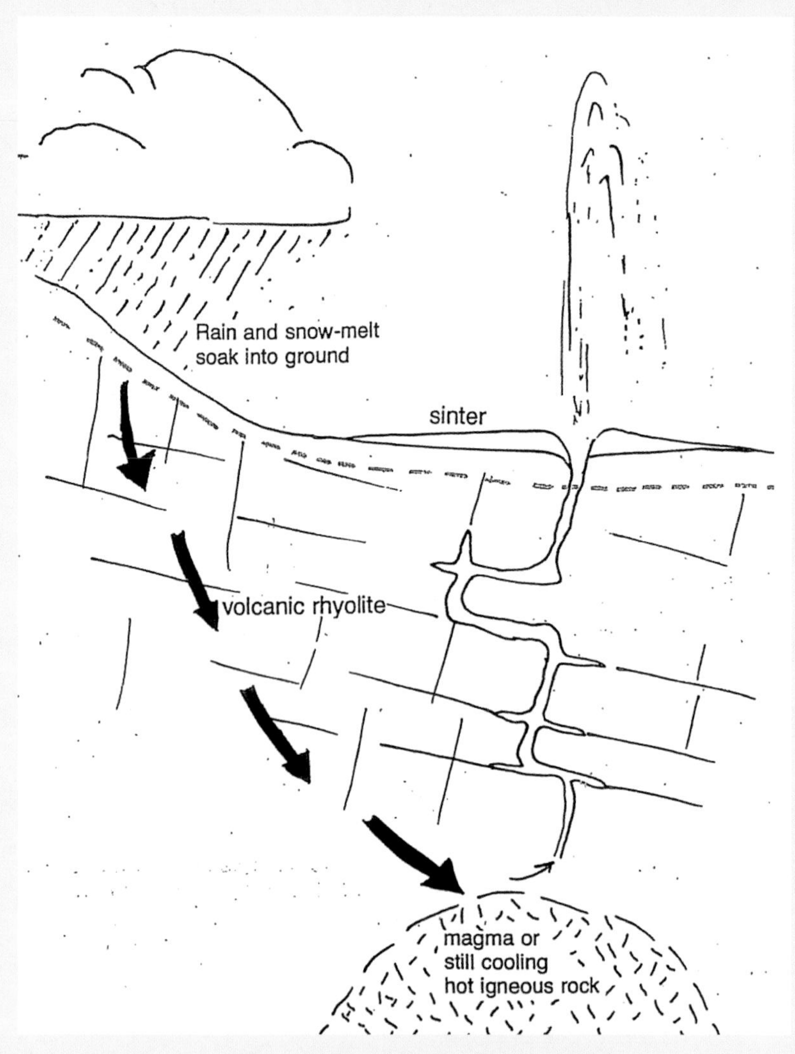

SB FIGURE 1: Cross section of a geyser system - For a geyser to operate there must be water, heat, and a plumbing system. Note the high water table (dashed line) or the top of the zone of water saturation in the ground. The column in the diagram is full of water below the WT, which means that when the water at the base of the column above the magma flashes to steam, the warm-to-hot water above is blown out (geysering occurs). The sinter mound around the base of the geyser builds from the precipitation of silica or calcium carbonate with each geyser eruption (modified from Bryan, 2005).

At least 99.9 per cent of visitors to Old Faithful Geyser in Yellowstone National Park do not understand how geysers work. While the principle is quite simple, conditions necessary for a geyser[1] to operate are rarely met. That's why there are so few geyser fields in the world. Other notable geyser fields include Larderello in Italy, El Tatio in Chile, the North Island of New Zealand, Iceland, and the Kamchatka Peninsula of eastern Russia.

The ingredients needed to create geysers are water, heat, a plumbing system of constricted vertical cracks or holes, and a near-surface groundwater table.[2] To fathom how geysers work, it's crucial to first understand the behavior of water.

Water is an unusual substance in many ways. The important characteristic in this discussion is that when liquid water converts to steam during boiling, it instantaneously expands in volume up to 1,600 times. That is, a cubic centimeter of water expands to up to 1,600 cubic centimeters of steam. This rapid, massive expansion is key to understanding how geysers work.

At Old Faithful Geyser (as in other geyser fields), the water table (the planar surface below which the ground is saturated with water[2]) is only meters below the ground surface. Also, the Yellowstone magma chamber several kilometers below heats the surrounding rock and water at depth. While water boils at 100°C (212°F) at sea level, it boils at only 93°C (193°F) at the 2,150-m (7,000-foot) elevation of Old faithful). This difference is due to atmospheric pressure. At sea level, the added weight of the atmosphere suppresses boiling until 100°C is reached. Drill holes to a depth of 332 m (1,088 feet) in the nearby Norris Geyser Basin have encountered liquid water at 241°C (465°F) due primarily to the increased rock load pressure at that depth, which suppresses boiling until even hotter temperatures are reached.

Old Faithful erupts because deep within the restricted, mostly vertical plumbing system of cracks, the water is heated until it boils at the base of the water column. There it converts instantaneously to vapor. The many-fold expansion at the base of the water column explosively forces the slightly cooler water above out of the restricted vent. This eruption or geysering continues until expansion of the basal boiling water is complete, with a major portion of the water in the column above forced out.

We now must wait 45 to 90 minutes at Old faithful for the process to repeat itself: water flows back into and fills the vent plumbing, basal water in the plumbing system heats to boiling, and then, instantaneously converting to vapor, it explosively expels water from the column above. Each geyser has its own repose or rest period between eruptions. Repose periods vary from minutes to years; Giantess Geyser recently rested for 2.4 years before erupting, then spouted and sputtered for more than 40 hours. Some geysers spray 3 m to 9 m high, while others blow water to 90 m above their vents.

NOTES

1 The word geyser comes from Geysir ("the gusher"), a geyser in Iceland's Valley of Haukadal.

2 The water table is a planar surface below which the ground (soil and/or rock) is saturated with water. It ranges from the surface in swampy areas to depths of hundreds to thousands of feet beneath the land in desert regions. In other words, if you are drilling a water well, in order to extract water you must penetrate this surface. The rate of flow to the well pump is a function of the permeability of the aquifer or 'bearer of water." Permeability is defined as the capability of a substance to allow the passage of a fluid. Porosity is defined as the percent void or space in the water-containing medium. Materials may have porosity but not permeability.

These deposits can be seen in Yellowstone Canyon and at Obsidian Cliff. Mantle-derived basalt flows also occur.

The great amount of heat continuously being liberated by the magmatic system heats the water that spouts from hot springs and geysers within and next to the Yellowstone Caldera. The hot water is not derived from the magma beneath, but is almost entirely recycled and heated precipitation--snow and rainwater. For more about Yellowstone, see the DVD *Supervolcano* produced by the British Broadcasting Corporation (BBC).

◆ Future Yellowstone Activity

Will the Yellowstone caldera erupt again? If so, how large an eruption might we expect? To date, caldera-forming events have been 600,000 to 800,000 years apart, with smaller, intracaldera eruptions between. Using that recurrence interval and other parameters as guides, most geologists believe Yellowstone will erupt again. Indicators of possible future eruptions include (1) the enormous amount of heat liberated in geysers, hot springs, and heated ground, together with (2) the geophysically determined low seismic velocity under the caldera. This indicates magma at depth, with the top of the magma chamber 6 to 7 km (3.7 to 4.4 mi) below the surface.

In addition, since formation of the Yellowstone Caldera, the Mallard Lake area (site of Old Faithful Geyser) and the Sour Creek region have risen. This is probably due to recharging of the magmatic chamber or injection at depth by new magma. Interestingly, uplift in parts of the caldera accelerated starting in 2004, with velocities up to three times faster (7 cm per year) than previously observed.

◆ Source of Rhyolite

But where does the erupted rhyolite come from if hot spots produce mantle-derived basalt? Geochemical and isotopic evidence show how this caldera system works, and the rhyolite source. At Yellowstone, basalt rises to the base of the continental crust, where it ponds. And because the temperature of basaltic magma (at least 1,200°C) is much higher than the melting temperature of the granitic continental crust (700 to 900°C) above, melting forms rhyolitic magma. This rhyolite magma then rises into upper crustal level magma chambers.

Eventually, gases, derived from the cooling rhyolitic near the upper part of the magma chamber, pressurize to where the crustal rocks above rupture, thus allowing the highly gas-charged rhyolitic magma to vesiculate as it surges upward. Plinian-type eruptions ensue up these vertical cracks, generally curvilinear in map view.

How long might eruptions last? Clearly, with volumes on the order of hundreds to thousands of cubic kilometers, eruptions probably last days to possibly weeks. Fortunately, modern man has never experienced an event of this scale. Such an eruption would catastrophically affect huge regions, altering the Earth's climate, food production, and the global economy for years (see Chapter 11).

ICELANDIC HOT SPOT

Not only does the Icelandic hot spot track depict an irregular path of the North American lithospheric plate (or the hot spot), but both earthquake data and seismically determined crustal structure show it presently lies near—but east of—the axis of the Mid-Atlantic spreading center where it built a huge island, Iceland.

The hot spot may have been active for 130 million years. In the beginning, Ellesmere Island in the Canadian Arctic was situated over it. The hot spot may also have built the north-trending Mendeleyev Ridge in the Arctic Ocean. But deciphering this hot spot track has been more difficult than deciding where the linear Hawaiian hot spot was at times in the past (see Figure 5.3).

About 40 million years ago, the hot spot would have been beneath Greenland's east coast. But determining its exact position is made difficult by today's 2-km- to 3-km-thick ice sheet covering 1.7 million km^2 (660,000 mi^2) of Greenland. Also note on Figure 5.3 that two 70-million-year-old hot spot locations exist. They are really one in the same because the crust spread apart here between previously connected Baffin Island and Greenland, creating Baffin Bay. Even more confusing, by 36 million years ago, the North American plate had changed from a northwesterly trajectory to a more westerly path.

Why, then, is the hot spot centered 240 km (146 mi) east of the spreading center? Apparently the axis of the Mid-Atlantic Ridge shifted westward, as also occurred in the South Atlantic Ocean 30 million years ago, leaving the Tristan da Cunha hot spot east of the spreading center axis. Why this shift happened in the North Atlantic Ocean is not known.

AFRICAN HOT SPOTS

The African lithospheric plate has more hot spots—a total of at least 20—than any other plate (see Figure 5.2). But upon examining the plate's tectonic history, between 90 million years ago and about 30 million years ago, Africa apparently had few hot spots. This change in "hot spot personality" is because prior to 90 million years ago, and from 30 million years ago to the present, the African plate was moving very slowly, if at all. When at rest, even "weak" mantle-derived hot spots can penetrate or burn through lithospheric plates. This produces doming, intrusions, and, in some places, volcanism.

When plates move, especially at rapid rates, only vigorous hot spot systems are capable of penetrating the lithosphere. Presently, hot spots are abundant on the very slow-moving African plate. Conversely, the fast-moving South American plate has just two hot spots, with one on the thin oceanic lithosphere.

The African situation, then, reinforces the conviction of how powerful the Yellowstone, Icelandic, and Hawaiian hot spots are as they burn through their moving lithospheric plate hosts.

CHAPTER 5 – NOTES

1 Flotational equilibrium of crustal material on the plastic mantle is called isostasy. To visualize isostasy, consider an iceberg floating in the sea. The folks on the *Titanic* in 1912 found out, if they didn't already know, that nine-tenths of the berg was below the sea surface. As bergs melt, they maintain the same ratio or flotational equilibrium in water: a cubic block of ice 100 feet on a side would expose only 10 feet of ice above sea level, whereas a block 10 feet on a side would protrude only one foot above the water surface.

If ice was denser than liquid water, it would sink and we would not be around to discuss this topic. This is because the world would be a much colder planet, and the oceans would be filled with ice.

2 The change in direction from north to northwest of the Pacific Plate 43 million years ago coincides approximately in time with the start of the collision of the Indian Subcontinent, moving northward, with Asia. It is theorized that this collision reoriented Pacific Plate motion from north to northwest.

3 The Hawaiian-Emperor hot spot track is linear rather than curvilinear, which supports the contention of linear motion of the Pacific Plate. This simple straight-line motion is not true for many lithospheric plates. For example, Figure 5.3 shows that the North American Plate has undergone erratic motion.

4 Danish meteorologist Alfred Wegener (1870 -1930) proposed the concept of continental drift in 1912 after carefully studying the configuration of present-day landmasses, especially South America and Africa. He coined the term Pangaea in 1927. Support for the continental drift theory includes glacial, fossil, rock composition, and other geological data.

5 St. Helena, in the equatorial South Atlantic Ocean, is British-controlled, and the location of Napoleon's exile starting in 1815, after his defeat in the Battle of Waterloo.

6 The failed rift arm was named an aulacogen or "born of furrows" by Russian geologist Nickolas Shatsky in the 1940s. Because aulacogens are failed rift valleys or depressions, they now contain major rivers such as the Amazon along the northeast coast of South America and the Niger River on the west coast of Africa. In addition, they fill with sediment and so may contain petroleum. One classic example is the major oil-producing Niger River delta of Nigeria, which happens to be the aulacogen associated with the St. Helena hot spot.

7 Crustal extension, or rifting, results from tension or the lateral pulling apart and thinning of crust. Long, linear valleys, which initially result from tension, widen and deepen over time through down-drop faulting. The deepest valleys on Earth such as Death Valley in California and the Dead Sea between Israel and Jordan formed by rifting. The rift valleys of East Africa will eventually drop below sea level and fill with water. This happened to the Red Sea rift to the north, separating the Arabian Peninsula from Africa. Rifting also occurs when the land surface is domed upward, and for other reasons.

[8] One thousand cubic kilometers of ejecta would cover about 30,000 km^2 varying in thickness from a few to more than 100 meters.

CHAPTER 5 – REFERENCES

Abbott, 2004; Anderson, 2005; Anderson and Natland, 2005; Anderson and Schramm, 2005; Anderson et al., 1992; Burke and Wilson, 1976; Chang et al., 2007; Chernioff et al., 2002; Christiansen et al., 2002; Clague and Dalrymple, 1987, 1989; Clouard and Bonneville, 2001, 2005; Courtillot, et al., 2003; Geist and Richards, 1990; Gudmundsson, 2000; Hamblin and Christiansen, 2004; Huffman, 1990; Humphreys et al., 2000; Koppers, 2011; Lange, 2002; Lawver and Muller, 1994; MacDougall, 1988; Moore and Clague, 1992; Moore et al., 1989; Morgan, 1971, 1972; Murphy et al., 1999; O'Conner et al. 2012; Perkins and Nash, 2002; Pitt, 1989 (Yellowstone USGS seismic map); Rampino and Stothers, 1988; Sharp and Clague, 2002; Skinner and Porter, 1995; Storey, 1995; Vink et al., 1985; Wang and Liu, 2006; Wilson, 1963.

CHAPTER SIX

Calderas—The Big Bananas

Introduction

Calderas (the Spanish word for cauldrons) come in different shapes, sizes, and types, but most have one thing in common: they are volcanic basins generally having steep interior walls. Most calderas form by subsidence into a partially emptied magma chamber. Subsidence occurs either during or following an eruption. We know that many calderas form during an eruption because we've found thick intracaldera pyroclastic fill associated with major eruptions.

Calderas also form by the enlargement of a preexisting crater during eruptions, by crater wall collapse, and by cone collapse (land sliding). The latter mechanism initiated the 1980 eruption of Mount St. Helens by lowering the overlying containment pressure on the magma chamber below.

The Earth is pocked with 225 large calderas that are 2.5 million years or less in age and with diameters of 5 km (3 mi) or greater. At least 138 of these have been restless—they exhibited some change from dormancy during historical times. And in any given year about 19 of the 225 are restless, with 5 erupting on average each year. Furthermore, a striking correlation exists between caldera area and the volume of eruption ejecta: the larger the caldera, the greater the amount of material that has been or may be subsequently erupted.

Mount Mazama—Crater Lake, Oregon

To better understand what calderas are and how they form, let's examine a classic: Crater Lake, Oregon, which also happens to be one of the most beautiful places on Earth. The lake resulted from the destruction of ancient Mount Mazama.[1] Geological reports in the early 20th Century postulated that Mount Mazama volcano was about the size of Mount Rainier, or about 4,267 m (14,000 feet) tall, prior to blowing its top and forming Crater Lake.

More recent research indicates its former elevation was about 3,658 to 3,960 m (12,000 to 13,000 feet), or similar to that of Mount Adams in Washington State, and that it didn't "blow its top." Instead, we now know that Mount Mazama experienced a huge pyroclastic eruption from an upper level magma chamber(s), with subsequent cone collapse into the partially emptied chamber(s) below. Prior to that eruption, Mount Mazama had been growing for 400,000 years, accumulating lava, ash, and mudflows of andesitic and more silica-rich materials.

Rather than resembling the symmetrical single cones of Mount Hood or the former Mount St. Helens, Mount Mazama was a complex assemblage of cones. It had a large basin glacially carved into the north side of the massif. The basin formed during the late Pleistocene ice age that ended about 10,000 years ago. This period of glaciations coincided in part with Mazama's dormancy, which lasted between about 30,000 and 7,700 years before the present day.

Then just before 7,700 years ago, a series of rhyodacite eruptions emanated from two vents on the north flank of the mountain. The huge, thick lava flow seen at Llao Rock that fills the U-shaped, glacially carved valley resulted from one eruption (see Figure 6.2A). It not only preceded the mountain-destroying eruptions, but is also thought to have blocked the vent, thus allowing pressure to build for the final cataclysmic events.

The climactic eruptions leading to the destruction of the strato volcano began 7,700 years ago. According to studies by Charles Bacon, the first event was violent, explosive Plinian-type venting of ash into the stratosphere from a source north of the main summit of the volcano. Prevailing winds carried this material away from the mountain to

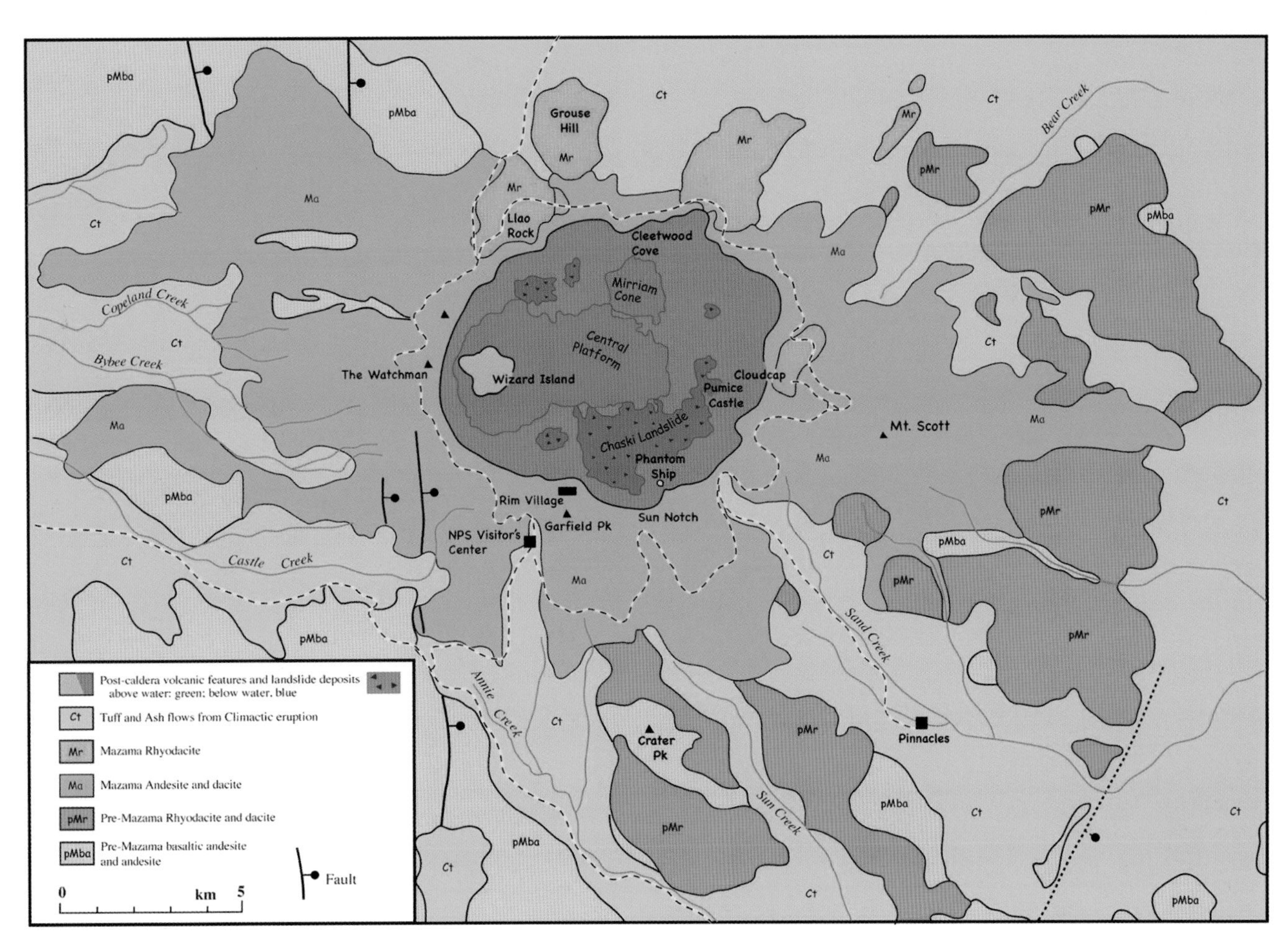

FIGURE 6.1: *Generalized geologic map of Mount Mazama and vicinity (modified from Bacon, USGS SIM-2832 by Marli Bryant Miller, University of Oregon).*

the northeast, covering at least 1.3 million km^2 (500,000 mi^2) with ash. Deposits up to 6 m (20 feet) thick are present around the base of the mountain while air fall thicknesses of 1 foot are found 113 km (70 mi) to the northeast.

Next, the Plinian tephra/gas column collapsed, resulting in the Wineglass pyroclastic flow that raced northward. This was followed by numerous pyroclastic flows as new vents opened on the mountain. One pyroclastic flow traveled 64 km (40 mi) down the Rogue River to present-day McLeod. And so the eruption proceeded from various vents on the mountain. Pumice blocks as much as 2 m (6 feet) across were swept 32 km (20 mi) away from their source, and lowlands around the mountain buried with as much as 80 m (250 feet) of hot tephra. Some pumice even reached Klamath Lake, more than 45 km (26 mi) away, after being transported by the Williamson River from its initial eruptive location in Klamath Marsh.

Probably coincident with these happenings was the collapse of the cone of Mount Mazama, along a steeply oriented ring-shaped fracture, into its emptying magma chamber below. What was once a large, complex 3,960-m (13,000-foot) tall strato volcano became a mammoth, steaming hole, or caldera 1,200 m (4,000 feet) deep and 10 km (6 mi) in diameter.

Estimates of the amount of erupted material vary from 46 to 58 km^3 (11 to 14 mi^3). With time, the caldera filled with water, making Crater Lake the deepest lake in North America at over 580 m (1,900 feet). Volcanism, however, didn't cease with the formation of the caldera.

Andesitic outpourings, apparently erupted from different places along the now submerged ring fracture (see Figure 6.3), have partially filled the caldera. Some of the small intracaldera volcanoes do not breach the surface of Crater Lake except for the 6,000-year-old cinder cone and lava flow of Wizard Island.

Is the volcano extinct? A manned Oregon State University submersible several years ago explored the deep, lightless bottom of the lake and found hot springs with strange looking white, organic material growing around them. This situation is analogous to what has been found around thermal vents in mid-oceanic spreading centers described

FIGURE 6.2A: *View to the northeast across the caldera to Llao Rock, a large, dark colored rhyodacitic lava flow that filled a glacially carved, U-shaped valley just before (perhaps 200 years) the 7,700-year-old, caldera-forming eruption. Wizard Island is on the left (Lange photo).*

FIGURE 6.2C: *Lava flow resting on top of an earlier lahar or mudflow, Crater Lake (Lange photo).*

in Chapter 2. Clearly heat is still being generated beneath the caldera lake.

If we look at the history of the mountain and other Cascade volcanoes, we see that they commonly have long periods of dormancy, sometimes measured in tens of thousands of years. Mount Mazama, then, may sometime again erupt, fill in Crater Lake, and build a new cone. We shall just have to be patient.

Caldera Generalizations

Besides actual eruptions, calderas can exhibit restlessness in a number of ways. These include (1) magmatic- or tectonic-induced tremors, (2) changes in the gas chemistry of fumaroles, (3) changes in

FIGURE 6.2B: *Crater Lake Wineglass ash flow outcrop seen from Godfrey Glen turnout above Annie Creek Canyon. Note color gradation upward due to chemical changes as upper, more silica-rich rhyodacite portion (pumice clasts with 70.4 percent silica) of the erupting magma chamber (light yellow) was succeeded by lower silica crystal-rich hornblende-bearing andesite magma erupted from deeper portions of chamber (composition of clasts range from basalt to andesite with about 20 percent silica-rich pieces) (Lange photo).*

FIGURE 6.2D: *North wall of Crater Lake caldera wall showing a dark gray-black vertical dike cutting the composite layering of lavas, tephra, and mud flows (Lange photo).*

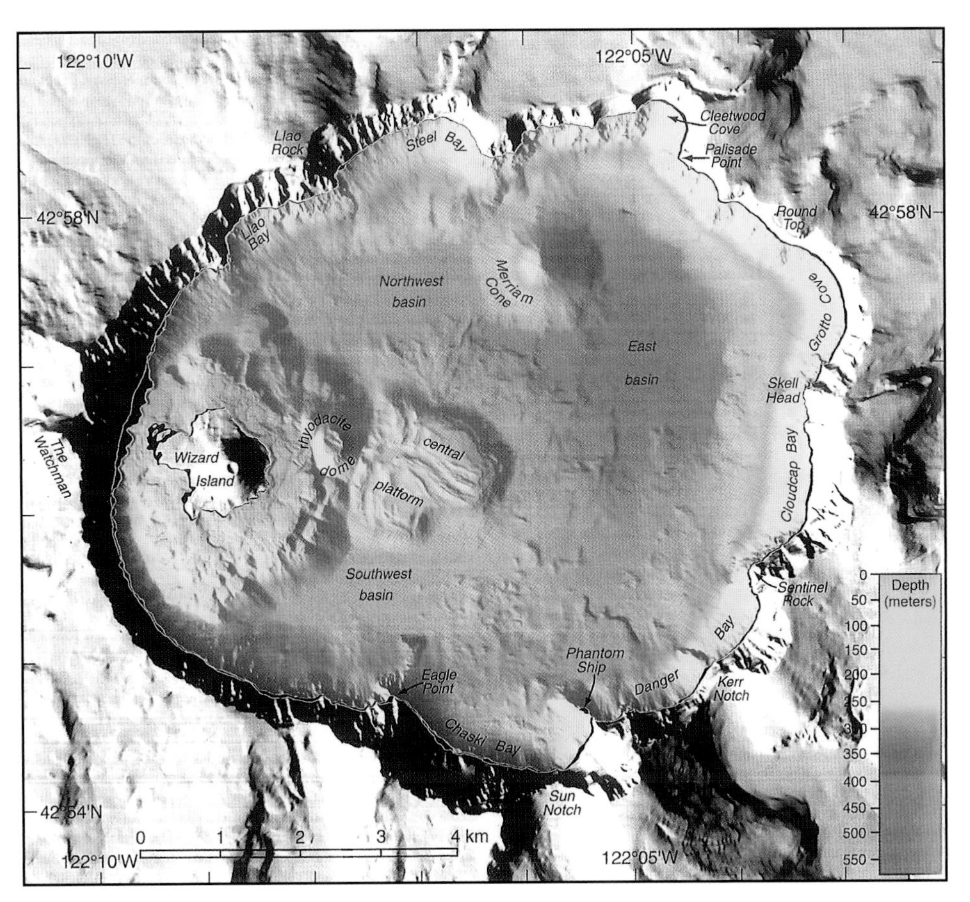

FIGURE 6.3: *Bathymetry of Crater Lake. Note Wizard Island Volcano of andesite composition and the other smaller, post-caldera volcanoes below lake level. These small volcanoes probably arose along the major ring fracture along which the cone collapsed into the partially emptied magma chamber below (Bacon et al., 2002, USGS).*

electrical conductivity, magnetism, and gravity, (4) caldera surface uplift, subsidence, or tilting with or without ground fissures, and (5) temperature changes in soil, water, or gas on or emitted from the volcano (see Chapter 12 for details).

The world's largest recently active (in geologic terms) caldera, Toba, is found near the northwest end of the island of Sumatra in Indonesia. Subduction zone-related but also in a tensional environment, Toba measures 100 km (60 mi) by 30 km (18 mi) and exists in a tectonic or fault-bounded depression. It last erupted about 73,500 years ago when between 2,500 km^3 (600 mi^3) and 3,000 km^3 (720 mi^3) of pyroclastic material was blasted into the atmosphere (see Famous Calderas sidebar, page 104). While most calderas don't erupt such huge amounts of material, some much smaller volcanoes, such as Mount Mazama in Oregon, have produced impressive amounts of ejecta.

Calderas have been divided into three types based on their volcanic features and tectonic host environment. About 70 percent form in stratovolcanoes, mostly associated with destructive or subduction zone volcanic arcs. Another 15 percent are hot spot-related, and the remaining 15 percent develop in tensional environments where the crust is being stretched or pulled apart.

Katmai in Alaska, Mount St. Helens in Washington, and Mount Mazama (Crater Lake) in Oregon are examples of arc-related calderas. Summit calderas on Hawaiian volcanoes and the Yellowstone system are hot-spot examples on oceanic and continental lithosphere, respectively. Long Valley Caldera in east-central California developed in a tensional regime.

Compositionally, calderas range from silica-rich (producing ejecta with more than 57 percent silica) to those of mafic[2] composition (between 50 and 57 percent silica). Neither compositional type appears to erupt more frequently than the other.

Arc-related calderas are generally smaller than hot-spot and tensionally produced calderas. They form by an enlargement of the summit crater of their stratovolcano hosts or by cone collapse. Commonly andesitic to dacitic in composition, arc calderas have diameters between 2 and 13 km (1.2 and 8 mi).

Tensional and hot-spot calderas, except basaltic shield volcanoes such as those found in Hawaii, are usually not tall constructional features like stratovolcanoes, and so it has been difficult to discover the source caldera for some, huge, ancient ash flow deposits. Satellite imagery of the Earth's surface, however, has helped with the discovery some of these obscure, huge calderas. Two of the most exciting recent discoveries include the gigantic La Pacana and Cerro Galan Calderas in the Andes Mountains of northern Chile in South America.

A striking example of a subtle-appearing caldera is Long Valley on the east side of the Sierra Nevada in east-central California. U.S. Highway 395, the main road linking Reno and Carson City, Nevada, to the north with Bishop and southern California to the south, crosses this 30-km (18-mi) by 15-km (9-mi) east-west trending structure. But most tourists who use the road are unaware as they drive over the dormant but restless and very much alive, flat-floored beast because there is no high volcanic mountain or strikingly steep caldera walls. What is apparent are just a few steaming areas along the highway and a wonderful hot spring used by Mammoth Mountain skiers and locals for soaking. Examine an aerial photograph of the region, however, and the caldera is readily apparent!

Caldera Geometries

The diversity of caldera structures may be seen in Figure 6.4, with gradations between the main types. The simplest geometric structure is the plate or piston type, common in large calderas. The roof or plate that subsides during or after a Plinian eruption into a partially emptied magma chamber

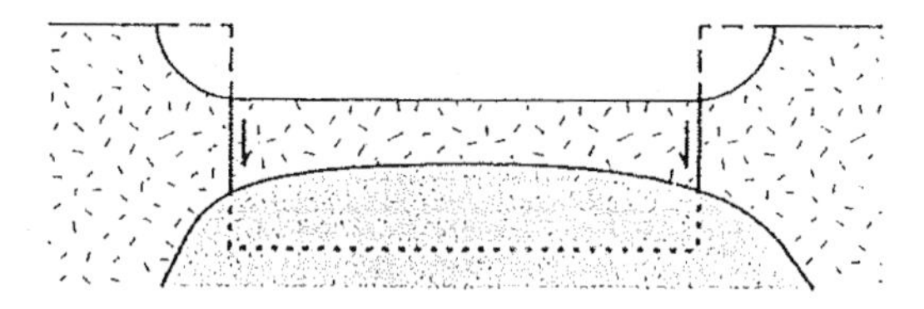

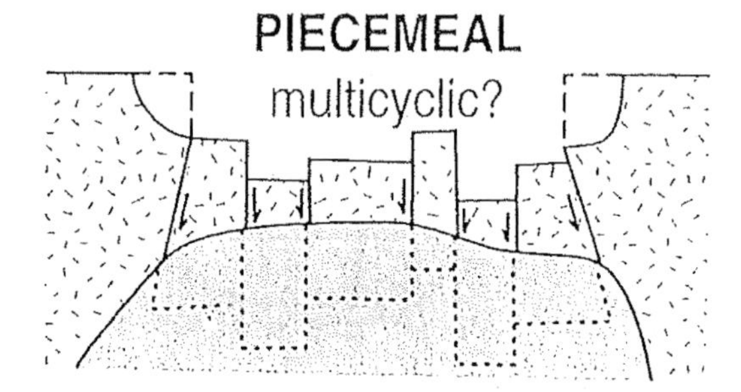

FIGURE 6.4: *Models of caldera geometries (modified from Lipman, 1997, 2000).*

below slips down a ring fracture fault. There may be several kilometers of vertical offset along the fault. Long Valley Caldera is a classic example of this type of structure.

Plate-type caldera structures also form above smaller volcanic systems, such as Crater Lake (Mount Mazama), Oregon. As typically occurs in big plate-type structures, the oversteepened caldera walls well outboard of the eruption-produced ring fracture sloughs off, enlarging the original depression (see bathymetric map of Crater Lake, Figure 6.3). How do we know where the ring fracture is in Crater Lake? The best evidence is the distribution of volcanoes within the caldera, including the Wizard Island volcano, whose location was apparently governed by the ring fracture.

During some Plinian eruptions, the caldera floor, rather than subsiding as a single plate, breaks into pieces. It has been theorized that this piecemeal plate subsidence may be due to eruption dynamics, successive small eruptions, the influence of basement structure below, and/or the interference of overlapping calderas.

Determining whether the caldera is of the piecemeal type is difficult without drilling, structural exposure after a significant amount of erosion, or geophysical studies including seismographic surveys. Examples of possible piecemeal calderas include Campi Flegrei in Italy, Rabaul in Papua New Guinea, calderas in southwestern Nevada, and the San Juan volcanic field of southern Colorado.

With trap door-type subsidence, collapse is incomplete due to the plate being only partially surrounded by a ring fracture. Trap door-type structures may form due to small eruptions, asymmetrical magma chamber geometry, and regional structural influences. The Silverton caldera in the San Juan volcanic field of southern Colorado may be an example.

Down sag-type subsidence appears to be common in large caldera systems. This structural type is characterized by inward sloping topography and strata, and lack of a bounding ring fracture with large offsets. Down sag subsidence, rather than plate collapse, may happen due to ash eruptions from a deep magma chamber. It may also result, in some cases, from the draping of tuff over deeply buried, previously active plate-type calderas.

Funnel-shaped calderas are typically small structures with diameters of less than 4 km. They may develop from subsequent sloughing inward of oversteepened caldera walls. It has also been suggested that they may form (1) from recurrent subsidence during a series of eruptions, (2) from explosive eruptions from a central vent rather than through ring fractures, and (3) above small magma chambers.

Geophysical data show that some large calderas are funnel-shaped. This may be due to upward vent flaring during pyroclastic eruptions from volcanoes situated in structurally weak crust. Other large calderas may take on an overall funnel shape after repeated eruptions coupled with subsidence along nested ring faults. The huge, 26-million-year-old San Juan volcanic field of Colorado may contain some calderas of this type.

Caldera Eruption Products

Calderas erupt the same materials that other volcanoes do: gases, tephra, lava, mud, and ash flows. But the catastrophic Plinian eruptions from some produce huge ash flows. Plinian eruptions with their resulting ash flows, described in Chapter 3, represent one of the greatest threats to humankind because of the speed, temperature, volume, and distance the products may travel from their source. They may also affect Earth's climate. Those who lived in Pompeii, Italy, in 79 A.D., or St. Pierre, on the Island of Martinique in the West Indies in 1902 would, if they could, collaborate these findings.

How much of a magma chamber can empty in a single catastrophic eruption such as the one that changed Mount Mazama into the Crater Lake caldera? Estimates place the amount at up to 10 percent. If true, the 35 km^3 of ejecta from Mount Mazama may have come from a magma chamber beneath the volcano that had a volume of 350 km^3 or more. Likewise, the 2,500 km^3 of material that erupted from the Yellowstone Caldera 2.1 million years ago may have come from a magma chamber with a volume of at least 25,000 km^3. Some single eruptions have produced over 3,000 km^3 (>700 mi^3) of ejecta, with cumulative totals from a caldera complex over several eruptions exceeding 10,000 km^3!

In some eruptions, ash flow material is so hot when it reaches the ground that the particles fuse together. These, called welded ash flow tuffs, are quite resistant to erosion and so form cliffs. Many have columnar jointing similar in appearance to some basaltic and andesitic lava flows. This phenomenon results from the same process—contraction during cooling of the hot material.

As the degree of welding increases, the specific gravity of the tuff increases and can reach 2.2 or more (water and granite have specific gravities 1.0 and 2.65, respectively, for comparison). This change is most remarkable considering that the erupted material starts as a hot, particle-bearing gas blast with a specific gravity of between 0.1 and 1.0.

We can estimate the amount of welding by looking at the degree of flattening of the contained pumice fragments. If we start with a spherical piece of erupted pumice, flattening following ash flow or tuff emplacement resulting in a 4:1 length-to-width ratio would produce a partially welded tuff. The specific gravity of this piece would range from 1.4 to 2.2. A length-to-width ratio of 8:1 results in a densely welded tuff and a specific gravity of greater than 2.2. Some welded ash flow tuffs, however, have length-to-width ratios of as much as 15:1 or 20:1. These result from secondary flowage following emplacement and are called reoignimbrites (see Figures 3.11B and 3.11C of unwelded and welded tuff or ignimbrite, page 45).

Why Such Large Calderas?

How do the very large, hot-spot and tensional calderas such as Yellowstone and Long Valley, respectively, develop? No one knows for sure because, fortunately, such really large calderas have not formed (or erupted) in historical times for us to witness. The current theory is that magma with a composition between andesite and rhyolite forms from the melting of silicic crustal rocks. The melting is driven by heat released by intruded, and much hotter, mantle-generated basaltic magma. The melted, mostly crustal-derived magma, rises to within perhaps 6 or 7 km (3 or 4 mi) of the surface. This ponded silicic magma so near the surface causes regional doming of the crust above. The intruded magma, if at rest for some time, chemically separates compositionally top to bottom. This separation is known as magmatic or chemical differentiation.

Differentiation occurs because the first minerals to form are rich in magnesium and iron or calcium. They then sink because their density is greater that that of the remaining magma. This process effectively enriches the upper part of the chamber in silica, which forms dacitic and rhyolitic magma. The lower part of the chamber, then, due to the settling crystals becomes more iron, magnesium (mafic[2]), and calcium rich (see Chapter 8). Volatiles—mostly water and carbon dioxide—also concentrate in the upper part of the magma, thereby increasing pressure on the confining magma chamber walls and roof rocks.

Vertical to steeply dipping cracks then form in the rocks above the roughly cylindrical chamber due to (1) doming, (2) increasing bulging of the magma chamber below, and (3) the brittle nature of the cooler crustal roof rocks. Finally, pressure within the magma chamber exceeds the strength of the confining roof rocks, and the eruption commences up the crack(s).

As with the uncapping of a shaken bottle of beer or soda, there is a rapid expansion up the fractures, commonly circular in map plan, of the gas-charged mixture of magmatic particles and pieces of rock ripped from the chamber walls. The latter are called lithic fragments, the former particles are termed juvenile components—fresh pumice, crystals, and glass shards.

Eruptions may go on for hours, days, or perhaps even weeks. We can only speculate as to how long the really large eruptions last, like those from the Yellowstone and Toba Calderas. Recent work, however, suggests that really large eruptions may last up to 100 hours.

Plinian eruptions result in air fall and then ash flows that can be devastating, as we know from examining the buried remains of Pompeii located near Mount Vesuvius, and Akrotiri on Thera in the Aegean Sea. Then, as the magma chamber empties, the central rock plug that forms the top of the magma chamber within the circular ring fracture structure collapses into the partially emptied chamber below. The resulting caldera may be partially or totally filled with erupted material, so a giant depression or collapse structure (caldera) may not be apparent.

After the eruption, fresh magma may reenter the magma chamber from below, thereby repeating the whole process at some later time. This phenomenon, termed caldera resurgence, is known to have occurred as many as four times, with each cycle culminating in a catastrophic Plinian eruption. The Yellowstone complex has witnessed three major eruptions over the past 2.2 million years.

Frequency of Caldera Eruptions

Of paramount importance to volcanologists, geologists, and the general public is not only the size but also the frequency, or recurrence interval, of volcanic eruptions. We know the interval for such famous volcanoes as Kilauea in Hawaii and Mount St. Helens in the Cascade Mountains thanks to either witnessing eruptions or doing detailed geology and radiometric age dating of past eruptions. Materials that can be dated include radioactive uranium- and potassium-bearing minerals and, if less than about 50,000 years old, carbonized plant material found buried around the volcanoes.[3] Estimating the recurrence interval of large calderas, however, is more difficult because the time span is usually hundreds of thousand of years or more. The same techniques, however, are useful—careful mapping and radiometric dating of the deposits associated with caldera eruptions.

For good reason, much attention has been paid to caldera systems in or near populated areas and to other really large calderas. Studies indicate that big caldera systems generally erupt catastrophically several times. In general, the bigger the system, the longer the repose period between large eruptions. The Yellowstone system erupted approximately 2.1 million years ago, 1.3 million years ago, and 640,000 years ago. Between these giant eruptions, new magma moved into the caldera system and bulged the surface above.

Resurgence is presently occurring in two places within the 640,000-year-old Yellowstone caldera. These surface bulges or domes are named for the prominent features near them: Sour Creek and Mallard Lake. In addition to bulging, these parts of the caldera have intense geothermal activity, including geysers and hot springs. Another caldera whose recurrence interval is known is Toba, Indonesia, which has erupted catastrophically four times in the last 1.2 million years, roughly every 340,000 to 430,000 years, with the last being about 73,500 years ago.

In addition, small eruptions of ash and lava and dome formation within calderas often take place between major Plinian eruptions. For example, there has been much rhyolitic intracaldera volcanism at Yellowstone in the last 640,000 years. These erupted materials fill the caldera and in some places obscure caldera walls.

While the interval between major eruptions varies between caldera systems, it is believed that mantle-derived basalt injections into the basal parts of the systems recharge, reheat, and eventually lead to catastrophic Plinian outbursts.

Recent research on the oxygen isotopic composition of the mineral zircon ($ZrSiO_4$) within ash deposits is shedding new light on the eruption process.

It is now believed that the ratio of the two most abundant oxygen isotopes,[4] ^{16}O and ^{18}O, in zircons (which are quite resistant to temperature and pressure changes), reflect the environment in which zircons crystallize. If the oxygen is mantle-derived, the zircons are ^{18}O-enriched (or ^{16}O-depleted). If the oxygen comes from near surface rocks that contain meteoric water (rain and snow), the zircons are ^{16}O-enriched and thus ^{18}O-depleted.

FIGURE 6.5: *Aerial satellite false color photograph of Long Valley Caldera lower part of photo and Mono Lake in upper right corner. Grass vegetation is red and forests are green in the image. Note the striking northeast wall of the subtle, 760,000-year-old caldera, with Lake Crowley cutting the southeast corner (USGS photo).*

After determining this isotopic information about zircons, researchers have found that if small Plinian eruptions contain ^{16}O-enriched zircons, new magma has not moved into the base of the system from the mantle. Rather, these small eruptions are produced by magmas formed by the remelting of some of the earlier, big eruption products that have been contaminated with oxygen from surface water. Major eruptions are therefore not anticipated in the near term. But if the zircons have a mantle signature of ^{18}O enrichment, then the system is probably recharging for possibly another huge eruption.

Caldera Stories

LONG VALLEY CALDERA, CALIFORNIA

No one would deny that the eruption of Mount Mazama 7,700 years ago was a cataclysmic event. But it pales in comparison with the eruption of Long Valley Caldera 760,000 years ago. About 650 km^3 (155 mi^3) of rhyolitic magma vented violently and covered more than 1,500 km^2 (580 mi^2) of east-central California and southwestern Nevada with air fall and ash flow deposits. One pyroclastic lobe of the Bishop Tuff alone flowed 65 km (40 mi) south, down Owens Valley, burying the landscape in places with hundreds of meters of welded tuff.

The pre-eruption magma chamber was estimated to be about 19 km (12 mi) across, with the top of the magma body within 5 km (3 mi) of the surface. Following the Plinian eruption of tephra and ash flows, the chamber roof dropped 2 km into the partially emptied chamber. Rather than producing a deep caldera, however, the hole was filled with ash flow ejecta. Later, much younger tephra and obsidian domes and flows spread within the caldera.

Today the 32-by-24-km (20-by-15-mi), east-west-oriented caldera is restless and resurging, having had swarms of earthquakes coupled with deformation starting in 1980. Additional activity occurred in 1983, 1989-1990, and 1997-1998. These tremors, some with magnitudes of 6.1 on the Richter earthquake scale, were felt as far away as San Francisco. The quakes were attributed in part to magma rising

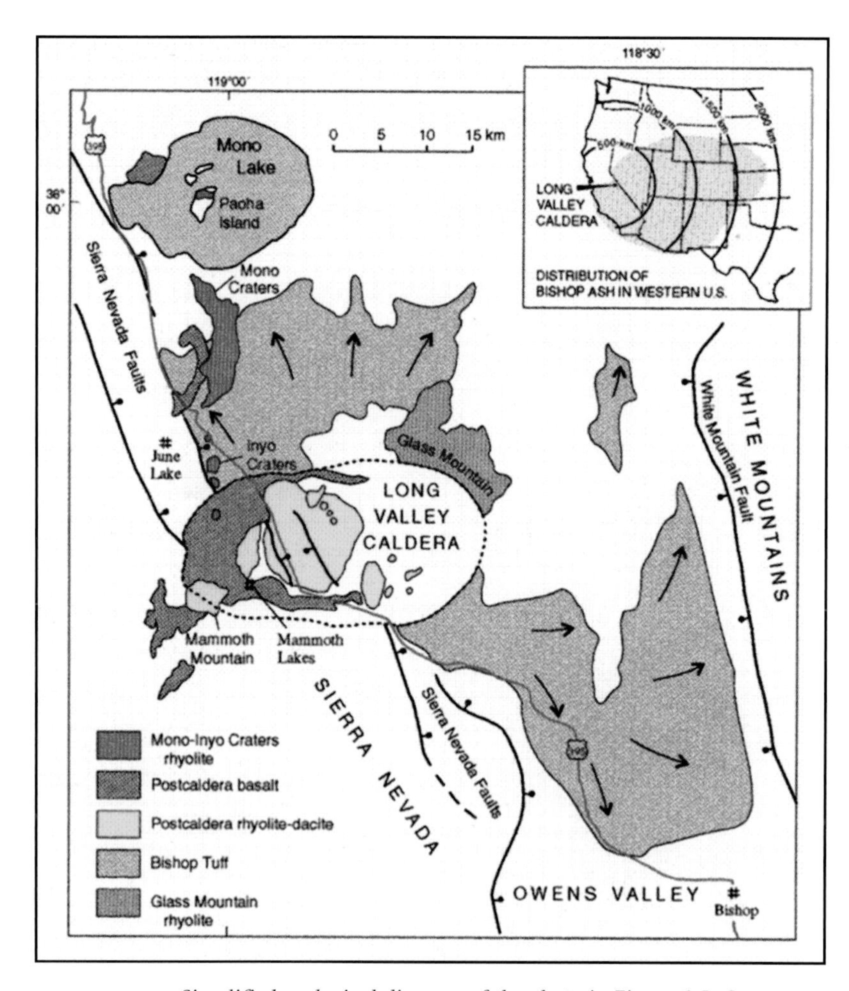

FIGURE 6.6: *Simplified geological diagram of the photo in Figure 6.5. Orange is Bishop tuff while yellow within the structure is intracaldera, post-eruption volcanic deposits. Glass Mountain in red is between 2.1 and 0.8 million years old. Drilling within the caldera indicates that more than 1,500 m (4,900 feet) of unexposed Bishop tuff lies below the intracaldera fill (USGS diagram).*

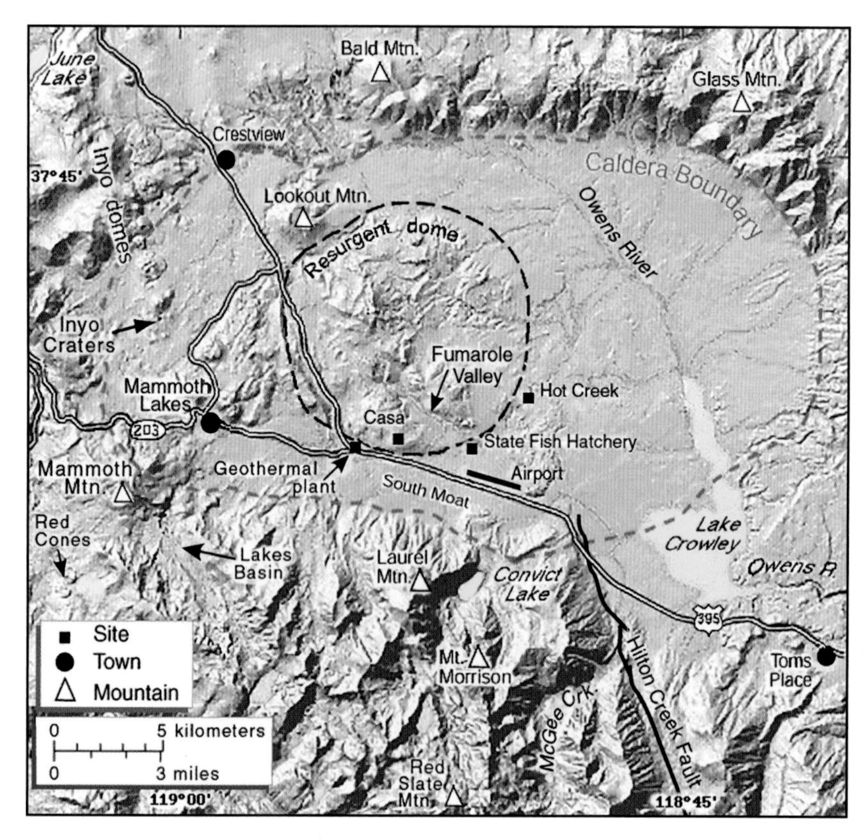

FIGURE 6.7: *Topographical relief map showing aspects of the Long Valley Caldera (USGS map).*

up within ring fractures along the southwest margin of the caldera to within a few kilometers of the surface.

Two magma chambers are believed to exist beneath the caldera. The top of the upper level magma chamber is about 10 km (6 mi) in diameter and between 7 and 10 km (4 to 6 mi) below the surface. The deeper chamber, located below the South Moat, is about 15 km (9 mi) beneath the surface. Research suggests that silicic magma and hydrothermal fluids fill the upper chamber.

Long Valley Caldera is just part of the Mono Lake-Long Valley volcanic-tectonic depression that witnessed basaltic eruptions starting some 3.5 million years ago. These were followed by andesitic and rhyolitic outpourings. For example, the northeast margin of the caldera cuts the Glass Mountain rhyolite flow complex dated at between 2.1 and 0.8 million years ago.

Volcanic activity, however, did not cease with the formation of the Long Valley Caldera. Some 200,000 years ago, columnar-jointed basaltic lavas extruded at today's Devils Post Pile National Monument to the southwest and at other locations to the north. Next, 3,362m (11,027-foot) Mammoth Mountain Volcano, site of a world-class ski resort, began rising along the southeast margin of the caldera. Mammoth Mountain is made up of at least 20 overlapping silica-rich domes and last erupted 57,000 years ago, just 10,000 years before the initiation of the north-south, Mono Lake-Inyo chain of rhyolitic eruptive centers to the north. Interestingly, episodic seismic unrest began under Mammoth Mountain in 1979, coupled with ground deformation and magmatically derived carbon-dioxide emissions. The gas, along with hydrogen chloride, escapes through fumaroles on the ski hill. Tragically, in 2006

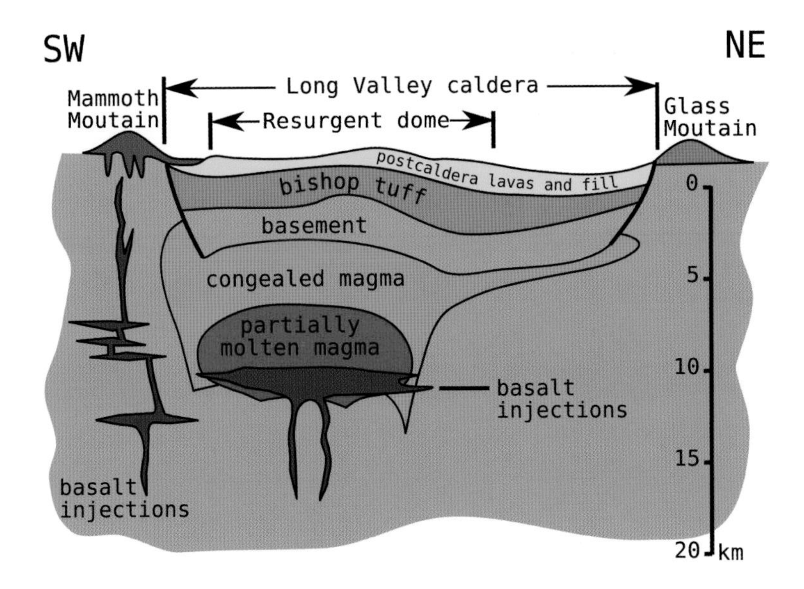

FIGURE 6.8: *Cross section oriented southwest to northeast through the Long Valley Caldera, California, depicting the area of 1980 uplift and 1980-1982 intrusions, which caused much seismic activity (USGS diagram).*

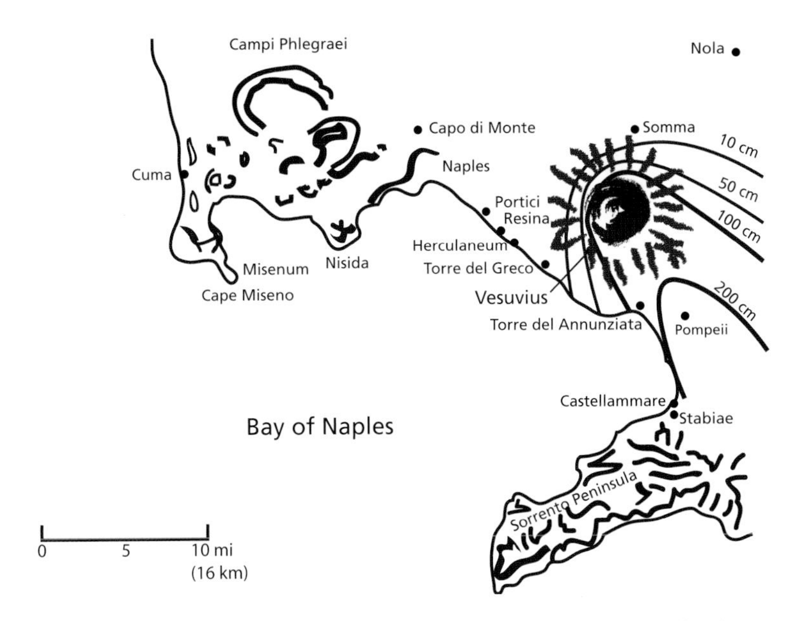

FIGURE 6.9: *Map of the Bay of Naples area, west-central coast of Italy, showing Mount Vesuvius, Campi Flegrei Caldera, Naples, and other cites in the area. Contours around Vesuvius are 79 A.D. eruption pumice fall-out depths (modified from Abbott, 2004).*

three ski patrollers died from inhaling the gases while trying to fence off a fumarole. Also, trees on about 100 acres at the base of the mountain have died due to high concentrations of carbon dioxide in the soil.

This most recent regional activity started about 40,000 years ago with the inception of the 45-km (28-mi) long volcanic chain of Mono-Inyo rhyolitic domes and short flows (see Figure 3.10A, page 44). These pristine-looking outpourings rose along a north-trending fracture system that extends from the center of Mono Lake south and through the western margin of Long Valley Caldera. In fact, at least 30 vents have been active in just the last 2,000 years, including Panum Crater (just a few hundred years old) immediately south of Mono Lake (see Figure 3.6A, page 37). Furthermore, in 1890, there apparently was an underwater eruption in Mono Lake.

With the documented increasing frequency of eruptions during the last 10,000 years as our guide, there is good reason to believe we will witness another eruptive event some time soon—in geologic terms. It will probably entail the formation of another rhyolitic plug dome, with accompanying tephra eruptions from one of these contiguous but discrete volcanic systems within the Mono Lake-Long Valley basin.

PHLEGRAEAN FIELDS, ITALY

Long Valley Caldera, should it erupt violently soon and without warning, would be catastrophic for those folks unlucky enough to be skiing on Mammoth Mountain or living in nearby Bishop, California. Possibly as many as a few thousand lives might be lost. The area around the caldera, however, is sparsely populated, with perhaps 10,000 living in the greater Bishop-Mammoth Lakes region near the ski resort.

The relatively small population of east-central California contrasts starkly to that in and around the Campi Flegrei or Phlegraean Fields (Burning Fields) Caldera in southern Italy. The caldera itself, with a diameter of 12 km (7 mi), contains at least 650,000 people, and is within an urbanized region home to more than 1 million people. Situated 8 km (5 mi) west of Naples on the coast, the caldera is also only 20 km (12 mi) from another killer, Mount Vesuvius.

Volcanism in the region began during the late Pleistocene, or about 34,000 +/- 3,000 years ago, with the formation of a 12-km (7-mi) wide collapse caldera. About one third of the caldera is submerged and forms Pozzuoli Bay. Caldera formation either accompanied or followed the eruption of the 80-km^3 (19-mi^3) welded Campanian tuff or ignimbrite. The 10-km^3 (2.4-mi^3) Neapolitan yellow

FIGURE 6.10A: *View of Roman Serapi market in Pozzuoli, within the caldera. Built in 105 B.C., it was used until the 4th century when land subsidence inundated the market. Eleven m (36 feet) of subsidence occurred until about 1000 A.D., when reemergence commenced. Note the dark rings on the columns marking submergence, the result of saltwater mollusk burrows of Lithodomus (Lange photo).*

FIGURE 6.10B: *A close up of a market column and mollusk burrows formed when the market was partially under water (Lange photo).*

tuff composed of ash and ash flows erupted about 11,000 years ago. Since then, and especially in the intervals between 10,000 and 8,000, and 4,500 and 3,700 years ago, extensive volcanism within the caldera resulted in the formation of cinder cones and phreatic explosion craters.[5]

What makes this caldera especially both interesting and dangerous is its documented restlessness over the last 2,000 years. Not only has seismic and fumarolic activity been abundant since Roman times, but there were explosive emanations of pumice in 1538, and decades to centuries of slow but steady uplift and subsidence of parts of the caldera floor.

Lest there be any doubt of changes in caldera floor elevation, one has only to view the famous Roman Serapi market in Pozzuoli, situated on the coast near the middle of the caldera. The market, built in 105 B.C., was used until the middle of the 4th century A.D. It has two floors, with the second having been built after the first sank below sea level. The 11 m (36 feet) of alternating subsidence and uplift is dramatically marked by the Lithodomus lithophagus marine mollusk bore holes in marble columns now exposed well above high tide.

The first mention of submerged buildings was about 840 A.D. By 1000 A.D., the area had subsided a total of 11 m (36 feet) at an average rate of 1 to 2 cm per year. From 1000 to 1198 the area rose 2 m (6.5 feet), and by 1538 the total uplift was about 12 m (39 feet), rising more than half, or 6 m (19 feet) after 1500 alone! Many earthquakes, culminating in the eruption of Monte Nuovo on September 29, 1538, accompanied this uplift. The size and number of earthquakes preceding the eruption, especially the quakes occurring on the Friday and Saturday before Easter of 1538 ". . . greatly increased the devotion of the people."[6]

Between 1538 and 1969, the Pozzuoli area subsided at an average rate of 1.5 cm per year. Since 1969, however, the area has been rising at variable rates commonly accompanied by swarms of earthquakes. Between 1969 and 1985, total uplift was 3.3 m (11 feet). But in January 1985 the region once again started sinking.

The geological future of this heavily populated region will probably mirror the past; the land will rise and fall, fumarolic and seismic activity will vary, and there will be intracaldera eruptions. Much of the activity will be triggered by the interaction of water and magma or hot rock at depth.

What makes this region especially dangerous is the submerged nature of a major portion of the caldera—there is no shortage of water available to interact with magma and cause violent phreatomagmatic explosions. Compounding this situation is the ever-growing population in the region.

CHAPTER 6 – NOTES

1 See the Famous Calderas sidebar for Mount Mazama/ Crater Lake legends, page 104.

2 The term *mafic* is derived from the elements magnesium (Mg) and iron (Fe). Another term that means mafic is basic.

3 These radiocarbon ages were determined on vegetation affected by the eruption and have a precision of about +/- one per cent of their ages.

4 Oxygen has three stable isotopes or species, ^{16}O, ^{17}O, and ^{18}O, each differing only in their number of neutrally charged particles or neutrons. ^{16}O, ^{17}O, and ^{18}O each have 8 positively charged protons, and 8, 9, or 10 neutrons, respectively. ^{16}O constitutes 99.738 percent of all oxygen isotopes, ^{18}O = 0.205%, and ^{17}O = 0.038 percent. Because of their different atomic weights, physical processes such as evaporation and diffusion can fractionate or separate one from another. (For example, it is easier for the lightest isotope, ^{16}O, to evaporate from water, and diffuse into leaves than ^{18}O).

5 Phreatomagmatic explosions occur when water comes into contact with magma. The water immediately flashes to steam and undergoes up to a 1,600-fold expansion at sea-level pressure, which causes a very powerful explosion, which in turn fractures rock, producing ash and craters.

6 From unpublished work by Gola Aniello Pacca.

CHAPTER 6 – REFERENCES

Bacon, 1983; Bailey, 1987; Bailey et al., 1976; Bindeman, 2006; Bindeman and Valley, 2001; Chapin and Elston, 1979; Dvorak and Gasparini, 1990; Francis, 1983; Harris, 1988; Hildreth, 2004; Hildreth and Mahood, 1986; Hill and Prejean, 2005; Lipman, 1997, 2000; Mandeville et al., 1996; Miller et al., 1982; Moos and Zoback, 1993; Ross and Smith, 1961; Sheridan, 1979; Smith, 1960a, 1960b, 1979; Sorey et al., 2003; U.S.G.S. Bulletin 1855; Wark et al., 2007; Williams, 1942.

Some Other Famous Calderas

Toba, Sumatra, Indonesia

The world's largest known caldera, the Toba complex, is found in northern Sumatra, Indonesia (SB Figure 1). This 100-km (60-mi) by 30-km (18-mi) oblong structure is composed of four overlapping calderas parallel to the curved volcanic front. Its most recent major activity produced one of the world's largest eruptions 73,500 +/- 4,000 years ago. This cataclysmic event enveloped three older calderas, and since that gigantic eruption, a large resurgent dome, Samosir Island, has risen within the huge, water-filled caldera (Lake Toba) complex (SB Figure 1).

This eruption produced a minimum of 2,800 km^3 (670 mi^3) of quartz latitic to rhyolitic magma (68 to 76 percent silica) over an estimated 9- to 14-day period. The top of the Plinian-type eruption may have reached heights of between 50 and 80 km above the vent.

At least 800 km^3 of the 2,800 km^3 erupted total was air fall, which presently covers 4 million km^2 (1.544 million mi^2). The layer of tephra was as thick as 10 cm up to 2,000 km (1,220 mi) away from the vent in the Bay of Bengal in the Indian Ocean. Non-welded ignimbrite deposits from this eruption cover 20,000 km^2 (7,720 mi^2) around the caldera. They vary in thickness from less than 100 m to more than 400 m.

Little is know about the oldest Toba eruption, which produced the Haranggaol dacite tuff. It has been radiometrically dated at 1.2 million years +/- 160,000 years. The next eruption, called the Oldest at 788,000 +/- 2,200 years, produced a densely welded tuff that can still be observed around the caldera. Between 800 and 1,000 km^3 of rhyolitic magma issued forth, with extensive air fall deposits now found in sediments in the South China Sea 2,500 km to the east. The third major eruption, named Middle, dated at 501,000 +/- 5,000 years, also produced a densely welded ignimbrite.

SB FIGURE 1: *NASA Landsat false color photograph of the Toba caldera complex situated near the northwest end of Sumatra Island, Indonesia. Dark green is forest, pink little or no vegetation, dark gray is water.*

The dates of the caldera-forming Toba eruptions show repose periods of 360,000, 340,000 and 426,000 years, respectively. Interestingly, these repose periods are similar to the 350,000-year average of another major caldera complex, the Valles Caldera in New Mexico. However, the period is roughly half of those of the Yellowstone system of 600,000 to 700,000 years.

The effects of the massive eruptions of the Toba caldera were both far-reaching and long-lasting, likely lowering the Earth's average annual temperature for many years (see Chapter 9).

Tambora, Indonesia - 1815

On April 10 and 11, 1815, Mount Tambora on the Island of Sumbawa, Indonesia, produced the largest and deadliest volcanic eruption during the last 10,000 years. It was ten times larger than the more famous 1883 eruption of Krakatau.[1] The giant shield volcano, believed to be extinct or even non-volcanic, showed signs of life in 1812 with earthquakes, summit smoke, and some erupted ash. Then on April 5, 1815, a moderate-size eruption occurred, including explosions heard as far away as Jakarta, 1,260 km to the west, and Ternate in the Molucca Islands 1,400 km to the northeast (See SB Figure 2).

At about 7 P.M. on April 10, the low-level, continuous eruptions of ash intensified as three great columns of fire rose from the crater. By 8 P.M., pieces of pumice 20 cm in diameter began raining down on Sanggar 30 km to the east. Between 10 and 11 P.M., violent winds blew through the village, uprooting trees and destroying roofs.[2]

Explosions continued into the evening of April 11, some being heard as far as 2,000 km (1,220 mi) to the west in Mukomuko in central Sumatra. Skies remained dark for up to two days in some locations within 600

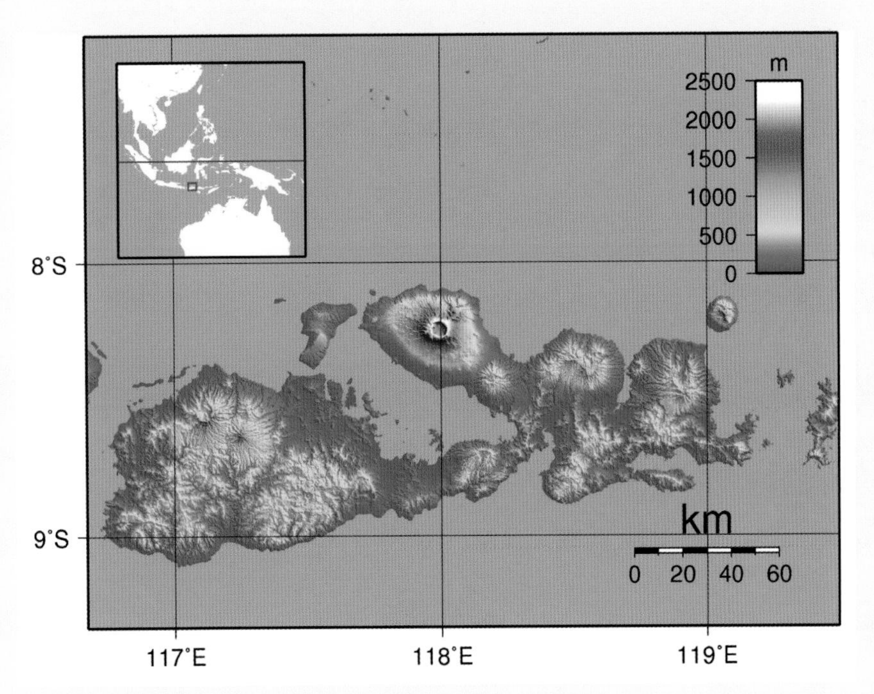

SB FIGURE 2: Map of Tambora caldera and 1815 eruption deposits located on the Sanggar Peninsula, Sumbawa Island, Indonesia (USGS diagram).

km (360 mi) of the volcano. Ash finally stopped falling on Java between April 14 and 17, and on the Island of Celebes to the north by April 15. In total, an estimated 150 km^3 (36 mi^3) of tephra was ejected.

Explosions ceased on July 15, with summit smoke not seen after August 23. But earthquakes and flames from the caldera were again recorded in August of 1819.

When visibility around the volcano finally improved, observers noted that the formerly twin-peaked mountain had lost 30 km^3 of it's original volume, and about one third of its original height of 4,300 m (14,100 feet). In place of the big mountain was a huge caldera 6 km across and 600 to 700 m (1,970 to 2,300 feet) deep, leaving the highest part of the rim 2,722 m (8,930 feet) above sea level.

This huge Plinian eruption not only produced large ash flows but also tsunamis that raced away from Sumbawa at speeds estimated to be 70 m (230 feet) per second. These were probably generated when ash flows reached the sea some 20 km from the volcano and displaced seawater.

Sanggar was struck with a 4-m-high wall of water at about 10 P.M. on April 10, and later Besuki in eastern Java was battered with a 2-m-high wave. Meanwhile, huge rafts of pumice and burnt trees several kilometers across and up to 1 m thick floated west, north, and east of the island.

On the islands of Sumbawa and Lombok alone, at least 88,000 people lost their lives due to direct and indirect effects of the eruption. But the whole world was affected for several years afterward due to the volcanic effects on the Earth's atmosphere. Famine affected places as far away as India and Europe, and 1816 was called the year without summer in eastern North America and western Europe (See Chapter 9).

Krakatau, Indonesia - 1883

Without doubt, 1883 is one of the most notorious historic years of volcanic catastrophes.[3] Krakatau Volcano, located in the Sunda Straits separating Java from Sumatra, erupted violently on August 26 and 27. Day turned into night and several killer tsunamis swept adjacent shores (see SB Figure 3, tsunamis heights in meters).

Reports of the event are good because the straits are on a major trade route between Europe and Asia, and the Dutch who administered Indonesia kept

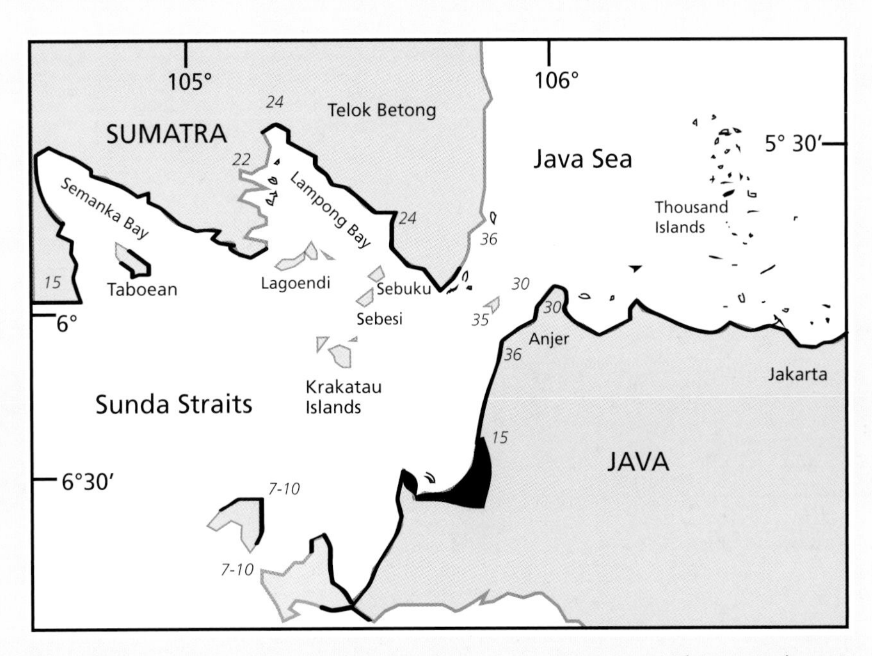

SB FIGURE 3: Map of the Sunda Straits separating Sumatra and Java, Indonesia, showing the location of Krakatau islands and extent of inundation by the 1883 killer tsunamis produced by the eruption (modified from Carey et al., 2000).

careful records. Also, during the long history of use of the straits, there has almost always been at least one ship transiting the narrow passage. This time was no exception—several British ships were in the region.

Volcanic-related activity in the area appears to have begun in the late 1870s with the occurrence of frequent but minor earthquakes. Then on September 1, 1880, a powerful quake shook the region, followed on May 20, 1883, by emissions of steam and ash from Perbuwatan crater on Krakatau Island. This was apparently the first volcanic activity on Krakatau since pumice erupted in 1680. A party of 86 who boated from Jakarta to see the volcano observed subsequent eruption phenomena on May 27. They discovered that Perbuwatan Crater on the north end of Krakatau Island was producing minor explosive eruptions at five- to ten-minute intervals.

This level of activity subsided after about one week, but on June 9 the tempo increased again and after August 23, gradually intensified. August 26 brought a loud explosion at 1 P.M. The noise was heard 120 km (75 mi) away by sailors aboard a British ship that also recorded a black cloud rising to about 25 km (15.5 mi) above the horizon. Another ship 20 km from the vent recorded ash and pumice up to 10 cm in diameter falling on its decks.

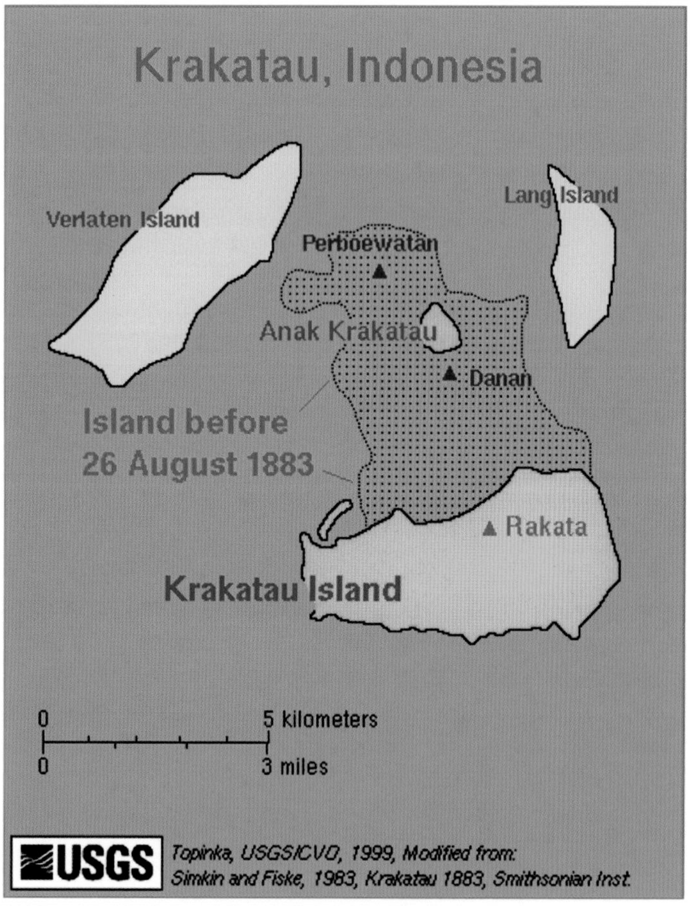

SB FIGURE 4: Map of the Krakatau volcanic islands remaining following the great 1883 eruption with an outline of the pre-eruption island (USGS diagram).

By 3 P.M., explosions occurring 10 minutes apart were heard at least 240 km (150 mi) distant. By 5 P.M., all of Java could hear the eruptive activity; the sound in Jakarta was likened to artillery barrages.

Intense lightning discharges were recorded starting about 7 P.M. by one of the ships in the strait as the eruption intensified.[4] By 10 A.M. the next morning, the top of the Plinian eruption cloud of dacitic material had risen to at least 60 km (37 mi) according to observers.

The largest of the numerous explosions during the eruption were heard at 5:30, 6:44, 9:58, and 10:52 A.M. on the morning of August 27. The biggest of these, at 9:58 was heard 3,224 km (2,000 mi) away in Elsey Creek, South Australia, and on Rodriquez Island in the Indian Ocean 4,811 km (2,980 mi) away. In addition, pressure waves from the explosions were detected throughout the world. This blast was also accompanied by the largest tsunami generated during the eruption (see SB Figure 3).

Tephra fell as far as 1,850 km (1,150 mi) away in the Cocos Islands, while chunks of pumice the size of pumpkins rained down on the decks of the Sir Robert Sale 40 km to the east-northeast. But the real damage and loss of life was wrought by the numerous tsunamis that swept nearby islands during the eruption.

More than 36,000 people perished due to these giant waves that crested up to 36 m (115 feet) high as they washed up to 10 km (6.2 mi) inland on the southern coast of Sumatra to the north, and the western and northern coasts of Java (SB Figure 3). Another 2,000 people were incinerated on the southeast coast of Sumatra, 40 km away from the volcano, by pyroclastic flows or surges that sped over the sea surface.

Since 1883, scientists have pondered the origin of the killer tsunamis.[5] Proposed mechanisms include (1) lateral blasts, (2) pyroclastic flows displacing sea water, (3) collapse of the northern part of Krakatau Island into the emptying magma chamber during caldera formation, and (4) shallow, phreatomagmatic submarine explosions.

While the answer(s) are not definitively known, many researchers believe that pyroclastic flows entering the sea and collapse of the island may be closest to the truth, based on investigations of the voluminous pyroclastic deposits found both on land and on the sea bottom. Some of these deposits, up to 40 m thick, were emplaced up to 15 km (9.3 mi) from the volcano.

In addition to widespread death and destruction, the eruption obliterated 23 km^2 of the northern end of the island, including the 450-m (1,475-foot) peak. In its place was a caldera 5 to 6 km in diameter, open to the sea, and 290 m deep at its south end (SB Figure 4).

Estimates place the release of ejecta during the eruption at 20 km^3 (5 mi^3), with most of the ash flow ignimbrite deposited in the straits to the north and northeast. Unusual optical phenomena were observed in the sky around the world between 1883 and 1886, while North Hemispheric average annual temperatures dropped between 0.5 and 0.8°C.

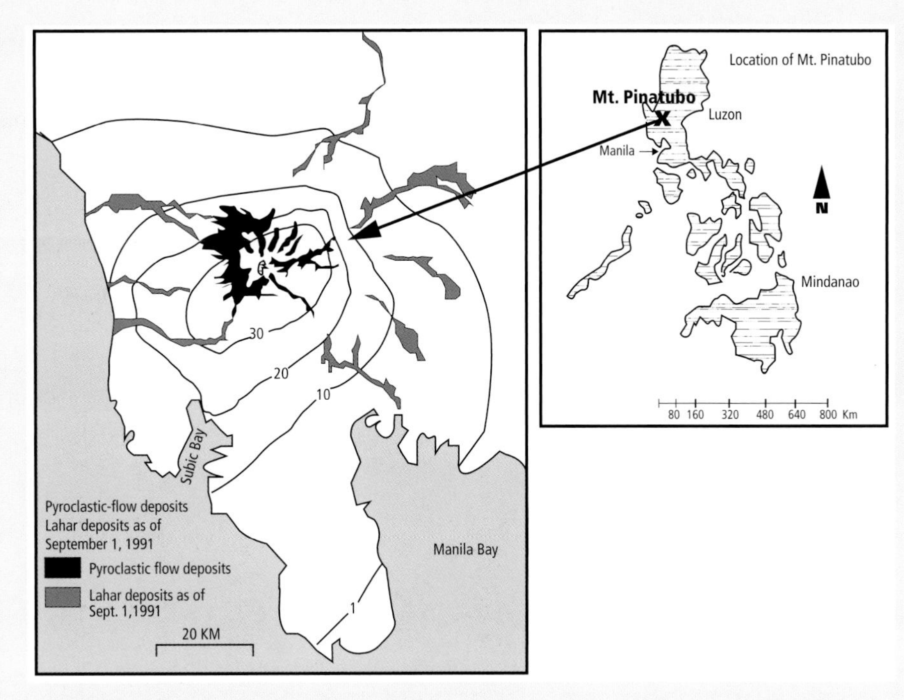

SB FIGURE 5: Map of Mount Pinatubo depositional deposits resulting from the 1991 eruption. North is to the top of the page. Note the different types of erupted deposits: air fall, lahars, and pyroclastic flows (after Decker and Decker, 2004).

Anak Krakatau or "child of Krakatau" was first recognized growing on the seafloor in 1927. Starting in 1972, an island formed which is presently 2 km in diameter.

What caused this catastrophic 1883 eruption? Pumice fragments may hold the answer, as some of the predominantly light-colored dacitic pieces contain streaks of darker, basaltic material.[6] One plausible scenario is that a quiescent dacitic or silica-rich magma chamber was intruded by a batch of very hot, volatile-rich basaltic magma. As the basaltic magma rose into the lower pressure upper portion of the chamber and mixed with the more silica-rich dacitic magma, water- and carbon dioxide-rich volatiles exsolved from the basaltic magma under this lowering pressure. Finally the volatile pressure exceeded the strength of the chamber walls and the violent eruptions commenced.

Pinatubo, Philippines - 1991

Most folks living on the highly eroded, jungle-covered slopes of 1,700-m (5,575-foot) high Mount Pinatubo in the Philippines were unaware that it was a volcano until 1991. That all changed in March when, after at least 500 years of dormancy, seismic tremors began shaking the mountain. One month before the June 15 cataclysmic eruption—the largest of the 20th century anywhere in the world—phreatic explosions began in the summit region, together with sulfur dioxide emissions as the amount of released sulfur increased with time.

On June 7, a lava dome was observed in the summit crater, and scientists noted continuing seismic activity. But just prior to the climatic eruption, sulfur dioxide emissions decreased. The volcanically induced seismic activity alerted the Philippine Institute of Volcanology and the U.S. Geological Survey to the potential danger to lives and property, and they installed monitoring devices including seismographs on and around the mountain.

Concern mounted for both the people living on and near the volcano and service personnel on the U.S. Clark Air Force Base upon the discovery of the volcano's past: Pinatubo had witnessed explosive eruptions approximately 500, 3,000, and 5,500 years ago. The air base, in fact, had been constructed on extensive, flat-lying laharic, pyroclastic, and thick air fall deposits on the east side of the mountain (SB Figure 5).

By June 10, more than 14,000 service personnel had evacuated the air base, as had 80,000 people living within 30 kilometers of the summit. The result was that only 300 people were lost during the climatic eruption 5 days later. Between the morning of June 12 and the afternoon of June 14, vertically directed eruptions occurred together with pyroclastic surges and long-period earthquakes (see Chapter 12).

On June 15 at 1:42 P.M., the 9-hour climatic event began, resulting in (1) a Plinian mushroom-shaped eruption cloud 400 km (240 mi) in diameter reaching an elevation of 34 km (21 miles), (2) pumice spread over 2,000 km^2 (770 mi^2) and pyroclastic flows that

SB FIGURE 6: NASA photograph of the caldera created during the great 1991 eruption.

reached 16 km from the vent impacting about 400 km^2, and (3) summit collapse forming a caldera 2.5 km in diameter (See Figure 3.11A, page 45, and SB Figure 6 which shows the resulting, huge caldera now filling with water). Estimates place the volume of ejecta at approximately 10 km^3 (2.4 mi^3).

The main eruptive phase was 3 hours long. Observations were difficult due to the simultaneous arrival of Typhoon Yunya. Heavy rains intensified the amount of material swept off the volcano, burying some valleys in up to 200 m (660 feet) of mudflow debris. Deaths resulted mainly from roof collapse due to the weight of water-soaked ash.

Mount Mazama/Crater Lake, Oregon Legends

The beginning of this chapter described the eruption of Mount Mazama in Oregon, which resulted in the formation of Crater Lake. Interestingly, like the legend of the lost continent of Atlantis, an ancient legend also tells of how Crater Lake formed, based upon witness accounts. However, as with other legends, the interpretation of the actual observations has changed with the passage of time.

This legend was told to and recorded by army soldier, William M. Colvig, stationed at Fort Klamath in 1865, by an old Klamath Indian Chief named Lalek. Chief Lalek began by saying that once upon a time, long ago, his ancestors lived in rock houses in the region. And when spirits of Earth, sea, and sky spoke to his people, Chief Llao of the Below World would come above ground and stand on top of a very high mountain, presumably Mount Mazama.

One day, Chief Llao observed a beautiful, young woman, Loha, and asked her to accompany him down into his Underworld Lodge. In return, the Chief offered Loha eternal life. Evidently the offer was not sweet enough for Loha because she declined his gracious offer. Furthermore, her tribe's ruling council would not force her to go live with Chief Llao in his digs.

Chief Llao became furious and tried to destroy Loha's people with fire. But this action provoked Chief Skell of the Above World to drop on to the summit of Mount Shasta, over 100 miles to the south, to help Loha's people.

War then broke out between the earth and sky gods. Day turned into night, and sheets of fire roared down the slopes of Llao's mountain, igniting the surrounding forests and driving Loha's tribe to the shelter of Klamath Lake to the southeast.

The future looked bleak for the tribe at this point until two old medicine men stepped forward to save their people. They carried torches to the summit of the flaming mountain and hurled themselves down into the fiery mouth of the Underworld. Chief Skell of the Above World, aware of the bravery of these two old medicine men, shook the earth, causing the mountain to collapse onto Chief Llao below.

When the sky cleared and the sun came out, the curse of the fires was gone. All that was left of the mountain was a huge hole or crater, which subsequently filled with snowmelt and rainwater, thus forming Crater Lake.

As pointed out by author Stephen Harris, this legend was built upon an actual event and contains a number of truths as well as myths. For example, there once was a large mountain, called Mazama, that witnessed a gigantic Plinian-type eruption. This changed day into night, and glowing ash flows roared down the slopes, igniting forests below. Concurrently with the eruption, or shortly thereafter, the mountain collapsed, creating a huge caldera, which subsequently filled with water.

Interestingly, the legend is more accurate than some of the early geologic investigations that concluded that Mount Mazama blew its top. It wasn't until 1942 that geologist Howell Williams wrote that the upper part of the stratovolcano collapsed into its partially emptied magma chamber situated under the cone, thus forming a huge, 1,280-m (4,000-foot) deep caldera.

NOTES

[1] The eruption was witnessed by the few British subjects and officials who lived in the region.

[2] The violent winds probably resulted from the collapse of the massive Plinian eruption column.

[3] Other memorable years include 79 A.D. when Vesuvius erupted destroying the Roman towns of Pompeii and Herculaneum; Tambora erupted in 1815, while Mount Pelee on the French Island of Martinique in the West Indies in 1902 destroyed St. Pierre, killing 30,000 people. That same year, Santa Maria volcano in Guatemala blew up.

[4] Lightning is caused by the build up of electrostatic electrical charges as millions of erupted particles hit each other during ejection from the vent.

[5] Tsunamis are usually caused by a sudden up or down motion of the seafloor, which causes a displacement of water. Ground motion may be generated by faulting, as in the great 2005 Sumatra quake, but also by volcanic explosions, caldera formation, landslides, and ash flows into water bodies. Gauges showed that sea level rises preceded most of the Krakatau tsunamis.

[6] Dacitic magma is much more silica-rich and more viscous (greater resistance to flow) than the darker, iron and magnesium-rich basaltic magma (65 percent vs. 50 percent silica).

REFERENCES

Carey et al., 2000; Chesner et al., 1991; Christiansen, 1979; Dartevelle et al., 2002; Decker and Decker, 2006; Fisher et al., 1997; Francis, 1983; Francis and Oppenheimer, 2004; Francis and Self, 1983; Harris, 1988; Lee et al., 2004; Mandeville et al., 1996; Rampino and Ambrose, 2000; Rose and Chesner, 1987; Self and Rampino, 1981; Self et al., 1984; Sigurdsson et al., 2000; Stothers, 1984; Verbeek, 1885; Wiesner et al., 1995; Yang et al., 2004.

Large Igneous Provinces—Flood Basalts and Silicic Volcanism

Introduction

Since time immemorial, people have gazed at the Moon and wondered what is it made of, why does it glow, and why the dark and light areas. Some thought the Moon was made of green cheese. Skeptics, however, believed that it was a stony celestial body. But obvious to the cheese lovers and skeptics alike were the Moon's dark and light areas. Some saw the face of an old man; the Mayans of Central America saw the profile of a rabbit.

Telescopes provided more detailed views of these regions. It became apparent that the lightest areas were meteorite-cratered, dust-covered uplands, whereas the dark areas, called lunar maria or seas, were lower elevation, relatively smooth plains. These latter regions also contained fewer craters than the uplands.

FIGURE 7.1: *The full, frontal side of the Moon, showing dark, smooth maria, covered with lava flows. The rough, bright areas are the cratered, ancient lunar highlands. Note the circular form of the maria. The edges of these circular structures are the rims of early, immense meteoric craters. Cratering caused instantaneous pressure-release melting of the lunar mantle (see Chapter 8, Magma Generation). This resulted in massive basalt eruptions that filled the craters.*

Prominent flood basalt-filled impact craters include 1. Mare Crisium, 2. Mare Serenitatis, 3. Mare Tranquillitatis, 4. Mare Fecunditatis, 5. Mare Imbrium, 6. Mare Nubium, 7. Oceanus Procellarum, 8. Mare Humorum, and 9. Mare Nectaris. The younger, light-colored crater with prominent radiating debris deposits, near base of photo, is Copernicus (NASA photo).

Starting in 1969, astronauts obtained lunar rock samples, and we learned that basalt eruptions had flooded the lunar surface early in the Moon's history. These eruptions had immediately followed huge impacts that occurred as our solar system was forming, leaving large craters on the Moon.

More recent discoveries show that Mars, Venus, and Mercury also have extensive flood basalt-veneered regions. These other planets are discussed in Chapter 11.

In places, remnants of crater rims not totally engulfed by basalt allow us to observe just how large some of the early lunar impacts were. These dark regions of the Moon, then, were covered by basalt flows following huge meteoritic impacting between about +4 billion and 3.0 billion years ago. These impactors so wounded the Moon's still warm interior that they caused instantaneous, pressure-release melting of the lunar mantle below the crust. Basaltic magmas were thus formed (see Chapter 8, Magma Generation).

We don't have to venture to the Moon to see flood basalt-covered regions. On Earth they include the area between Spokane and the Cascade Mountains of Washington state, the Mumbai (formerly Bombay) region in India, and the Paraná region in Brazil that contains Iguazu Falls. These are just a few of the huge areas on Earth covered by thick sequences of basalt. In fact, the vast expanse between Spokane and the Cascade Mountains, and south into Oregon, is one of Earth's smaller (but best exposed) flood basalt provinces.

Why Study Flood Basalts?

Flood basalt provinces are geologically important

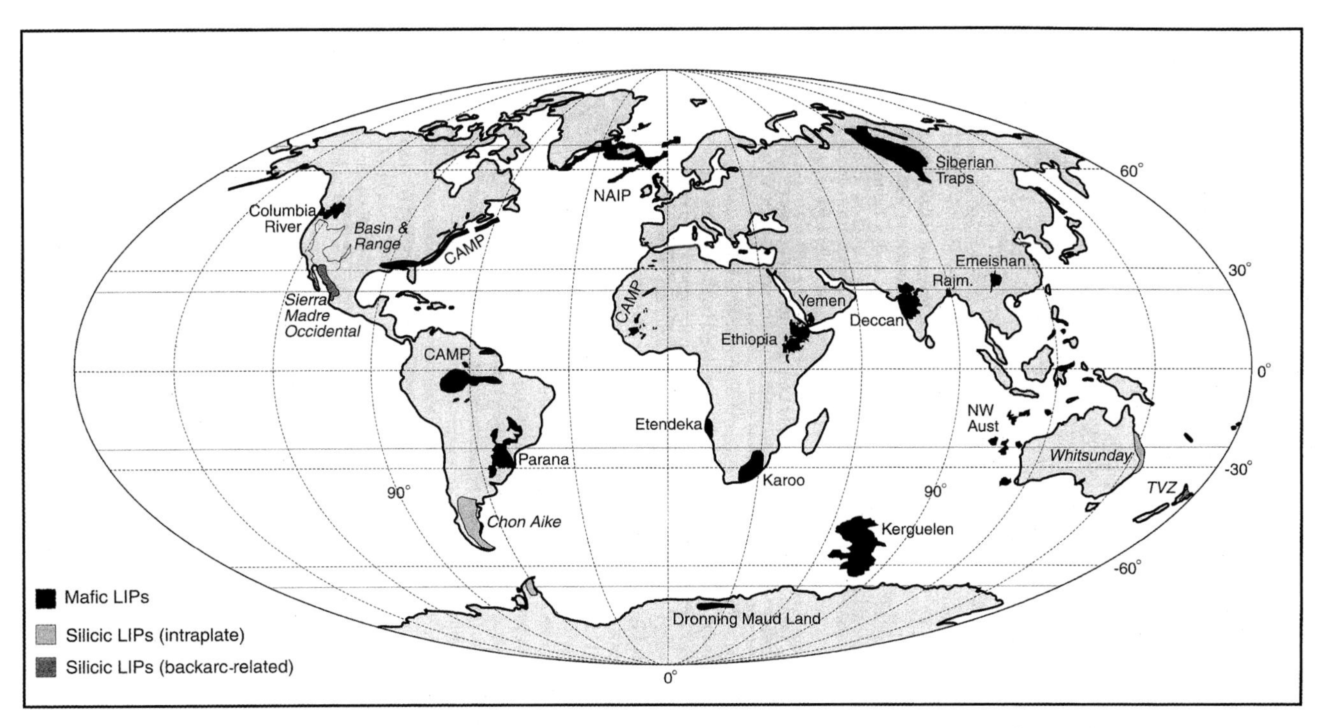

FIGURE 7.2: *Earthly flood basalt provinces are found on five continents (permission from Bryant et al., 2002).*

for many reasons. Some—perhaps most—of these areas are related to continental breakup. They include the opening of the North Atlantic region where Greenland separated from Europe 57 million years ago, the separation of Antarctica from South Africa about 184 million years ago, the opening of both the mid and South Atlantic Oceans starting about 200 and 132 million years ago, respectively, and current rift spreading in the Afar region in the Horn of Africa (Table 5.2, page 85).

Flood basalts have affected how and where we live as they cover huge areas of the Earth's surface. But more important is how these huge eruptions may have affected Earth's climate: many scientists believe these ancient eruptions resulted in major global extinctions due to severe and sudden climatic changes (See Chapter 9). Flood basalt provinces also provide information on major mantle events, and some host valuable mineral deposits of copper, nickel, gold, and platinum-group metals. Some basalt flows even contain hydrocarbons—oil and gas—introduced into these rocks later.

Flood basalts occur on oceanic and continental lithosphere. The "younger" ones range in age from 258 million years ago (in southwest China) to 17 to 6 million years ago (in the Columbia River basin). Currently, volcanism is occurring in the Afar region of East Africa and elsewhere. Many of Earth's basalt provinces appear to derive from hot spots (see Chapter 5).

What cannot be gleaned from Figure 7.2 is that:

1. Provinces apparently form over geologically very short time periods. For example, 96 percent of the lava of the Columbia River flood basalts erupted in about 2.5 million years, starting 17 million years ago.
2. Some flood basalts originate from long-lasting hot spot sources.
3. The rate of basaltic output varies considerably over time from long-lived hot spots.
4. Lavas, when compared with normal mid-ocean ridge basalt (MORB) or spreading center tholeiitic basalts, range from 45 to 55 percent silica, and some erupt very high-temperature, magnesium-rich lavas that travel great distances from their vents.
5. Individual flows are typically 2 to 3 m (6.5 to 10 feet) thick but can exceed 30 m (100 feet) and cover extensive areas. For example, the Roza flow in the Columbia River Province contains 1,300 km^3 (312 mi^3) of basalt and covers 40,300 km^2 (25,025 mi^2).
6. Accumulated thicknesses of basalt flows within provinces can exceed 7,000 m (22,960 feet).
7. Lava flows issue primarily from vertical cracks or fissures, rather than from central vents or cones.
8. Fissures can be very extensive, and the congealed basalt within them forms vertical wall-like, cross-cutting structures called dikes. These may be seen in southeastern Washington and

northeastern Oregon because dikes are more resistant to erosion than the lava flows they fed (see SB Figure 2, CRFB sidebar, page 116).

9. Lava flows are of the pahoehoe type, but in places they are pillowed, indicating deposition under water.
10. Basalt is generated by partial melting of mantle found below the crust (see Chapter 8).
11. Earth's crust "floats" on the plastic mantle below, and the general flotational equilibrium of areas of greater and lesser crust density is called **isostasy.** Thus, basalt-heavy regions that initially form plateaus eventually sink, forming basins. These basins subsequently fill with sediments. Thus, flotational equilibrium on the mantle is eventually achieved.

No one witnessed (fortunately) the formation of the gigantic flood basalt provinces. The closest analog to that type of event was the Icelandic Lakagigar eruption in 1783-1784. The 14 km^3 (3.3 mi^3) Laki flow issued from a 25-km-long fissure between June and November in 1783, covering 500 km^2 (195 mi^2). Even this relatively minor eruption, compared to the creation of flood basalt provinces, resulted in some dramatic climatic effects (see Chapter 9). This Icelandic eruption showed us how fluid lavas could be, and the great amount of climate-altering gases, especially carbon dioxide, fluoride, and sulfur compounds, that could be emitted.

Why Do Flood Basalt Provinces Exist on Earth?

Researchers have proposed at least three major theories to explain how these provinces form: meteoric impacting, tectonic rifting, and hot spot (mantle plume) eruptions. Table 5.2, page 85, lists most of the major flood basalt provinces younger than 300 million years. Table 5.2, page 85, also shows at least four are still active: Reunion Island in the Indian Ocean (which created the Deccan Traps of western India), Tristan da Cunha in the South Atlantic Ocean (see Chapter 5), Afar in east Africa, and Iceland. All but Iceland are now producing much smaller amounts of basalt. The other systems were large, intense, but relatively short-lived phenomena.

Large impactors from space, the results of which we have observed on the Moon, no doubt also produced flood basalt provinces on the Earth billions of years ago. This was when the solar system, due to gravitational attraction, was cleansing itself of debris. Confirming that this mechanism created flood basalt provinces on Earth is difficult. This is because (1) some of the largest flood basalt provinces are located in ocean basins (for example, the Kerguelen in the southern Indian Ocean and Ontong Java in the southwestern Pacific Ocean), and (2) no "smoking gun" impactor features have been found near terrestrial provinces. In ocean basins, provinces are eventually destroyed by plate tectonic forces (it's the rare ocean floor that is older than 200 million years).

The Earth does contain some relatively small meteoritic craters filled with igneous rocks. The Sudbury, Ontario, Canada, crater is an example. This oval structure, roughly 60 km (37 mi) northeast by 30 km (19 mi) northwest, formed about 1,850 million years ago. Not only did the crater fill with basalt, but also copper, nickel, gold, and platinum-group element deposits formed at the base of the spoon-shape structure. The deposits have been mined since the late 19th century.

Definitive signs of an impact structure at Sudbury include impact-forming shatter cones.[1] Within the crater is a fall-back layer of breccia (angular rock fragments) blasted into the air during impact. The impactor was not large enough, however, to form a flood basalt province.

Another province-forming mechanism is tensional thinning (pulling apart) of the Earth's crust. Crustal thinning decreases pressure on the mantle below, thereby lowering the melting temperature of mantle rock. Geologists disagree as to whether rifting precedes or follows the formation of flood basalts. What's understood is that rifting, unless aborted, establishes basalt-producing spreading centers, such as the Mid-Atlantic Ridge.

The third mechanism, the eruption of hot spot mantle plumes, is favored by many as the most likely source for flood basalt provinces (see Chapter 5). With precise radiometric age dating, we know that there is a striking temporal relationship between flood basalt-sized eruptions and subsequent continental breakup (see Chapter 5). But the question remains of how to separate true causes and effects.

A review of Table 5.2, page 85, shows the timing of the final phases of the opening of oceans and separation of continents. Note that the separation of northern Africa and Europe from eastern

North America, resulting in the opening the central Atlantic Ocean, was preceded by enormous eruptions of basalt.

Much of the lava produced during these events has been destroyed. We can still find, however, basaltic remnants, together with dikes and sills[2] that fed the flows. Thickness and volume estimates of the flood basalt provinces, then, are probably low.

The 250-million-year-old Siberian plateau and the 258 million-year-old basalt of the Emeishan Traps of southwest China apparently are related to rifting, which, for some reason, ceased following vast basalt eruptions (Figure 7.2). Some scientists speculate that the huge 115-million-year-old Kerguelen submarine basalt plateau preceded the separation of India from the then united continents of Australia and Antarctica. Also, the Balleny hotspot probably resulted in the separation of New Zealand from a still united Australia and Antarctica.

The rate of basalt production during flood basalt province formation evidently was truly astounding! For example, the Reunion Island volcano, now in the Indian Ocean, erupted on average between 2 and 8 km^3 (0.5 and 2 mi^3) per year to produce the Deccan Traps 66 million years ago. The Ontong Java field in the Pacific Ocean produced between 12 and 15 km^3 (2.9 and 3.6 mi^3) per year! By way of comparison, it is estimated that the total magmatic output of the entire 40,000-km (25,000-mi) spreading system, including the East Pacific Rise and the Mid-Atlantic Ridge, is about 21 km^3 (5 mi^3) per year. Subduction zones, however, are estimated to produce a paltry 9 km^3 (2 mi^3) of magma per year, of which only 0.65 km^3 (0.16 mi^3) is extruded.

Flow Emplacement

Most flood basalt provinces contain thick, aerially extensive sequences of flows. Much speculation exists as to how these eruptions occur. It is known from radiometric age dating that the bulk of each basaltic province formed in about one million years, a geologically very short time. How, then, do the eruptions differ from large modern basaltic eruptions in Hawaii or Iceland? Other than much greater flow rates and amounts of lava erupted, they probably don't differ much.

Province-forming eruptions apparently vary in intensity, with lavas flowing primarily from fissures (dikes) and less so from circular vents. Therefore, there would be some enormous eruptions producing large flows, such as the Roza in Washington State, sandwiched between thinner and much less extensive flows.

How Can Basalt Flow so Far without Solidifying?

Geologists also debate about how single flows can go great distances, covering huge areas, without solidifying. Early researchers envisioned the flows traveling very rapidly. They estimated that placed outpourings were at least 0.6 km^3 per hour per linear km from fissures three m (10 feet) or wider. This mechanism is now believed to produce turbulent flow that would promote cooling, and thus inhibit great flow distances. Also aa, or rough-surfaced lava, not common in flood basalt flows, would form.

Self et al. (1997) proposed the inflation model, in which slow-moving, more laminar flow produces pahoehoe-type lavas. Here, lobate-shaped sheet flows develop a crust that inflates and deflates according to the dynamics, or flow rates, of venting. The crust insulates lava flowing below, allowing it to go much farther before solidification. For example, a thermal study on the Columbia River Basin Ginkgo flow concluded that over its 550-km length, lava temperature dropped only 20°C. During the journey, fresh lava breakouts occur at flow fronts, with the ensuing formation of new crusts.

Support for the inflation model comes from recent Kilauea eruptions in Hawaii. There, pahoehoe flows initially 20 to 30 cm thick may inflate into flows 3 to 5 m thick. Also, lobes coalesce to form wide lava sheets.

Confirmation that this model is correct, perhaps, awaits the next really large basalt eruption. But lava tubes also transport lava great distances without much thermal loss, only at a much smaller scale. Lava tubes show fluctuating strand or flow lines within their now empty tubes. These depth markers, frozen into the interior of the walls, demonstrate the variability of lava flow rates with time.

Finally, pahoehoe flow requires that lava must be very fluid, possess low strain rates, or have both characteristics. Because the Columbia River Flood Basalt (CRFB) viscosities are estimated to have been 3 to 5 times greater than typical Hawaiian lavas, low strain rates probably prevailed (see Chapter 3 for strain explanation). This has led researchers to estimate flow velocities for CRFB of about 1 km per day. Flows on slopes averaging less than 0.1 percent, then, would take at least one year to travel 300 km (180 mi).

Associated Silicic Volcanic Rocks

Most continental flood basalt provinces, except the Siberian province, also contain silica-rich volcanic rocks. These rocks, including rhyolite, are mostly ignimbrites that erupted primarily from caldera complexes 10 to 30 km (6 to 18.6 mi) in diameter. Some ignimbrite units are actually larger than the biggest basalt flows. For example, the Paraná and Etendeka regions in South America and southwest Africa, respectively (Figure 7.2), each contain an enormous unit of at least 8,000 km^3 (1,920 mi^3).

The restriction of silicic volcanic rocks to continental flood basalt provinces is related to their genesis: granitic crust was melted by mantle-sourced basalt (see Chapters 5 and 8). The amount of silicic volcanism varies from province to province. And silicic volcanic units are found throughout stratigraphic sequences, including near the base.

Finally, large continental igneous provinces that contain only minor amounts of basalt relative to silicic rocks also occur. These regions of relatively long-lasting volcanism include the Chon Aike province of southern South America and the Antarctic Peninsula (188-153 million years ago), the Whitsunday province of eastern Australia (132-95 million years ago), and the Sierra Madre Occidental region of Baja and western Mexico (34-27 million years ago) (Figure 7.2).

Geochemical and isotopic evidence shows that magmas of these large silicic provinces, restricted to continental margins and interior settings, were generated primarily by partial melting of lower water-bearing continental crust. The impressive amounts of ignimbrites and pyroclastic rocks found in these areas were erupted from large caldera complexes.

Both the Chon Aike and Whitsunday provinces are spatially and temporally related to continental breakup. The Sierra Madre Occidental province in Mexico, however, is within a convergent continental margin setting.

Chapter 7 – Notes

[1] Shatter cones form in structurally sound rock around craters during meteoric impacting. They also may develop around calderas and can be caused by dynamite and nuclear blasts. The cone-shaped rock breakage phenomenon points away from the blast or impacted zone as intense shock waves travel through rock.

[2] A sill is a sheet-like igneous body intruded conformably between layers or strata of sedimentary and/or other igneous rocks. The Palisades Sill found in northern New Jersey along the west side of the Hudson River is a classic example. Here it forms a cliff due to its resistance to erosion. It was intruded when the Atlantic Ocean was opening in that region. Dikes are tabular igneous injections that cross cut rock strata. They may be sills in places, or conformable to layered rock strata.

Chapter 7 – References

Anderson, 2005; Bryan et al., 2002; Courtillot, et al., 1999, 2003; Decker and Decker, 2006; Hooper, 1997, 2000; Jerram, 2002; Reidel et al., 1994; Self et al., 1996, 1997.

Columbia River Flood Basalt Province

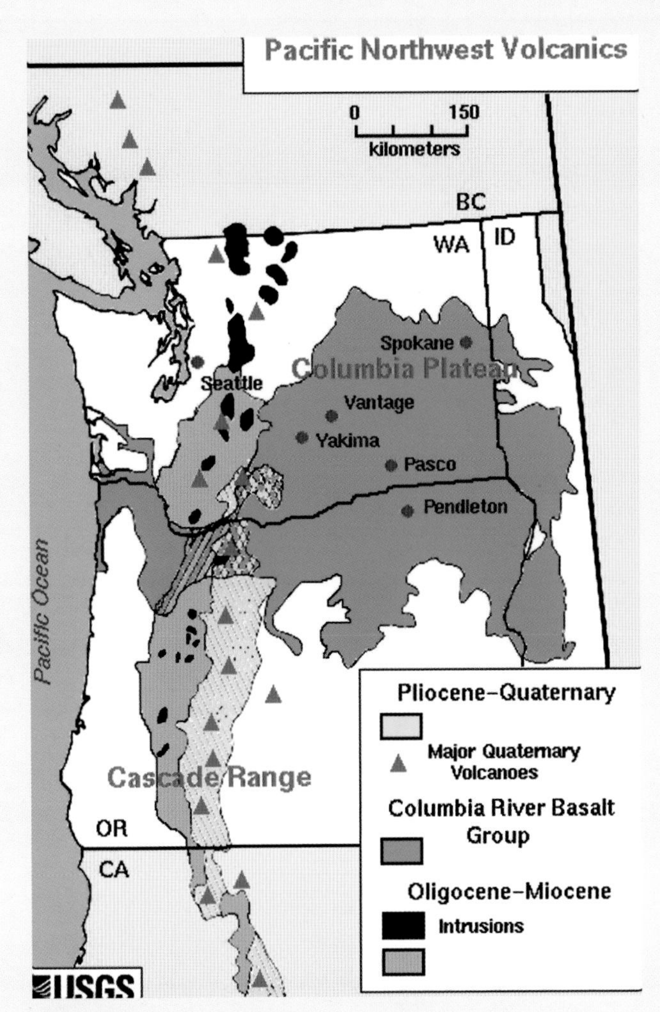

SB FIGURE 1: Map of volcanic provinces of the Pacific Northwest, including Cascade volcanoes and off shore plate tectonics (USGS map).

Stratigraphic Subdivision of Columbia River Basalt Group (CRBG)

SERIES		GROUP	SUB-GROUP	FORMATION (Age, Volume, % of CRBG)	MEMBER	MAG*
Miocene	Upper	Columbia River Basalt Group	Yakima Basalt SubGroup	Saddle Mountain Basalt (14-6 Ma, 2,400 km3 volume, 1.5% of CRBG)	Lower Monumental Member	N
					Ice Harbor Member	N,R
					Buford Member	R
					Elephant Mountain Member	R,T
	Middle				Pomona Member	R
					Esquatzel Member	N
					Weissenfels Ridge Member	N
					Asotin Member	N
					Wilbur Creek Member	N
					Umatilla Member	N
				Wanapum Basalt (15.5–14.5 Ma, 10,800 km3 volume, 6.0& of CRBG)	Priest Rapids Member	R3
					Roza Member	T,R
					Frenchman Springs Member	N2
					Eckler Mountain Member	N2
	Lower			Grande Ronde Basalt (17–15.5 Ma, 151,700 km3, 87%)		N2
						R2
				Picture Gorge Basalt		N1
						R1
				Imnaha Basalt (17.5–17 Ma, 9,500 km3 volume, 5.5% of CRBG)		R1
						T
						N0
						R0

* Magnetic Polarity:
N, normal; R, reversed; T, transitional; subscripts denote magnetostratigraphic units

USGS Topinka, USGS/CVO, 1997, Modified from: Swanson, et.al., 1989, AGU Field Trip Guidebook T106

SB TABLE 1: CRFB stratigraphic column (USGS diagram).

The Columbia River Flood Basalt province (CRFB) (SB Figure 1), seemingly large, is one-tenth the size, or less, compared with flood basalt provinces elsewhere in the world. But the CRFB, situated between the Cascade Range on the west and the Rocky Mountains to the east, has all the characteristics of the largest continental flood basalt provinces. And because of its relatively young age of 17 to 6 million years and excellent exposures partly due to climate aridity, the region has provided much information about flood basalt provinces.

The CRFB covers 164,000 km^2 (63,300 mi^2), has a volume of 175,000 km^3 (42,000 mi^3), and is composed of more than 300 lava flows. These flows average between 500 and 600 km^3 (120 and 144 mi^3). Most flows, ranging from 20 m (66 feet) to more than 100 m (330 feet) in thickness, are distinguished over this vast region on the basis of mineralogy, chemistry (particularly elements in trace amounts of less than 100 ppm), and by magnetic polarity (see Paleomagnetism sidebar, page 13).

Some of the oldest and most observable CRFB flows built Steens Mountain in south-central Oregon. This mammoth uplifted block contains 17- to 16.5-million-year-old lavas that correlate with the Imnaha Basalt Formation (see SB Table 1 CRFB stratigraphic column). Ninety percent of the CRFB, however, erupted between just 16.5 and 14.5 million years ago. This 2-million-year period also produced the most silica-rich flows of the CRFB, the Grande Ronde basaltic andesite. They contain between 53 and 57.5 percent silica.

The immense volume of the Grande Ronde Basalt Formation, which is 149,000 km^3 (35,600 mi^3) and 85 percent of the CRFB, was erupted from north-north-west-oriented fissures of the Chief Joseph dike swarm (SB Figures 2 and 4) east of Pasco, WA. These linear vent systems are 50 km (30 mi) to 200 km (120 mi)

SB FIGURE 2: Dikes of the St. Joseph dike swarm – see SB FIGURE 1 for location (Don Hyndman photo).

SB FIGURE 3: Morphology of a series of stacked CRFB flows. Note the prominent columnar jointing of the colonnade (Don Hyndman photo).

long and in zones a few kilometers wide. Large, black, vertically oriented dikes, meters to tens of meters wide, fed the flows. They can be seen in southeastern Washington and northeastern Oregon.

SB Figure 3 shows the morphology (structure) of the most common type of lava flow. A basal and very prominent colonnade is capped with an entablature (finer grained and more glassy than the colonnade), which is in turn topped with a scoracious and vesicular zone. Some flows, however, have the entablature sandwiched between a basal and upper colonnade. Flows typically were emplaced on slopes of less than 0.1 percent. Pillowed lavas formed where flows entered into lakes on the basalt plateau. Many of these lakes formed where flows blocked river drainages.

Some flows are separated by soil horizons. This shows that: (1) vegetation covered the region and (2) in places, flows were separated by thousands of years that allowed for soil development. The preserved soils containing fossilized vegetation between the flows also demonstrate that the climate was wet and mild.

At Vantage, Washington, where Interstate 90 crosses the Columbia River, a wonderfully preserved petrified forest lies between flows. The remains of the Gingko forest exist because water-saturated logs at the bottom of a shallow lake were buried by a lava flow. With time, silica impregnated and partially replaced the logs,[1] thus preserving some of the original wood.

The enormity of individual lava flows may be seen in SB Figure 4. For example, the Roza flow of the Wanapum Basalt Formation had a volume of 1,300 km^3 (310 mi^3) and covered 40,300 km^2 (15,555 mi^2). Its flow lobes extended more than 300 km (185 mi) from its vent(s). At an eruption rate of about 4,000 m^3 per second (10m^3/second/vent meter length), it took 6 to 14 years to form this immense field of pahoehoe lava.

The 12-million-year-old, 700-km^3 (178-mi^3) Pomona flow of the Saddle Mountain Basalt Formation actually reached the Pacific Ocean, as did the 15.6-million-year-old Ginkgo Basalt unit of the Wanapum Basalt Formation (SB Figure 4). In reaching the ocean, the 600-km (375-mi)-long Pomona flow, one of the longest known on the Earth, had to travel down the Snake River Canyon from west-central Idaho, fill the Pasco Basin of south-central Washington, and then proceed down the ancestral, pre-Cascade Range Columbia River channel.

The source of the CRFB was the mantle, with magmas generated by partial melting of mantle (peridotite). But why did the province form? The Yellowstone hot spot has been held responsible by some. They reason as follows. North America has been moving at 4 cm per year to the southwest over the stationary Yellowstone hot spot for tens of millions of years. About 16.5 million years ago, the hot spot was under the Nevada-Oregon-Idaho border near McDermott. The main area of basaltic venting moved rapidly north to the Washington-Oregon-Idaho border following the eruption of the Steens Mountain basalts. During this time, the hot spot-forming calderas remained stationary.

Subsequent eruptions came from north-to-northwest-oriented fissures 300 to 400 km (185 to 250 mi) north of the hot spot. These formed near the tectonic boundary between the thinner, but denser, accreted (attached) oceanic lithospheric terrains to the west with

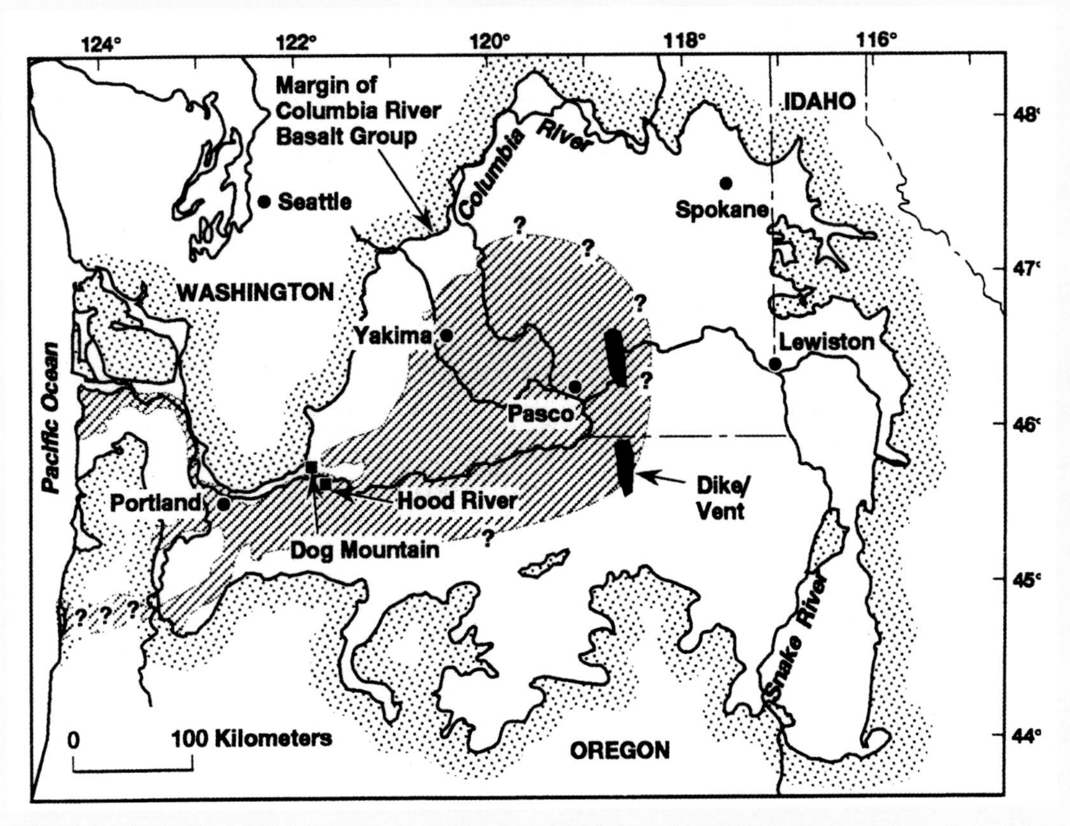

SB FIGURE 4: Map of the extent of the Ginko flow (Reidel et al., 1994).

the thicker, more competent (stronger) continental lithosphere to the east. The fissures, then, developed in weakened and/or thinned lithosphere.

Will basalt issue forth again? Probably not in Washington State but most likely in the Snake River Plain of Idaho in the wake of the Yellowstone hot spot. You can see fresh-looking, 2,000-year-old basalt flows, lava tubes, vents, pahoehoe lava, and a wonderful visitors' center at Craters of the Moon National Monument.

NOTES

1 Trees recognized in Ginkgo Petrified Forest State Park include maple, Douglas-fir, spruce, walnut, elm, and ginkgo.

REFERENCES

Carey and Bursik, 2000; Hooper, 1997, 2000; Reidel and Tolan, 1992; Reidel et al., 1994; Self et al., 1996, 1997.

CHAPTER EIGHT

Magma Generation

Introduction

Scientists believe our 4.6-billion-year-old Earth was very hot during its early years. In fact, the planet was probably entirely molten because of the intense heat generated by giant meteorite bombardment, early gravitational contraction, and heat released during radioactive isotope decay (see Chapter 2). This one-billion-year-long intense phase of early solar system "debris cleansing" resulted in the growth of planets through accretion of "space junk" gravitationally attracted to these large, growing bodies. The early molten state also resulted in early differentiation, or segregation, of elements. A heavy, metal-rich, liquid core (iron and nickel), and a lighter, basaltic lava surface were separated by a zone—the mantle—of intermediate composition. However, remnants of the very oldest surface rocks no longer exist due to subsequent and continuing meteoric impacting, plate tectonism including magmatism, and erosional processes.

The oldest terrestrial rocks discovered on Earth, 4.03-billion-year-old Acasta gneissic rocks,[1] are located in the Northwest Territories of Canada. These rocks record magmatic processes apparently similar to what occur in our presently active subduction zones. Also, the Jack Hills of western Australia contain +4.404-billion-year-old zircon minerals ($ZrSiO_4$) derived from even older igneous rocks. Because of these and other data, earth scientists believe plate tectonic processes have operated since the Earth was young and much hotter.

Each year, about 30 km^3 (7 mi^3) of volcanic and intrusive rocks form on Earth's surface. Sites of magmatism include divergent or constructional plate boundaries, convergent or destructional plate boundaries (subduction zones), along transform fault boundaries, and hot spots located within plates and on spreading ridges. Seventy-five percent of this igneous activity is related to constructional plate boundaries, twenty percent is subduction zone-related, and five percent is involved in hot spot activity.

Generation of Basalt

About 60 percent of the Earth is covered by basalt derived from the mantle. Most is found in ocean basins. Melting of mantle peridotite[2] occurs when upward convecting rock crosses a solidus—the tem-

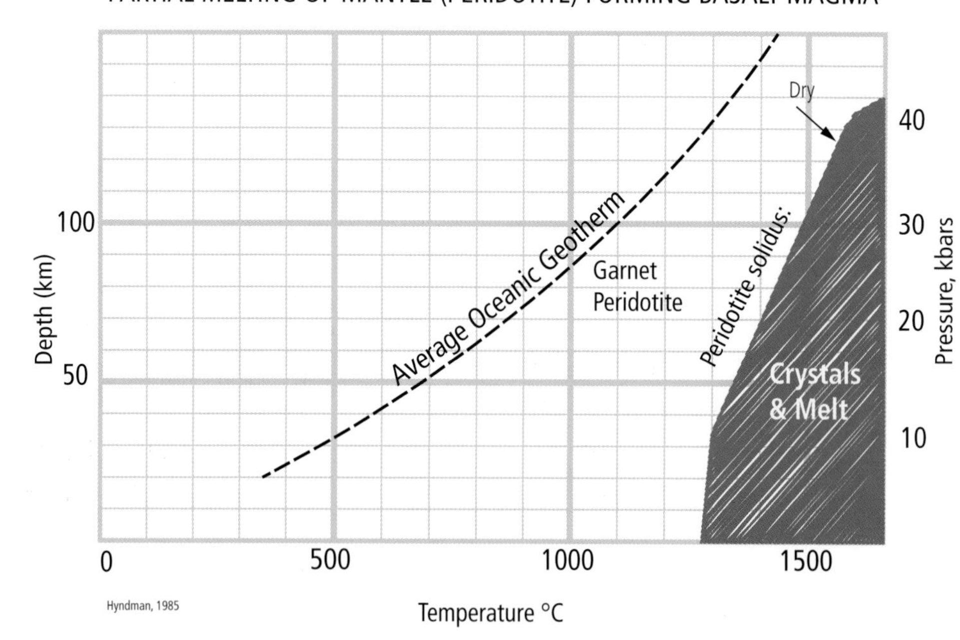

FIGURE 8.1: *Pressure (depth) increasing upward and temperature (increasing to right) diagram showing the generation of basalt from mantle peridotite.[2] The gray field right of the sloping line boundary contains crystals and liquid of basaltic composition—the mantle melting zone. Note how mantle melting takes place at continually lower temperatures as pressure decreases right area in Figure 8.1). This is a dry melting system. The addition of water lowers the melting temperature further. Also note the average rate of temperature increase (geotherm) with depth under oceans. With depth, the Earth gets hotter but at rates that differ according to the tectonic province (modified from Hyndman, 1985).*

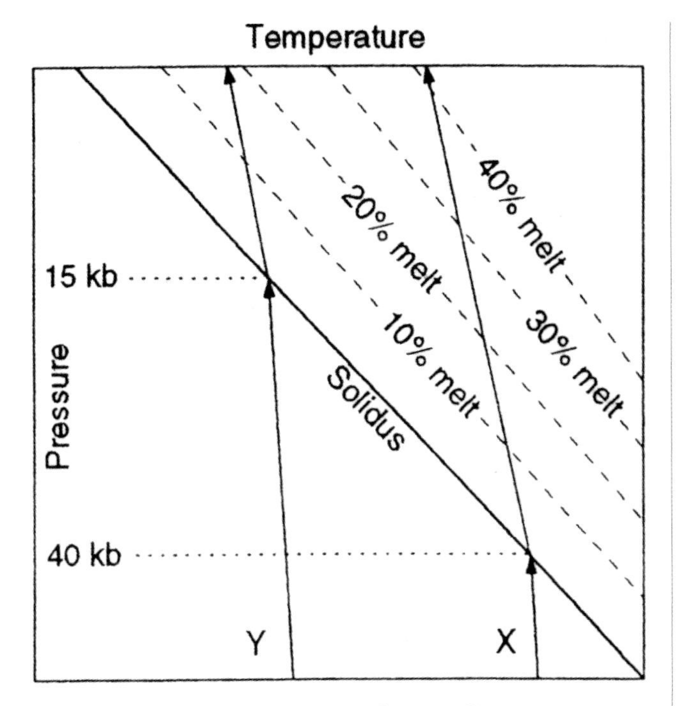

FIGURE 8.2: *Pressure (increasing downward)-temperature (increasing to the right) diagram depicting the melting of upward convecting hot mantle peridotite as the rock crosses the solidus (boundary surface between solid and liquid). Note that the amount of partial melting is greater, up to 40 percent, along the higher temperature path "X" than along the lower temperature path "Y" (modified from Grove, 2000; Klein and Langmuir, 1989).*

perature at which mantle rock starts to melt. This is due to lowering pressure.[3] And mantle rock undergoes convection because the lower mantle, heated by the molten outer core below, expands thus becoming less dense than peridotitic mantle above. This hotter, less dense, plastic mantle rock rises and displaces cooler, denser upper mantle that in turn is displaced slowly downward.

This type of mantle solidus melting is an essentially adiabatic (meaning no heat is exchanged),[4] decompression phenomenon. As hot peridotite rises into the upper mantle or asthenosphere (see Figure 8.1), pressure continues to drop, as does the temperature of mantle melting. So, the higher the temperature of convecting mantle, the deeper in the mantle melting occurs. That situation results in a greater percent of melted mantle and more basaltic magma. Figure 8.2, which depicts temperature plotted against pressure, shows this graphically. Note the effects of pressure on the solidus—the surface separating solid mantle from mantle where liquid and solid (crystals) coexist.

Of the major minerals in peridotite,[2] clinopyroxene and the aluminous phase(s) (plagioclase, spinel, and/or garnet) melt first.[2] Therefore, this basaltic melt is higher in silicon dioxide, aluminum oxide, calcium oxide, sodium, and potassium but lower in magnesium oxide than parent peridotite from which it melted. And because these minerals contribute in different proportions to the melt, and the amount of partial mantle melting varies, the resulting basaltic magma compositions may vary from location to location.

Other mechanisms can cause decompression melting of mantle rock, also forming basaltic magma. One process is crustal thinning through tension, or the pulling apart of the crust. This reduces pressure on the mantle below. Locations of major extension include the basin and range region between eastern Utah and the Sierra Nevada of California. The amount of extension for that region may be 100 percent. If correct, the distance today between Reno, Nevada, and Salt Lake City, Utah, is twice what it was 30 million years ago, prior to extension and thinning of the crust there.

Crustal tension produces steeply dipping fractures, some that penetrate the upper mantle. These fractures become conduits for decompression-type, mantle-generated magmas. Examples include the small basaltic cinder cones and lava flows found along such faults bounding the east side of the Sierra Nevada in Owens Valley, California.

These basalts may be primitive or primary. Primitive magmas form from mantle that has never undergone partial melting (anatexis). Primary basaltic magmas, however, are chemically unchanged since their origin in a mantle that has already witnessed partial melting or contamination from surface rocks. This is depleted mantle. Geochemical and isotopic analyses are used to determine whether mantle basalts are primitive or depleted.

Primary magmas of various compositions also form from preexisting rock, in addition to depleted mantle, in lower or subducted crust.

Subduction Zone Magma Generation

Subduction zones—that is, convergent plate margins—also generate basalt from mantle rock. But mantle melting here is in response to the release of fluids, mostly water, from the subducting slab as it descends into drier mantle. Fluids are released continuously from the water-saturated slab during the entire subduction process, starting on the seafloor. This process, known as flux melting, lowers the mantle solidus by 200°C to 300°C, resulting in the formation of basaltic liquids at temperatures of about 1,100°C.

At the deep ocean-trench interface, mud (fine-grained sediment) is 80 percent water. Spreading center-generated basaltic seafloor is also water-saturated due to the great porosity and permeability of the rocks (see Figure 2.6, page 21). As material descends into subduction zones, or is buried elsewhere, water is released from these materials and is also incorporated into new, water-bearing (hydrated), chemically and structurally complex metamorphic minerals.[5]

Hydrated minerals include chlorite, amphibole, and serpentine. But the release of water does not result in mantle melting until depths of 75 km to 120 km (47 to 75 mi) are reached. Here, temperatures in the mantle wedge above the subducting slab reach 1,100°C or more.

At these depths, water is liberated from the previously formed hydrous, metamorphic minerals listed above as they move out of their stability range into an even higher temperature-pressure regime. Water encountering mantle at such depths lowers the solidus or melting temperature of peridotite (see Figure 2.5, page 16). The basaltic melt formed here contains large amounts of water. For example, at a depth of 90 km (56 mi) , or under 30 kb of pressure,[6] the melt can contain 20 percent water.

Basalt, as we already know, is just one igneous rock type found in subduction settings. The most abundant rock within Andean-type subduction zones is andesite. Rhyolite, dacite, and other extrusive rocks—and their intrusive equivalents (see Table 3.1, page 31)—are also common.

Island arcs contain lesser amounts of the more silica-rich rock types; basalt and basaltic andesite are most abundant. This is understandable because mantle-generated basaltic magma rises through lithosphere of roughly the same composition. Little opportunity exists for chemical changes except through the processes of magma modification discussed next.

Modification of Magmas

How does mantle-generated basaltic magma change into more silica-rich magmas? Mechanisms include (a) magmatic differentiation, (b) mixing of magmas of different compositions, and (c) assimilation and/or melting of country (surrounding) rock of different, more silica-rich compositions.

MAGMATIC DIFFERENTIATION

Magmatic differentiation (magma separation) can be accomplished through **crystal-liquid fractionation, convective fractionation,** and **flow differentiation.** Howell Williams first described the process of **crystal-liquid fractionation** during his classic study of the cataclysmic destruction of Mount Mazama, which formed the caldera that now holds Crater Lake in the Cascade Range of southern Oregon. Williams noted color and mineralogical changes accompanying a chemical change from bottom to top in the thick, 7,700 year-old sequence of pyroclastic flow deposits around the volcano (see Figure 6.2B, page 94). Silica content of pumice in the ash flow decreased upwards in the deposits from 71 percent to 54 percent as the color darkened from buff to smoky gray.

Williams postulated that while the melt resided in the magma chamber before the climactic eruption, iron, magnesium, and calcium-rich minerals developed. And because their density exceeded that of the liquid, these early formed minerals settled, producing a chemically differentiated or vertically zoned magma chamber. This resulted in enrichment with silica, potassium, sodium, and volatiles above, and calcium, magnesium, and iron in the bottom of the chamber.

Initially puzzling was the fact that most of the silica-rich ash flow material occurred beneath the more mafic portion of the ash flow. This reversal in the usual magma/mineral chemistry happened because the chamber drained from the top down into the more mafic, deeper portions of the chamber (see Figure 6.2B, page 94).

Since the classic work of Norman L. Bowen in 1928, we've known that silicate minerals crystallize sequentially in a silicate melt. Bowen devised the "reaction principle," or Bowen's Reaction Series, whereby crystals form and chemically react with their host melt either continuously or discontinuously with dropping temperature (Figure 8.3). For example, plagioclase feldspars change composition continuously during formation. Crystallization starts with a calcium-rich core and ends with the incorporation of sodium (see the right side of Figure 8.3). Under ideal but rarely achieved conditions, the plagioclase crystals are a homogeneous

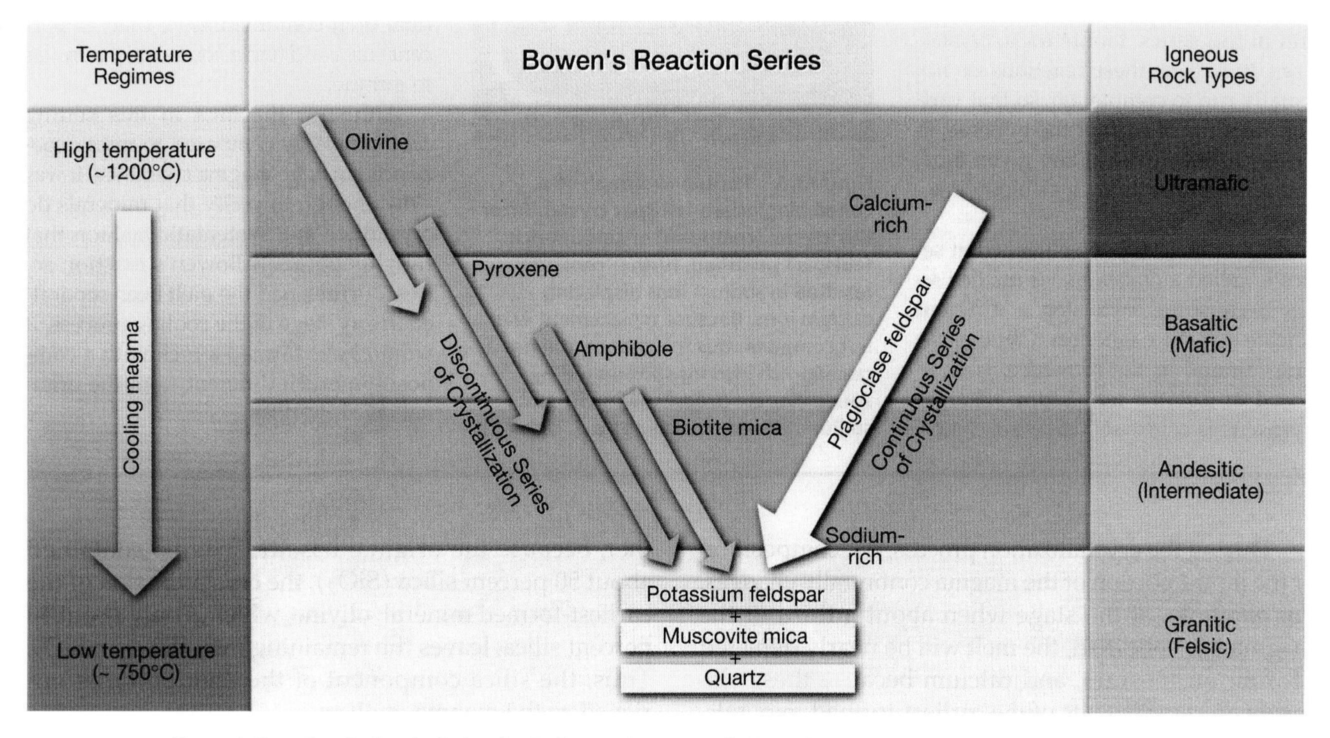

FIGURE 8.3: *Bowen's Reaction Series depicting both the continuous and discontinuous reactions and formations of minerals with decreasing tempeature. Note the resulting rock types listed on the right side of the figure (modified from Lutgens and Tarbuck, 2006).*

mixture of sodium and calcium. Most commonly, however, plagioclase crystals do not homogenize during formation, but rather show an optical as well as chemical zonation from a calcium-rich core to sodium-rich rims.

Minerals forming in the discontinuous series, as seen on the left side of Figure 8.3, are from highest to lowest temperature: olivine, pyroxene, amphibole, and biotite mica. Crystallization begins at about the same temperature as continuous series crystallization. Magma chemistry determines which mineral is in the final rock mineral assemblage, but the order of crystallization remains the same. If the magma is rich in magnesium and iron but poor in silica, olivine plus or minus pyroxene with calcium-rich plagioclase crystallize and constitute the resulting rock assemblage. If the melt is silica-rich and iron-magnesium-calcium-poor, some olivine will initially form. It is subsequently replaced by pyroxene that is in turn is replaced by amphibole, with or without biotite mica eventually forming.

Also, if the original magma was silica-rich, the small amounts of contained magnesium, iron, and calcium will be incorporated into amphibole and/or biotite well before total crystallization of the magma is complete. The remaining silica, aluminum, and potassium will combine to form potassium-feldspar. Any silica remaining after the formation of these minerals will form quartz: pure silica. And since quartz is the last mineral to crystallize, it fills space around the earlier formed mineral crystals.

Convective fractionation involves mineral crystallization from magmas that flow past surfaces where crystallization takes place. In **flow differentiation**, crystals concentrate in the center of vertical or horizontal conduits. During both processes, magma chemistry changes with solution passage, resulting in the development of more silica-rich liquids.

MIXING OF MAGMAS

This process entails magmas from different chambers, by coincidence or not, meeting and then mixing. For example, andesite or dacite magmas may originate through the interaction of basaltic and rhyolitic magmas. This may be common in large magmatic systems and can be detected through textural, geochemical, and isotopic examinations. It is visually apparent, however, when mixing is incomplete. For example, one might observe inclusions (xenoliths) of basalt in andesite, or olivine crystals (xenocrysts) within andesite.

ASSIMILATION OF COUNTRY ROCK

Rising magmas may (1) melt, (2) dissolve, or (3) react with the rocks they underlie or pass through. The result is a modified, final magma composition. For example, (1) magma contamination by **wall rock melting** can occur as the rising basaltic magma at 1,200°C encounters granite that melts at between 700°C and 900°C. Clearly, ample heat exists in the basaltic magma to melt the granite. This mechanism raises the silica and potassium content of the resulting mixture, leading to the formation of dacite or andesite. But our models for assimilation and magma mixing are not without problems. For example, granitic melt is very viscous compared with basaltic magma. That high viscosity impedes mixing and homogenization (see Figure 4.3A, page 60, of a xenolith—a partially digested rock fragment—that was incorporated in the granitic magma).

Some very large, silica-rich eruptions, such as from Yellowstone, result not from the melting and assimilation of lower-melting-temperature country rock, but are due to crustal rock melting because of the release of heat from basaltic magma below. Isotopic data support the generation of the rhyolitic rocks from the melting of ancient granitic crust by basaltic heat at Yellowstone. Minor amounts of basalt occur within Yellowstone caldera.

The **dissolution** process resulting in magma contamination requires that magma be undersaturated with respect to components that subsequently dissolve and are incorporated within the system. That this mechanism has operated is hard to substantiate.

The **reaction** mechanism produces changes in magma composition due to component exchange taking place along fracture conduits. It also occurs by diffusion of components along the contact between magma and wall rocks. Examples of these types of contamination mechanisms can be documented chemically (major, minor, and trace element gradational changes across intrusive-wall rock contacts) and isotopically. Oxygen isotopic studies have shown, for example, the presence of meteoric water well into the interior of otherwise pristine-appearing granite intrusions.

CHAPTER 8 – NOTES

1. Gneiss is a highly metamorphosed rock, commonly of granitic composition. It has a planar look or fabric because of the development of planar minerals, including micas.
2. Mantle peridotite rock is composed primarily of the minerals olivine, orthopyroxene, clinopyroxene, and an aluminous phase (plagioclase ($(Na,Ca)Al_{1-2}Si_{3-2}O_8$), and spinel ($MgAl_2O_4$) or garnet ($Mg_3Al_2(SiO_4)_3$).
3. The solidus is the melting temperature of a rock, or the boundary between rock (solid) and liquid. It varies with rock composition, pressure, and volatile content.
4. An adiabatic process is where a substance does not exchange heat with its surroundings during either compression or expansion. With compression the temperature of the substance increases. With expansion the substance temperature decreases. In the case of rising mantle rock into the lower-pressured upper mantle, the peridotite decreases in temperature but so does its melting temperature.
5. These minerals are metamorphic in origin. That is, they form under subsurface conditions where temperature and pressure, among several factors, are higher than on the seafloor. Composition of the material is very important, with contained and/or added fluids facilitating the processes and being incorporated into metamorphic minerals. And as temperature and pressure increase, these minerals in turn may change in the solid state into new metamorphic minerals stable under the higher metamorphic grade. Finally, the metamorphic assemblage, if originally derived from mud, for example, will melt when exposed to sufficiently high temperatures, forming granitic magma.
6. One bar of pressure is equal to atmospheric pressure at sea level or 14.5038 pounds per square inch. One kilobar is equal to 1,000 bars, so 30 kb is 30,000 times atmospheric pressure measured at sea level.

CHAPTER 8 – REFERENCES

Asimow, 2000; Bowen, 1928; Bowring, 2014; Bowring and Williams, 1999; Christiansen, 1979; Grove, 2000; Hyndman, 1985; Klein and Langmuir, 1989; Lutgens and Tarbuck, 2006; Raymond, 2002; Rogers and Hawkesworth, 2000; Spera, 2000; Wallace and Anderson, 2000; Wilde et al., 2001; Williams, 1942.

CHAPTER NINE

Volcanoes and Climate

Introduction

Can volcanic eruptions affect climate? Yes. We know that large eruptions in the 1700s, 1800s, and 1900s affected world climate for several years following each major event, some with disastrous consequences for life. Can they produce long-term climatic effects measured in tens to hundreds of years? Maybe.

The first person known to suspect that volcanoes could affect climate was Greek historian Plutarch (45 A.D. to 120 A.D.). He speculated that dimmed sun light, climatic cooling, and crop loss and famines in Italy and Egypt followed the 44 B.C. eruption of Sicily's Mount Etna.

Seventeen hundred years later, Benjamin Franklin (1706-1790) also suggested volcanism could cause climate changes. In 1784, while in Paris as the first U.S. diplomatic representative to the Court of Louis XVI, Ben theorized that the "dry" bluish haze and cold weather affecting Europe then was due to the huge basaltic eruptions occurring in Iceland. Located between Europe and Greenland, activity started in 1783 on that large island. The eruption from Laki Volcano produced 14 km^3 (3.35 mi^3) of lava, plus enormous amounts of ash, hydrogen fluoride, and sulfur dioxide gas. An estimated 190,000 sheep, 28,000 horses, and 11,000 cattle died from fluoridosis resulting from volcanic fluorine emissions.[1] Furthermore, 10,000 Icelanders, or one fifth of the island population, eventually died as a result of the demise of the animals and crops. Continental Europe also suffered crop failures, famine, and starvation.

And then in 1815, only 31 years later, came one of the most explosive, climate-altering eruptions in 10,000 years. Tambora, located on the island of Sumbawa in Indonesia, blasted 175 km^3 (42 mi^3) of ash plus huge quantities of gases into the stratosphere. Three days of total darkness followed for those within 200 miles of the volcano (see SB Figure 5, page 107). Worldwide, the average annual temperature decreased between 0.4 and 0.7°C for several years. (New England and western European temperatures may have dropped up to 2.5°C (4.5°F) per year.

In the northern hemisphere, 1816 became "the year without summer," a disastrous consequence of this eruption. Not only did 10,000 people die directly from the eruption, but at least another 70,000 subsequently died of starvation due to widespread crop failures. New England had killing frosts and snow in June, July, and August. Europe fared worse because the eruption followed the disastrous Napoleonic wars that had already ravished the continent. Ruined crops resulted in famine, starvation, civil unrest, and food-related crimes by fall.

In 1883, Indonesia was again the site of a major and more famous eruption of Krakatau. This time, more witnesses described the volcano-created havoc, and more deaths occurred. The series of eruptions from Krakatau, located between Java and Sumatra, destroyed most of the island. But more devastating were the tsunamis that drowned at least 36,000 people living on the neighboring coasts of southern Sumatra and northern Java (see SB Figure 4, Famous Calderas sidebar, page 106). While Krakatau produced less tephra than the Tambora eruption, Earth's climate was also affected; the average annual temperature in Europe dropped by 0.3°C (>0.5°F) for three years.

But there was another, earlier time when Europe and the Middle East witnessed an even dimmer sun coupled with an unusually cool period and significant crop failures. The dry, fog-clouded atmosphere of 536 and 537 A.D. was described by at least four observers, including Byzantine historian Procopius living in Rome, and Constantinople resident Lydus.

The culprit is suspected to have been Rabaul Volcano located on the island of New Britain, Papua New Guinea. The eruption, radiocarbon-dated at 540 A.D. +/- 90 years, spread at least 11 km^3

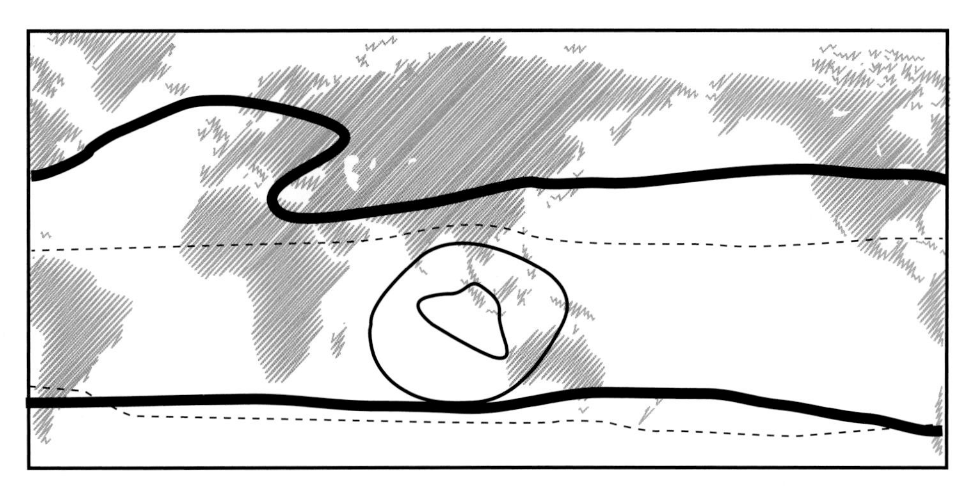

FIGURE 9.1: *Climatic effects of the 1883 Krakatau eruption. The thin line surrounding the volcano shows distribution ash fallout of over 700,000 km^2 (270,272 mi^2). The larger diameter circle shows the great area in which the explosion was heard. The dashed lines show the northern and southern atmospheric distribution of ash and aerosol particles before September 22, 1883, while the thick solid lines show the extent of atmospheric particles by late November (modified from Francis and Self, 1983).*

of tephra locally, mostly to the west of the island. Considerably more tephra fell more distally, and into the Bismarck Sea. This tephra distribution pattern correlates with an eruption occurring during the annual March to October monsoon season.

The "mystery" eruption is estimated to have been approximately 1.7 times larger than the great 1815 Tambora event. It blasted sulfuric acid aerosols into the stratosphere subsequently detected in Greenland ice cores. And if from Rabaul, the travel time to Europe would have been only two to three weeks.

The 20th century also witnessed large eruptions, with some affecting the climate. For example, the climatic effects of the 1980 Mount St. Helens eruption in Washington State were negligible, whereas the 1982 and 1991 eruptions of El Chichón in southeastern Mexico, and Pinatubo on the island of Luzon in the Philippines, respectively, lowered average annual world temperatures.

How Volcanoes Affect Climate

How, then, do eruptions affect climate? Much depends on the size of the eruption, rate of tephra ejection, height of the erupting column, and types and amounts of emitted gases. As we'll see, an eruption's location—particularly whether it's closer to the equator than the poles—also plays an important role in the scope of its effect on climate.

Large eruptions send ejecta higher into the atmosphere than small eruptions, with maximum heights attained in excess of 50 km (more than 165,000 feet). This places tephra well above the cloud- and precipitation-bearing troposphere and into the drier stratosphere. The lower boundary of the stratosphere varies with latitude and by season. At the equator, this boundary occurs at 15 to 20 km (48,000 to 66,000 feet), but it may be only 10 km (33,000 feet) at the poles, and as low as 7 km (23,100 feet) during polar winter.

In addition to the strong positive correlation between the height of an eruption column and mass of ejected material, an excellent correlation also exists between discharge rates of eject a and the amount erupted (see SB Table 2, Explosivity sidebar, page 43).

Clearly, size or volume of ejecta is important. But because some relatively small eruptions have

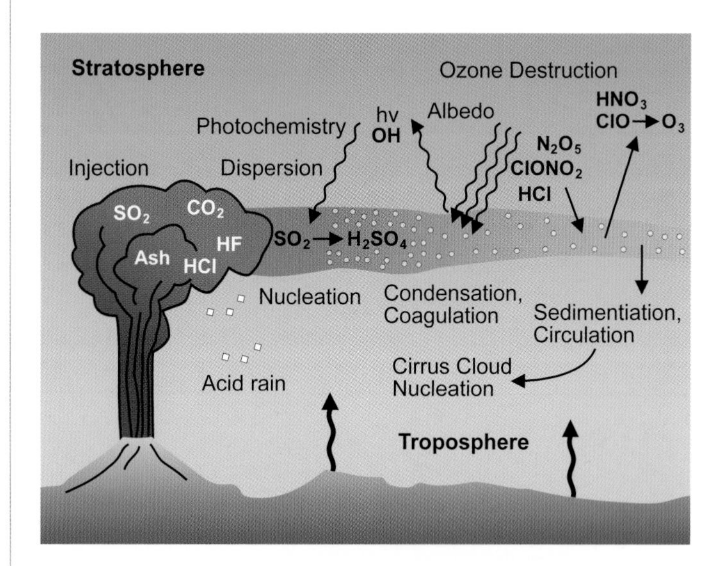

FIGURE 9.2: *Diagram of a large volcanic eruption that injects ash and aerosols including sulfur dioxide (SO_2) and hydrochloric acid (HCl) into the stratosphere. Some chemical reactions that occur include the formation of surface temperature-lowering sulfate aerosols such as sulfuric acid (H_2SO_4) from erupted SO_2 and destruction of ozone (O_3). This occurs by heterogeneous (complex) chlorine (HCl) and nitrogen pentoxide (N_2O_5) chemical reactions on aerosol surfaces (see down arrow on to aerosol layer. Theses reactions generate chlorine monoxide (ClO) and nitric acid (HNO_3) (up arrow), the former of which destroys ozone (O_3). Most of the erupted hydrochloric acid (HCl) and hydrogen fluoride (HF) dissolve in water droplets and fall as acid rain. Note while the Earth's surface cools, the stratosphere warms. Albedo is the percent reflectivity of a surface. It ranges from high values of fresh snow and silver to very low values of black surfaces (from the USGS after modification from R. Turco, 1992).*

had more profound effects on the Earth's climate than larger ones, other factors must come into play. Ash particles shield the Earth's surface from solar radiation, lowering surface temperatures, but those particles generally don't linger in the atmosphere more than a few months.

Some erupted gases also play a major role in surface warming and cooling. The gases, mostly water and carbon dioxide, are accompanied by much lesser amounts of sulfur dioxide, hydrogen sulfide, hydrogen, nitrogen, hydrochloric acid, and hydrogen bromide. Carbon dioxide, a green house gas, warms the Earth's surface by not allowing all wavelengths of solar radiation to re-radiate back into space. While volcanoes annually emit between 145 and 255 million tons (130 to 230 million tonnes) of carbon dioxide, anthropogenic or human-derived sources essentially mask the effect by releasing 30 *billion* tons (27 billion tonnes).

Sulfur dioxide (SO_2) and hydrogen sulfide (H_2S) emissions, however, are major global climate changers because they can produce profound, negative effects on surface temperature. This was determined By Hubert H. Lamb following the 1963 eruption of Agung Volcano in Indonesia. Lamb realized that dust and ash particles affected climate less than acid sulfur aerosols in the stratosphere. At these high altitudes, sulfur dioxide combines with oxygen and water to form sulfuric acid (H_2SO_4) aerosols. These are very tiny, less than 2-micron-sized suspended liquid particles.[2] Hydrogen sulfide (H_2S), in turn, oxidizes to sulfur dioxide (SO_2), which then also forms sulfuric acid aerosol particles.

At altitudes between 15 km and 30 km (9 and 19 mi)—in the lower stratosphere—the acidic aerosol particles join the dust- and sea salt-bearing layer. These acidic aerosol particles, then, increase stratospheric opacity.[3] Over time, aerosol particles collide, forming larger ones. Once a particle attains a diameter of about one micron, it eventually falls back to Earth. While in the stratosphere, however, these particles absorb solar radiation, and, more significantly, they reflect radiation back into space. The troposphere and land surface below are thus cooled, while the stratosphere above warms.

Rather than sheer volume, then, the chemical content of an eruption is often more significant. For example, the 1980 eruption of Mount St. Helens emitted 0.15 km^3 (0.04 mi^3) more tephra than the 1982 El Chichón eruption in southern Mexico. But the El Chichón blast produced almost 100 times more sulfur dioxide and rose 13 km (42,650 feet) high into the stratosphere. The Mount St. Helens eruption had little climatic effect, but worldwide cooling of about 0.2°C (0.36°F) followed the El Chichón event.[4]

Tools of the Trade

But how do we know the amount of sulfur released into the atmosphere? Since the 1970s, researchers have used instruments such as COSPEC (Correlation Spectrometer of ultraviolet light) to measure sulfur dioxide emissions. COSPEC, developed to quantitatively determine industrial smokestack emissions, has proved invaluable in monitoring volcanic sulfur dioxide releases as well. The device, usually mounted in an airplane, measures the intensity of ultraviolet radiation under blue-sky conditions. Additional absorption is then due to volcanically liberated sulfur dioxide.

Other airborne quantitative analytical tools include LIDAR (Light Detection and Ranging), which also quantitatively estimates sulfur dioxide discharges and aerosol concentrations. This optical technology uses laser light to analyze the back-scattered light off of airborne particles.

Differential absorption (DIA) measures specific gases such as ozone, carbon dioxide, water vapor, and sulfur dioxide released during eruptions. In addition, TOMS (Total Ozone Mapping Spectrometer) units housed in the Nimbus 7 and Meteor 3 satellites and operational since the 1980s also measure volcanic sulfur dioxide emissions.

In 1991, Mount Pinatubo erupted 17 to 20 megatonnes of sulfur dioxide, almost two times more than El Chichón's 12 megatonnes (a megatonne, abbreviated Mt, is one million tonnes, or one trillion—10^{12}—grams). The result was an even greater, longer-lasting cooling of the Earth. The Pinatubo eruption, ten times larger than Mount St. Helens, sent ejecta to 34 km (111,000 feet) and lowered average annual temperatures 0.4 to 0.5°C (0.7 to 0.9°F) during the second half of 1991 through 1992. The global cooling effect lingered into 1993, lowering average annual temperatures by about a 0.2°C.

Another common volcanic gas that adversely affects human health is hydrochloric acid (HCl).

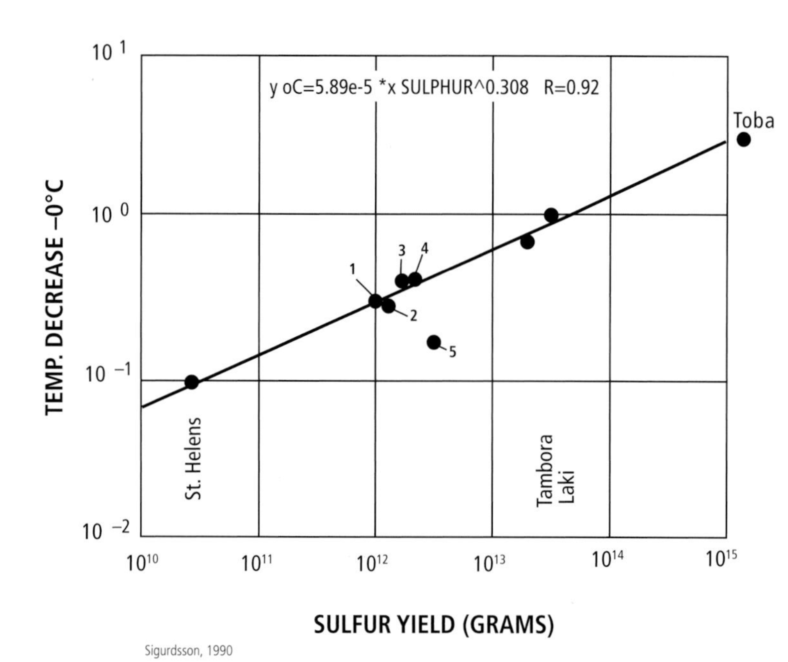

FIGURE 9.3: *Relationship between volcanic sulfur contribution to Northern Hemisphere atmosphere and temperature decrease during large volcanic eruptions. The 73,500-year-old Toba, Indonesia, eruption plots just to the right, off of the diagram. The 1991 Pinatubo, Philippines, eruption plots between the 1902 Santa Maria , Guatemala (3), and 1883 Krakatau, Indonesia (4) eruptions. (1) – Fuego, Guatemala, (2) – Agung, Indonesia, (5) – Katmai, Alaska (modified from Sigurdsson, 1990).*

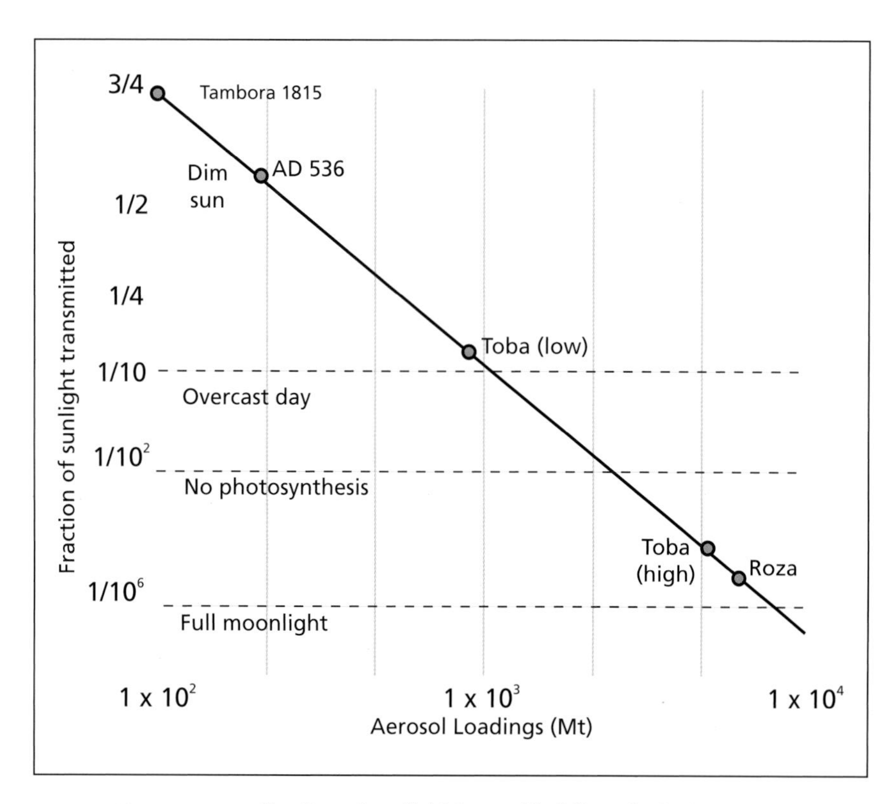

FIGURE 9.4: *Fraction of sunlight transmitted through stratospheric aerosols from some great historic eruptions. Note the high and low estimated values for the 73,500-year-old Toba eruption. The A.D. 536 value is thought to have come from Rabaul Volcano, Papua New Guinea, and the 14.5-million-year-old Roza flow within the Columbia River flood basalt province is estimated (modified from Francis and Oppenheimer, 2004).*

Rather than altering temperature, this gas breaks down to chlorine (Cl), which destroys our life-protecting, high-altitude ozone (O_3) layer. Ozone in the stratosphere helps shield us from harmful, cancer-causing ultraviolet solar radiation.

Ozone is also destroyed by the formation of sulfuric acid aerosols because one oxygen atom from the ozone molecule is incorporated into sulfuric acid (H_2SO_4) droplets. Scientists estimated, for example, that the 1991 eruption of Mount Pinatubo depleted stratospheric ozone by between 3 and 8 percent.

Finally, eruptions at low latitudes (generally no higher than 20 to 30 degrees north and south of the equator) cause greater atmospheric effects than similarly sized eruptions at higher latitudes. Not only do solar rays strike the surface more directly at equatorial latitudes, but atmospheric circulation disperses the aerosols in a wider belt around the Earth (see Figure 9.1).

Long-Term Atmospheric Effects

Let's now return to an earlier question. Can volcanic eruptions cause either long-term climate cooling or warming? Some researchers think so. For example, James Kennett and Robert Thunell detected more volcanic ash in deep sea sediment cores two million years old and younger, than in older sediment. This finding coincides with the Pleistocene Epoch ice ages—is the correlation merely coincidental?

As early as the late 19th century, Swiss scientists (and cousins) Paul and Fritz Sarasin suggested volcanic eruptions could affect global climate over long periods. Then, in 1913, U.S. meteorologist William Humphreys theorized that volcanic ash not only could reduce surface temperatures, but could disturb atmospheric circulation to such an extent that summers would be cooler and cloudier and winters colder and stormier than normal. Furthermore, Humphreys speculated that increased volcanism might even induce ice ages.

In 1952, U.S. Weather Bureau scientist Harry Wexler suggested that large

Tree Rings

The remains of a late Minoan settlement, Akrotiri, were discovered in 1967 on the Island of Thera, located in the Aegean Sea between Greece and Crete. Archeological excavations revealed that buildings had been wracked by earthquakes and partially repaired before being buried by tephra and ash flows from a mammoth volcanic eruption from Santorini. Research shows this volcano has had a very violent past Since discovery, absolute age daters have tried to determine when this event occurred. Since World War II, the radiocarbon (carbon 14) method has been widely used to date events younger than 50,000 years.[1] Because of analytical uncertainty, however, the margin of error for this method is plus or minus one percent of the actual age.

A more accurate method under ideal circumstances for dating major volcanic eruptions less than 5,000 years old is tree ring analysis. If huge eruptions like that from Santorini occurred during the growing season, significant worldwide climate cooling might follow. This is because of the stratospheric veil of aerosol and ash particles shielding the Earth's surface from solar radiation. Trees growing at high elevations might in turn suffer frost damage that could be detected in their growth rings.

Data show, in fact, that frost might affect tree growth up to two years after a huge eruption. This is due to the lag time between the eruption and fallout of stratospheric aerosols and particles circling high above, shielding the Earth's surface.

For damage to occur, trees must be actively growing when temperatures drop to freezing or below for two consecutive nights and the intervening day, at least. Freezing causes extra-cellular ice formation and dehydration within new growth, resulting in a distinctive tree ring.

Furthermore, if frost occurs early in the growing season, the inner part of the annual growth ring is damaged. If frost occurs late in the season, the late-developing, dark zone of thick-walled cells is affected.

For the tree ring technique to be applicable for dating eruptions, trees in widely separated regions need to simultaneously sustain frost-induced damage. This demonstrates regional, rather than local climatic effects. In the United States, subalpine bristlecone pines *(Pinus aristata)* and Great Basin bristlecone pines *(Pinus longaeva)* have proved valuable for tree ring studies because of their wide distribution and long life.[2] Bristlecone pines start radial ring growth in late June, with maturation complete by late August.

The technique was first tested on large, historically documented eruptions to determine whether it could be applied to dating major prehistoric volcanic events. Results are promising because damage was detected in growth rings from 1884 that followed the 1883 Krakatau, Indonesia, eruption. Damage was also detected in the 1902 tree rings following the early 1902 eruptions of Mount Pelée and Soufriere, West Indies, in 1912 tree rings following the early 1912 Katmai, Alaska, event, and in 1965 growth rings following the 1963 Agung, Indonesia, eruption.

For an additional check of the validity of the method, researchers compared acidity peaks in ice cores from Greenland (resulting from high concentrations of erupted sulfur) to tree ring data. While some historical events correlate, the fit is not perfect. The correlation is weak for major Southern Hemisphere eruptions after which particulates did not flow over Greenland, and for eruptions that did not produce large amounts of sulfur.

Dating old events is more problematic because of the decreasing availability of old trees coupled with their more restricted geographical extent. Also complicating the method are frost events present in old trees that appear unrelated to known volcanic eruptions.

Radiocarbon dating done on Akrotiri material including charred seeds and wood give average ages of 1675 B.C. +/- 57 years. While this age may appear older than Egyptian archaeological information suggests, the tree ring frost damage age of 1626 B.C. falls within the radiocarbon age range of 1732 B.C. to 1618 B.C. This age agreement supports, then, a 17th century B.C. date for the catastrophic Santorini eruption, the effects of which led at least in part to the precipitous decline of the Minoan civilization on Crete to the south.

NOTES

1 The radiocarbon age-dating method is based on the fact that the isotope carbon 14 (^{14}C), with an atomic mass of 14 (6 positively charged protons and 8 neutrons also with mass but no charge) has an unstable arrangement of these sub-

atomic particles. It is therefore unstable or radioactive. This unstable arrangement causes the ^{14}C parent isotope to decay into a stable, non-radioactive isotope called a daughter product. In this process, a neutron in the carbon-14 atom disintegrates into a proton with the release of energy.

The mass of the atom stays the same at 14, but the atom now contains a stable arrangement of 7 protons and 7 neutrons instead of the unstable arrangement of 6 protons and 8 neutrons in carbon14. The daughter isotope produced is nitrogen 14.

The rate of decay of carbon 14 is constant, having a half life of 5,730 years. So if we start with one pound of carbon 14 today, in 5,730 years we will have one half pound left, with only one-quarter of a pound left after 11, 460 years, or two half lives. And so it goes so that after about 50,000 years, or a little less than 9 half lives, there is not enough carbon 14 left to analytically obtain a trustworthy date.

Two other prerequisites necessary for valid carbon 14 ages are (1) that the carbon-containing material be shielded or isolated from atmospheric contamination from its time of formation until dated and (2), that the amount of carbon 14 has remained constant in the atmosphere.

Finally, since carbon 14 is a relatively short-lived isotope, why is there any carbon 14 left after 4.6 billion years of Earth history? The answer lies in the fact that carbon 14 is continuously regenerated in the upper atmosphere from nitrogen 14 because when a neutron from space collides with the nitrogen atom, it not only knocks out a proton but replaces it.

2 These trees occur from Colorado to California, a distance of 1,300 km, with some bristlecone pines being more than 4,600 years old.

REFERENCES

Filion et al., 1986; Lange, 2002; LaMarche and Hirschboeck, 1984; Rampino et al., 1988; Yamaguchi, 1985.

eruptions might upset atmospheric circulation by shifting normal westerly winds in the Northern Hemisphere to the south. The result would be a cooler northern Hemisphere climate.

Another Hubert Lamb contribution to our understanding of long-term volcanic climatic effects was his exhaustive compilation, back to 1500 A.D., of known volcanic eruptions. Lamb devised the Dust Veil Index (DVI). He noted a weak correlation between high DVI values and climatic cooling. This weak correlation resulted in his subsequent, very important discovery in 1963 that acid aerosol particles are a major climate-cooling mechanism.

But some of the largest volcanic eruptions in the past, such as from the Toba and Yellowstone Calderas, are orders of magnitude larger than what we have historically witnessed. What might their effect(s) have been? Even though smaller than the earlier, immense flood basalt eruptions (see Chapter 7), The 1783-1784 eruptions of Laki in Iceland, can give us some insight into the effects of those ancient cataclysms.

The Icelandic eruptions, while not blasting material into the stratosphere, still had profound climatic consequences in Iceland and Europe, as we know from Ben Franklin and others. The Laki eruptions lasted eight months and produced about 14 km^3 (3.35 mi^3) of lava and some tephra. Estimates place the sulfur dioxide release at 120 million tonnes, or comparable to the yearly release from human activities.[4] In addition, an estimated 15 million tonnes of hydrogen fluoride and 7 million tonnes of hydrogen chloride were released.

The late 18th century Icelandic basalt eruptions, however, were negligible compared to the ancient outpourings during flood basalt eruptions. Formation of the Columbia River Flood Basalt Province between 17 and 6 million years ago produced 175,000 km^3 (41,825 mi^3) of lava (see CRFB sidebar, page 115). While eruptive activity was not continuous, some huge flows emanated rapidly. For example, the 14.5-million-year-old Roza flow is estimated to have formed in about 10 years, with a volume of 1,300 km^3 (310 mi^3). Thordarson and Self calculated that the sulfur dioxide output might have been 1,240 million tonnes per year, or about ten times today's yearly emissions from all sources.

The worldly effect would have been dramatic, perhaps catastrophic![5] If the resulting atmospheric haze was distributed evenly, noontime would have appeared like night under a full moon, according to Rampino and others. But how long might this effect have lasted? The answer is unknown due to the low altitude of the haze and what happened to it. The haze might have rapidly precipitated compared with the long residence time of stratospheric aerosols.

There have been, however, some gigantic, plinian eruptions from calderas throughout history. The three from Yellowstone at 2.1 and 1.3 million years ago, and again 640,000 years ago, and from Toba 73,500 years ago come to mind. These monster eruptions blasted incredibly large amounts of tephra and sulfur dioxide high into the stratosphere in a matter of hours or days.

Ash and Aviation

Since the development of internal combustion engines early in the 20th century, scientists and transportation officials have been concerned with the effects of fine dust and tephra on the "health" of these motors. The May 18,1980, eruption of Mount St. Helens underscored their concerns, as voluminous amounts of fine ash deposited over central Washington State destroyed countless car, truck, and bus engines immediately following the eruption. The damage happened so quickly, and before most people were aware of the hazard.

The main factors responsible for engine destruction were the hardness and fineness of ash. Ash, considerably harder than steel, permeated engines and immediately abraded moving parts (see Figure 3.8C, page 40, of volcanic glass). This effect was compounded by the size of the ash particles. The best carburetor filtration systems commonly used in 1980 only removed particles 10 microns or greater in size. (A human hair is about 90 microns in diameter). As tephra spread eastward, finer and finer ash fell, much of it less than one micron in diameter. In all, about 150,000 km^2 were covered with ash. And much of it blew around central Washington state for years afterward.[1]

More recently, atmospheric scientists and air transportation officials have been concerned with the effects of large, tephra-bearing volcanic eruptions on air travel. This concern was highlighted during the December 14,1989, to August 31,1990, eruption of Mount Redoubt Volcano in Alaska, 177 km southwest of Anchorage International Airport, a major transportation hub.

Domestic and international airline flights above south-central Alaska were disrupted, with hundreds canceled. The situation was especially frustrating for airlines and travelers because it happened during the busy Christmas holiday season. Four commercial jet aircraft flew through the abrasive ash, resulting in sandblasted windows and wing edges. Engines of these planes were subsequently inspected and found to be undamaged.

The most serious, near fatal incident, however, happened on December 15, 1989. A Boeing 747 en route from Amsterdam with 244 passengers on board flew through an ash cloud at 7,500 m, about 240 km downwind of Redoubt. The encounter resulted in all four engines shutting down. This plane, flying just 20 minutes later than another 747 traveling through the same airspace, had entered a tephra cloud erupted just 90 minutes earlier.

The plane went into a steep, 4,000-m dive for the next 8 minutes, dropping to just 2,000 meters above the ground before the engines restarted. Engine failure was due to the melting of ash particles by engine heat, and the subsequent coating of turbine parts with glass that, in turn, restricted air intake. All four engines and some other plane parts were subsequently replaced at a cost of approximately $80 million.

But this was not the first time for catastrophic engine failure on an aircraft. On June 10, 1982, a British Airways Boeing 747 lost all four engines flying through an eruption cloud that issued from Mount Galunggung in Indonesia. Fortunately the pilot was able to restart three engines, and land safely in Jakarta. And then in July, 1982, a Singapore Airlines 747 also flying across Indonesia lost two engines to volcanic ash. It also landed safely in Jakarta.

Other volcanoes that have affected air travel or caused airplane damage include the 1991 massive Mount Pinatubo eruption in the Philippines and the late summer 1992 Mount Spurr, Alaska, activity that disrupted both U.S. and Canadian airspace. Air traffic was also disrupted by Rabaul on the Island of New Britain, Papua New Guinea, in 1994 and Popocatépetl Volcano near Mexico City between 1996 and 1998.

Fast forward to 2010 and the eruption of Icelandic volcano Eyjafjallajökull (Eyjafjoll) in the southern part of the country that last erupted between 1821 and 1823. Here the situation was especially perilous because of the eruption dynamics and the location of this large, tephra-spewing volcano west and upwind of Europe, beneath major airline flight paths.

Scientists became aware that the volcano might erupt following seismic activity beginning in March and inflation of the volcano due to magma intrusion into the cone. On March 20, alkali olivine basaltic (47 percent silicon dioxide) flows erupted on the volcano's flanks below the glacier.

The eruption dynamics changed on April 14 when pyroclastic activity of andesitic composition

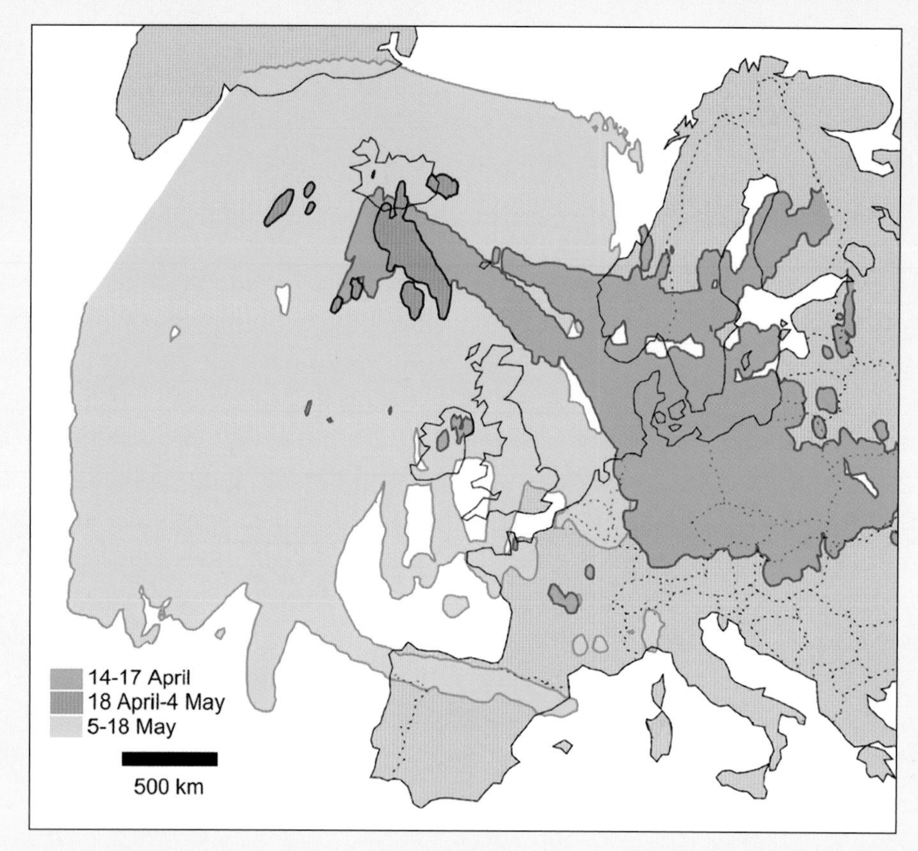

SB FIGURE 1: Three Eyjafjajokull Volcano, Iceland, Iceland, eruption ash fallout patterns during April and May, 2010 (modified from Gudmundsson et al., 2012).

(58 percent silicon dioxide) commenced under the ice cap. As volcanic fire met glacier ice and water, intense phreatomagmatic explosions sent fine, tephra-bearing blasts to 9 km above the volcano.[2] The ash was then carried by prevailing winds east-southeast toward Europe. One hundred thousand to 150,000 m^3 of ice, or 10 to 15 percent of the glacier subsequently melted, exposing a 1-km erupting fissure in the summit crater.

The eruption affected local people and agriculture,[3] and a minor jökulhlaup—a glacial meltwater outburst from under the glacier—eroded soils as it rushed to the sea. The largest problem associated with the eruption, however, was its effect on worldwide air travel to and from Europe and elsewhere. For more than six days, starting on April 15 when the eruption cloud reached Europe, many airports in the western part of the continent closed. More than 95,000 flights were canceled, stranding millions of travelers.

The impact spread worldwide as airports in New Zealand, San Francisco, Africa, and South America also canceled flights. Airline losses alone were estimated at $200 million per day. This monetary figure does not include the value of lost perishable agricultural products, including flowers, and nonperishable commodities, as well as the disruption of global tourism, trade, and manufacturing supply chains. Not surprisingly, ground transportation in Europe surged.

In conclusion, the tephra problem demonstrates once again not only how tied together countries and world commerce are, but how susceptible we all are to such naturally occurring, but fortunately infrequent, uncontrollable events. With world population, commerce, and air travel growing exponentially, we can only expect more disruptions that will cause human, property, and monetary loss.

NOTES

1. Soon after the recognition of how damaging ash was to engines, highway patrol cars (and commercial buses) in eastern Washington started sporting massive black boxes on their front bumpers--new air filtration systems.
2. In phreatomagmatic eruptions, water flashes to steam with an instantaneous, explosive volume expansion of about 1,600 times. (One cubic meter of water becomes 1,600 cubic meters of steam.)
3. People and animals that inhale very fine ash particles may develop respiratory problems such as emphysema. The inhalation of sulfur dioxide gas may be immediately toxic, whereas high fluorine concentrations can lead to fluoridosis. The 1783 fluorine-rich Icelandic eruptions resulted in the loss of thousands of farm animals due to fluoridosis poisoning.

REFERENCES

Brantley, 1990; Casadevall, 1994; Gudmundsson et al., 2012; Miller and Casadevall, 2000; 2010, *New York Times;* Segall and Anderson, 2014.

The three Yellowstone eruptions released 2,500 km^3, 280 km^3, and 1,000 km^3, respectively, of tephra. The eruption from the Toba caldera released 2,800 km^3 of material, which, according to Rampino and Ambrose, may have resulted in a six- to seven-year-long winter and 1,000 years of a colder than normal world climate. They also recently speculated that human genetic data indicate that the global cooling may have been responsible for almost exterminating our ancestors.

Unknown is the exact amount of sulfur released by this and other very large silicic eruptions. While basaltic eruptions may eject 2,000 parts per million (ppm) or more of sulfur, silicic magmas generally release much less sulfur (Table 9.1). Silicic eruptions, then, must be considerably larger in order to equal the sulfur output of a given basaltic eruption.

	Sulfur**	Chlorine	Fluorine
Basalt	800	65	100
Intermediate	560	920	500
Silicic	70	135	160

* in parts per million (ppm)
** The 1982 El Chichón eruption magma (55.9% SiO_2 - trachyandesite) contained between 6,000 and 9,000 ppm S making it very unusually S-rich! (Sigurdsson, 1990)

TABLE 9.1: *Average atmospheric yield values of sulfur, chlorine, and fluorine from eruption of different magma compositions.**

Volcano	Date	Estimated aerosol atmospheric loading*	Northern hemisphere summer temperature anomaly in degrees °C	Deaths
Toba, Indonesia	73,500 B.P.	1,350		
Laki, Iceland	1783-1784	215		
Tambora, Iceland	1815	>100 to 115	–0.5	>71,000
Krakatau, Indonesia	1883	-50 to 62	–0.3	>36,000
Santa Maria, Guatemala	1902	45	_	7,000 – 13,000
Katmai, Alaska	1912	20 to 40	–0.4	2
Agung, Indonesia	1963	14.4 to 20	–0.3	975
Fuego, Guatemala	1974	3 to 6		
Mount St. Helens, Washington	1980	2		57
El Chichon, Mexico	1982	12 to 14.4	–0.2	>2,000
Mt. Pinatubo, Phillipines	1991	30 to 40	–0.5	1,200

Sources: Francis and Oppenheimer, 2004; McCormick et al., 1995; Schmincke, 2004
*1 Tg=1 million tonnes. 10 (6) tonnes=10 (12) grams of H2SO4/H20 aerosol particles

TABLE 9.2: *Selected major sulfur aerosol-producing eruptions.*

While it might take 1,000 km^3 to 4,000 km^3 of erupted rhyolitic magma and 700 km^3 of intermediate magma to equal the sulfur dioxide output from 300 km^3 to 400 km^3 of basaltic magma, some silicic magmas contain—and would release—more sulfur than the averages found in Table 9.1. For example, the 1985 Nevado del Ruiz eruption in Columbia, South America, of rhyodactic magma (with intermediate silica content) contained 700 ppm sulfur. The great Toba rhyolitic eruption may have averaged close to 400 ppm sulfur.

Why do magmas vary so greatly in their sulfur content? The variance appears to be related, in part, to the iron content of the magma: higher amounts of sulfur occur in more iron-rich magmas, for example in basalt versus rhyolite.

It also appears that some explosive silicic eruptions, such as the 1991 dacitic Mount Pinatubo event, may be triggered by basalt injected into their magma chambers. Even if they don't erupt basalt, they release basalt-derived, sulfur-bearing volatiles.

Problems with Determining Long-Term Climatic Effects

We now know that some major volcanic eruptions influenced world climate. But how great and long-lasting are the effects of such infrequent but huge eruptions? The amount of sulfur released is usually a minimum estimate, and for many ancient eruptions even rough estimates are not available.

Furthermore, the timing of some eruptions, especially during the Pleistocene Epoch ice ages that ended 10,000 years ago, makes estimates of global effects very uncertain. For example, the Toba eruption 73,500 years ago could have decreased global surface temperature about 3.5°C (6.3°F). But because this eruption happened during major, Pleistocene climatic cooling, it is difficult at best to obtain a definitive answer to the relationship between that eruption and climate cooling.

Also important is the uncertainty in the absolute age dates

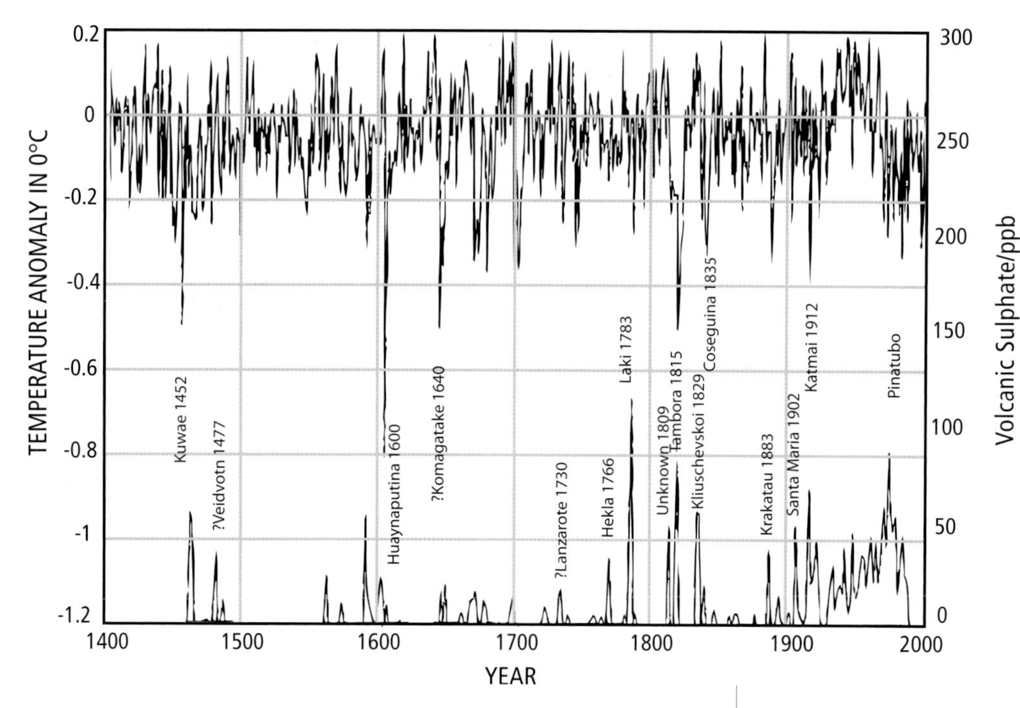

FIGURE 9.5: *Volcanic markers in GISP2 ice core from Greenland (lower curve) based on sulfate values in parts per billion (ppb) seen along the right vertical axis. Upper curve of Northern Hemisphere land and marine surface temperatures is derived from tree ring data. The temperature anomalies are relative to the 1881 to 1960 average or mean temperature for April to September (modified from the National Snow and Ice Data Center: University of Colorado, and Francis and Oppenheimer, 2004).*

of big, ancient eruptions. For eruptions in historical times (e.g., Vesuvius in 79 A.D.), the dates were accurately recorded by observers. Even for major eruptions less than 5,000 years old, we can date them fairly precisely using tree ring growth damage, such as occurs in bristle cone pines (see the sidebar on Tree Rings, page 127).

Similar to tree rings, ice cores from Greenland and Antarctica also record yearly events in the form of acid (sulfur) aerosol deposits. Of course, eruptions are recorded in such ice cores only if the aerosols reach these high latitudes. When they do, the ice core data can be very useful for dating events back hundreds of thousands of years.

The difficulty of obtaining accurate age dates increases with eruption age. Even with the best, most precise absolute age-dating techniques, dates are prone to uncertainties. For example, an error of plus or minus 3,500 years exists for the radiometric date for the Toba eruption of 73,500 years ago. This age uncertainty, then, gives a range of 7,000 years for that eruption. Only if distinct ash and acid aerosol particles can be found in Greenland and/or Antarctic ice cores, that in places record yearly events back almost one million years, can we get closer to accurate eruption dates.

In the case of the gigantic Toba eruption, recent ice core studies show that sulfuric acid was detected for only 6 years after the blast. This suggests that the cooling effect of sulfate aerosols may actually be short-lived.

In conclusion, more and better volcanic eruption data and precise age dating combined with greater knowledge of the factors that control both atmospheric circulation and climatic changes may lead to more definitive answers about the relationship between volcanic eruptions and climate effects. These advances are happening!

CHAPTER 9 – NOTES

1 Fluorosis in animals occurs when fluoride replaces calcium in bones. This condition can lead to brittle bones, joint pain, and death. Brittle bones eventually cannot support an animal's weight, and so bones fracture.

2 Aerosol particles may be either liquid or solid. While in the stratosphere they increase the albedo or reflectance back into space of incoming solar radiation. On Earth's surface, they absorb radiation. In contrast, snow has a very high albedo or reflectivity. The average hair is about 90 microns in diameter (one millionth of a meter). Blondes average 17, while hair may be 180 microns in diameter for folks with black hair.

3 Opacity of the stratospheric aerosol layer is measured by beaming upward on clear nights laser light from the National Oceanic and Atmospheric Administration's Mauna Loa Observatory. The observatory is located at more than 13,000 feet on the big island of Hawaii in the middle of the Pacific Ocean. This LIDAR (Light Detection and Ranging) technique measures the back scatter reflected from particles.

4 The bulk of yearly sulfur dioxide emissions comes from small volcanoes and fumaroles. For example, Arenal Volcano in Costa Rica releases approximately 200 tonnes per day of sulfur dioxide (200 X 10^6 grams per day) of the world yearly total of between 7.5 and 18.6 X 10^{12} grams. Much greater amounts of sulfur dioxide have been released at times from large volcanic systems such as Kilauea in Hawaii and Mount Etna in Sicily. Between 1956 and 1985, Kilauea released between 0.2 and more than 10 X 10^9 grams per day; Mount Etna emitted 4 X 109 grams per day between 1975 and 1987.

And then there was the incredibility sulfur-rich 1982 eruption of El Chichón Volcano in Mexico. The trachyandesite magma with 55.9 weight percent silicon dioxide is estimated to have had between 0.6 and 0.9 weight percent sulfur—so much sulfur that crystals of the mineral anhydrite ($CaSO_4$) formed on the erupted tephra. Mount St. Helens had only 0.007 weight percent sulfur.

In comparison, human activities of burning coal, gas, and oil release between 120 and 200 million tonnes of sulfur dioxide per year.

5 Huge flood basalt eruptions are believed by some to have caused massive extinctions including that of the dinosaurs 65 million years ago. That was when the vast Deccan Trap flood basalts formed, today found in west-central India around Mumbai. Others believe that huge meteoric impacts, such as the Chicxulub impactor that hit the Yucatan region of present day Mexico 65 million years ago, did the beasts in.

CHAPTER 9 – REFERENCES

Ball, 2008; Bindeman, 2006; Bluth et al., 1993; Decker and Decker, 2006; Fisher et al., 1997; Francis and Oppenheimer, 2004; Franklin, 1784; Halmer et al., 2002; Huffman, 1990; Humphreys, 1913; Kennett and Thunell, 1975; Lamb, 1970; Luhr et al., 1984; McCormick et al., 1995; McGee et al., 1997; Mills, 2000; Rampino and Ambrose, 2000; Rampino and Self, 1982, 2000; Rampino et al., 1988; Robock, 2000; Sanderson, 2009; Santer et al., 2014; Sarasin and Sarasin, 1901; Schmincke, 2004; Sigurdsson, 1990; Stothers, 1984a, 1984b; Thordarson and Self, 1996; Wallace, 2001; Wexler, 1952.

CHAPTER TEN

Volcanism—Related Resources

Introduction

Unless you live near an active volcano, you may not realize that many aspects of your daily life are tethered to volcanism and igneous processes. Consider some of the household products we use every day, for example—phones, computers, and automobiles. These contain metals such as copper, tungsten, iron, zinc, lead, gold, and silver.

Another vital industrial element, sulfur, is used to make rubber, soaps, medicines, matches, explosives, and agrichemicals, and even to dry apricots. These raw materials, and more, commonly develop in, on, or well under volcanoes. (See Volcanic Building Materials sidebar, page 53.) Some mineral deposits, now exposed at the surface by uplift and erosion, actually formed several kilometers (miles) below craters during active volcanism.

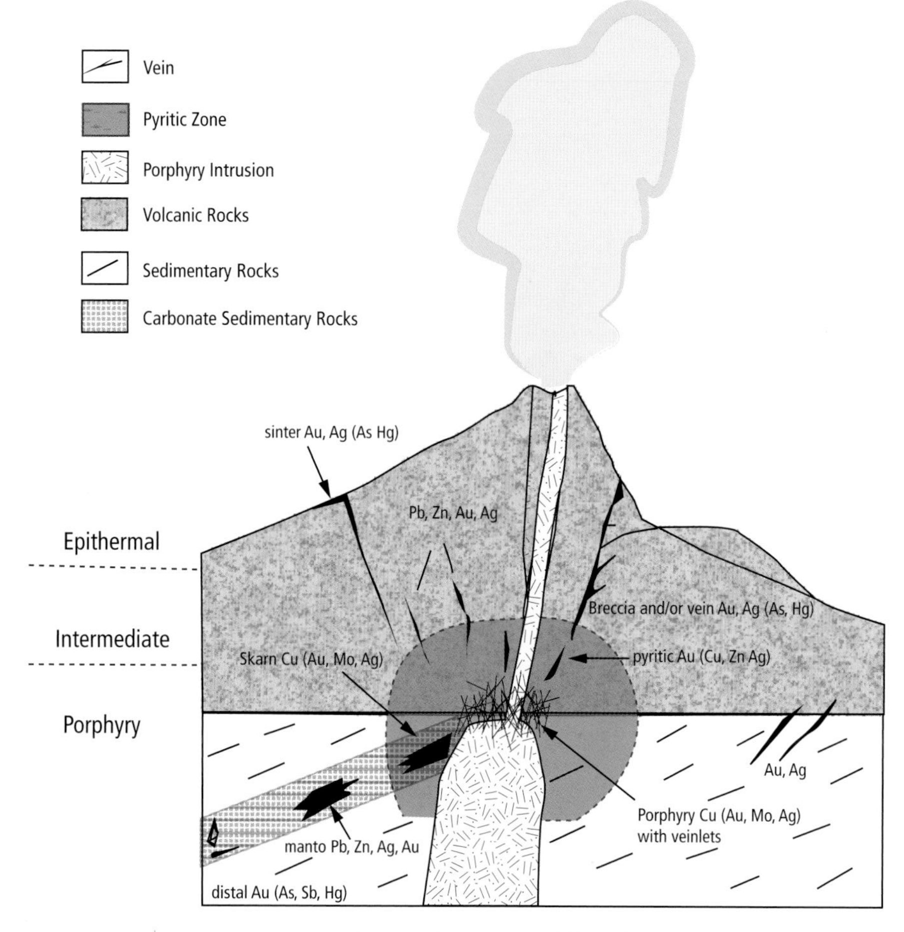

FIGURE 10.1: *Volcanic System with Mineral Deposits. Mineral deposits formed in an Andean-type arc volcano-intrusion environment. Note the different types of mineral deposits associated with different levels and rock types within this intrusive-extrusive system. Not shown are crater deposits mentioned in the text. Elements shown include Ag = silver, Au = gold, As = arsenic, Cu = copper, Hg = mercury, K = potassium, Mo = molybdenum, Pb = lead, Sb = antimony, Zn = zinc, alt = wall rock hydrothermal alteration. Argillic alteration is characterized by the formation of clays due to hot waters circulating through volcanic rocks where veins are forming at and near the surface. Pyritic (FeS_2) alteration occurs where sulfur from the hydrothermal solutions combines with iron found in original iron-bearing minerals. Skarn is Swedish for mixed rock* and manto *is Spanish for blanket (modified from Kirkham and Sinclair, 1996).*

Volcano Tops—Craters and Fumaroles

A good place to examine volcanic resources is at the top and on the upper flanks of active and dormant volcanoes. Fumaroles are common here, and they degas underlying magma chambers and their resulting mineral products. Yellow encrustations surrounding the vents may be sulfur. In fact, prior to about 1890, essentially all sulfur was derived from volcanoes; Mount Etna on Sicily was a major producer.[1] Sulfur was loaded into horse-drawn wagons and then shipped throughout the world. Mount Adams, in the Cascades of Washington State, also produced commercial sulfur from 1929 until about 1957.

In addition to sulfur, most fumaroles produce compounds such as aluminum chlorohydrate, the basic ingredient of antiperspirant deodorants. Fumaroles also emit trace amounts of other compounds in addition to many metals. While the concentrations of these metals are generally too low for exploitation, they suggest that metals may be found in much higher concentrations at depth. And they commonly are. For example, copper minerals precipitate (with or without sulfur) around vents on the dormant Aucanquilcha Volcano in northern Chile, volcanoes in the Kamchatka Peninsula of eastern Russia, and at Cerro Negro Volcano in Nicaragua.

Geologists recently estimated that about 16 ounces of gold per day are released into the air from fumaroles located within the crater of Galeras Volcano in Columbia, South America. In addition, at least another 640 ounces of gold are deposited within the volcano each year, proving that volcanic systems do produce ore deposits.[2]

Ore Deposits that Form within the Upper Parts of Volcanoes (<1km Depth)

A great place to see ore deposits formed by volcanic processes is in the extinct Farallon Negro volcanic complex in Argentina. There surface processes—running water and glaciers—have eroded deeply into the extinct volcano, exposing the guts of the igneous ore deposit-forming system.

EPITHERMAL GOLD-SILVER VEINS

Gold and silver deposits form in convergent plate margins or subduction zone volcanoes. Some of these deposits are enormously rich in precious metals, and so fueled the New World gold and silver rushes, and then colonization, following the conquest of the Aztec and Inca empires in Mexico and South America, respectively. Once the Spaniards saw the people adorned in precious metals, the rush was on to find this type of deposit—and the search continues to this day.

Famous U.S. bonanza deposits include the Comstock Lode at Virginia City, Tonapah, Round Mountain, and Goldfield in Nevada; Cripple Creek and Summitville in Colorado; and the deposits of Guanajuato and Pachuca in Mexico. Volcano-studded Indonesia and the Philippines and elsewhere where there is or has been recent subduction-type volcanism also host deposits. Other locations include the Dominican Republic, Spain,

	Type of Mineralization	Location	Metals, Elements*	Examples	Host	References
Strato Volcanoes (andesitic)						
	encrustations	Edifice crater	sulfur (S)	Mt. Etna, Italy	lavas, pyroclastics	
	breccia pipes, diatremes**	cutting crater and below	Cu, Au, Fe, S (Mo, Ag)	Montana Tunnels, Montana	Broken rocks	Sillitoe et al., 1985
	Acid-sulfate veins, replacements	below and into crater	Cu, Au, As, Zn, Fe, S		Lavas, pyroclastics	Silitoe, 1993
	Epithermal veins	below crater	Cu, Au, Ag, Pb, Zn, S Fe	Tui, Waiorongomai, New Zealand	flows, pyroclastics	Sillitoe, 1993
	Porphyry (disseminations)	3-5 km? below crater	Cu, Au, Fe, S (Mo, Ag)	Farallon Negro, Argentina	intrusive phases	Sillitoe, 1997
Flow Dome Complexes (dacite-rhyolite)* **						
	veins	in domes	Ag, Bi, Pb, Cu, W, Au, Sn, FS, Fe	Julcani, Peru	dome, flows	Petersen et al., 1977
	Acid -sulfate, Breccias, veins, replacements	Feeder pipes	Ag, Au, W, Fe, F, S	Summitville, CO; Goldfield, NV	dome, flows	Lipman, 1975 Sillitoe, 1993
	Veins	Below domes	Ag, Pb, Zn, Mn, Fe, FS	Chinkuashih, Taiwan	domes, below them	Wallace et al., 1968
	Climax Porphyry (disseminations)	kilometers below domes	Mo, W, Sn, Fe, S	Climax, CO	Intrusive phases	
Caldera Complexes (basalt - rhyolite)* **						
	Veins, disseminations	below caldera	Au, Ag, Zn, Cu, Fe, S	Lake City and Creede, CO	flows, pyroclastics	Stevens and Eaton, 1975
	Veins, disseminations	below caldera	Hg, U, S	McDermitt, NV-OR	flows, pyroclastics	Buchanan, 1981
	Veins	below caldera	Au, Te, S	Vatukoula, Fiji	flows, pyroclastics	Buchanan, 1981

* = Not all metals may by found in any particular deposit.
** = diatremes witness both up and down motion resulting in clast rounding; breccia pipes have angular clasts from primarily vertically directed explosions
*** = mineralization is post dome and caldera formation

TABLE 10.1: *Types of ore deposits that occur within three kinds of volcanic structures: strato volcanoes, flow dome complexes, and calderas. Note that some of the same types of deposits, e.g., veins, occur in different volcanic settings but may contain the same as well as different minerals.*

Japan, New Zealand, Peru, Bolivia, Ecuador, Argentina, and Chile.

Some deposits of this type, termed epithermal,[3] are so rich in precious metals that a ton of ore may contain more than 1,000 ounces of silver and many ounces of gold. Epithermal deposits have been divided into many subtypes. They include those that contain mercury or antimony with or without gold and/or silver, copper, lead, zinc, and other metals.

The most important precious metal-hosting systems are the high-sulfidation (acid sulfate) and the low-sulfidation (quartz-adularia-sericite) types. These two subtypes are distinguished on the basis of mineralogy and alteration of the enclosing wall rocks, and location within the volcanic system. The high- and low-sufiidation systems differ in composition, and due to the physical characteristics of the hot, mineralizing waters that coursed through the rocks for thousands, perhaps millions of years, altering the host rocks and depositing metals. These characteristics in turn are a function of where each deposit develops in the volcanic-magmatic system.

Picture a volcano that derives magma, periodically, from chambers located at various depths beneath the summit cone. These are water-saturated environments, whether they are island arc- or Andean-type. Island arc systems are bathed in sea water, whereas Andean systems derive most of their water from precipitation. Some water in both systems may be magmatic. The former, recycled water is termed meteoric (meaning from the sky), whereas the magmatic-type is called juvenile because it has never been on the surface. Oxygen and hydrogen isotope studies are used to differentiate the two water types.

This water may be heated by magma to temperatures of 500°C or more, and it can contain variable concentrations of salts, metals, sulfur, and gases that influence the solubility and solution carrying capacity of metals. These components may be derived from magma or leached from the rocks through which the ascending, volcanically heated waters flow.

As metal-bearing hot waters—hydrothermal solutions—move through fractures in the rocks, sulfur combines with metals—copper, lead, silver, and zinc to precipitate ore and gangue (waste) minerals, including, commonly, pyrite (see Table 10.2). Gold, however, forms its own mineral, called native gold, enters the structure of sulfide minerals, and alloys, in various amounts, with silver to form the mineral electrum.

Acid-sulfate systems are a product of mineral deposition and wall rock alteration from acid, oxidizing (oxygen-rich), sulfur-rich hot water. Typical vein mineral assemblages include abundant pyrite (FeS_2), with lesser amounts of copper-arsenic sulfides such as enargite (Cu_3AsS_4). Sphalerite (ZnS), galena (PbS), native gold (Au), and electrum, +/- tellurium and bismuth minerals also may occur (see Table 10.2).

Following mineralization,

	Formula	Low Sulfidation	High Sulfidation
Ore Minerals		Frequency of occurrence (abundance)	
pyrite	FeS2	ubiquitous (abundant)	ubiquitous (abundant)
sphalerite	ZnS	common (variable)	common (very minor)
galena	PbS	common (variable)	common (very minor)
chalcopyrite	CuFeS2	common (very minor)	common (minor)
enargite	Cu3AsS4	rare (very minor)	ubiquitous (variable)
tennantite	Cu12As4S13	common (very minor)	common (variable)
covellite	CuS	uncommon (very minor)	common (minor)
stibnite	Sb2S3	uncommon (very minor)	rare (very minor)
orpiment	As2S3	rare (very minor)	rare (very minor)
realgar	AsS	rare (very minor)	rare (very minor)
arsenopyrite	FeAsS	common (minor)	rare (very minor)
cinnabar	HgS	uncommon (very minor)	rare (very minor)
electrum	AuAg	common (variable)	uncommon (very minor)
native gold	Au	common (very minor)	common (minor)
tellurides	? + Te	common (very minor)	uncommon (variable)
selenides	? + Se	common (very minor)	uncommon (variable)
tetrahedrite	(Cu,Fe)12(Sb,As)4S13	uncommon (rare)	uncommon (variable)
argentite	(Ag,Cu,Au)2-4(S,Sb,S)	uncommon (rare)	uncommon (variable)
acanthite	Ag2S	uncommon (rare)	uncommon (variable)
proustite	Ag3AsS3	uncommon (rare)	uncommon (variable)
pyrargyrite	Ag3SbS3	uncommon (rare)	uncommon (variable)
sulfosalts	sulfide minerals with Sb, As	uncommon (rare)	uncommon (variable)
Gangue (waste minerals)			
quartz	SiO2	ubiquitous (abundant)	ubiquitous (abundant)
calcite	CaCO3	common (variable)	absent except post mineralizat.
adularia	KAlSi3O8	common (variable)	absent
illite (clay)	KAl2(OH)2{AlSi3(O,OH)10}	common (abundant)	uncommon (minor)
kaolinite (clay)	Al4Si4O10(OH)8	rare	common (minor)
pyrophyllite	Al2Si4O10(OH)2	absent	common (variable)
alunite	KAl3(SO4)2(OH)6	absent	common (minor)
barite	BaSO4	common (very minor)	common (minor)
chlorite	Mg4.9Al(Si3.2Al0.8)O10(OH)8	common	uncommon
fluorite	CaF2	common	uncommon
rhodochrosite	MnCO3	common	uncommon
hematite	Fe2O3	uncommon	uncommon
siderite	FeCO3	uncommon	uncommon
chalcedony	SiO2	common	uncommon
epidote	Ca2(Al,Fe)3(OH)(SiO4)3	common	uncommon
pyrrhotite	Fe1-xS	uncommon	uncommon
zeolite	Na and/or Ca (SiO3)2.H2O	common	uncommon

TABLE 10.2: *Common and less common ore and gangue (waste) minerals in epitherma-type deposits.*

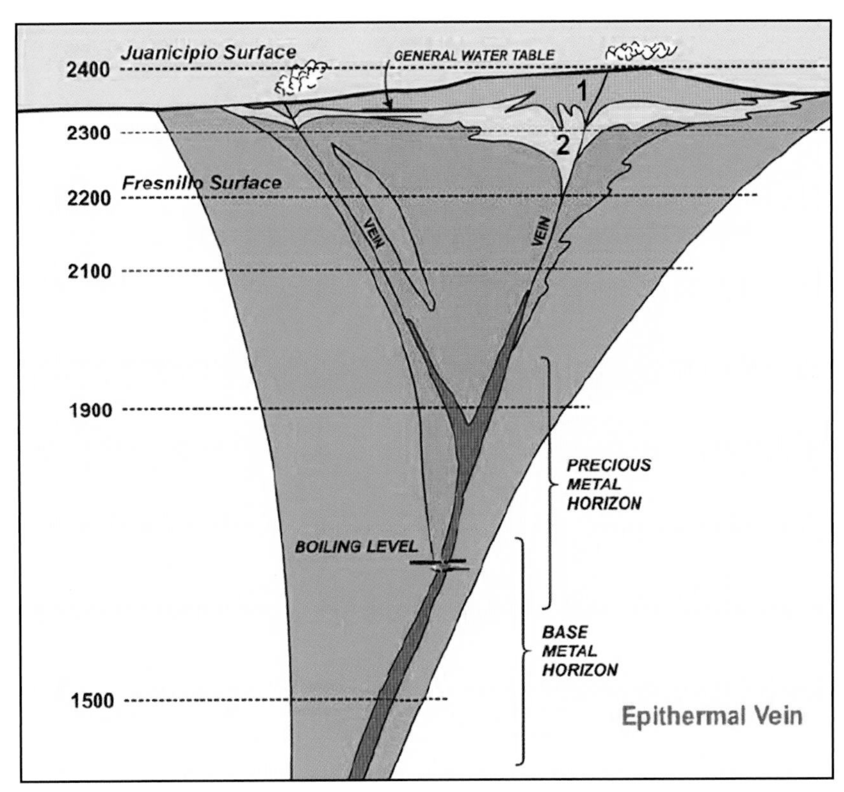

FIGURE 10.2: *Low-Sulfidation or quartz-adularia vein system. Idealized cross section of the structurally simple Juanicipio, Zacatecas, Mexico, low-sulfidation or quartz-adularia-type epithermal vein showing the present ground surface of both the Juanicipio and Fresnillo vein systems. Vertical and horizontal positions of ore and gangue minerals, and alteration mineral assemblages that develop in response to ascending hydrothermal solutions are shown. This type of vein usually forms at some distance from active volcanoes versus the high-sulfidation type. The volcanic wall rocks are altered by the mineralizing solutions as follows. A silica-rich cap (#1 - gray) forms if the system vents to the atmosphere. An acid water leached zone (#2- yellow) lies below and contains clay minerals. Below is intensely silicified rock (purple) containing quartz and adularia (potassium feldspar – $KAlSi_3O_8$). Surrounding the vein is a wide zone of propylitized altered rock, green in color due to the presence of green minerals including chlorite mica. Elevations are in meters, and the precious metal (gold and silver) and base metal (copper, lead, and zinc) bearing horizons are depicted. See Table 10.2 for mineral compositions (figure modified from Megaw, 2006; Buchanan, 1981).*

the original volcanic host rocks may look like very porous, vesiculated lava. This results from extreme acid-leach removal of essentially all original minerals but quartz. Alunite, a pink sulfate mineral, and clay commonly develop too (see Table 10.2).

Both epithermal ore deposit subtypes are structurally controlled—ore fills open spaces within fractures, fault zones, and in and around rock fragments from ascending and/or laterally moving mineralizing solutions. The acid-sulfate subtype generally develops spatially closer and temporally to magma. Sometimes these systems form within cooling lava domes.

The adularia-quartz-sericite deposits, not usually found in the same igneous systems with acid-sulfate deposits, develop from pH neutral, intermediate to reducing (oxygen deficient), and low sulfur-bearing solutions. Gold and silver are the principal economic commodities in these sulfide mineral-poor, open space filled-veins. Minor amounts of pyrite, chalcopyrite ($CuFeS_2$), galena, and sphalerite +/- arsenopyrite (AsFeS) usually occur, too.

Rarely do either of these two subtypes extend more than 1,000 meters (3,000 feet) below their paleosurface, or the ground surface during formation. Sharp metal gradients, or changes in metal concentrations, commonly exist within deposits: gold and silver values decrease with depth, while base metals (copper, lead, and zinc) usually increase.

Middle Level (Mesothermal) Ore-Forming Environments

PORPHYRY COPPER–MOLYBDENUM AND ASSOCIATED DEPOSITS

Epithermal veins in some deposits have been traced downward into large mineralized masses of shattered and broken porphyritic intrusive rocks[4] (see Figure 4.2D, page 59). These rock masses crystallized from magmas that fed the lavas above.

Inspection of the mineralized rock usually shows tiny veinlets and disseminated (isolated) grains of pyrite (FeS_2), chalcopyrite ($CuFeS_2$), +/- molybdenite (MoS_2), and bornite (Cu_5FeS_4) in a bleached-looking, silica-rich rock. One has to look carefully to see these tiny sulfide mineral grains and veinlets because their total amount rarely exceeds 10 percent of the rock by volume, with most of that being pyrite.

The pre-mineralized rock was composed of approximately 60 percent large and small feldspar crystals, 10 to 20 percent quartz, and variable amounts of dark minerals that include biotite mica and amphibole.

These systems, called porphyry copper ore deposits, are the world's number one source of copper and molybdenum. They also contain major quantities of gold and silver. Examples include the Bingham Canyon deposit[5] southwest of Salt Lake

Geothermal Systems

Those who have been in a mine know that Earth's temperature increases with depth. Wall rocks in South Africa's 12,000-foot deep gold mines, for example, are too hot to touch; air conditioning is required in all deep mines.

It is estimated that Earth's core temperature is at least 4,500°C and may be 7,000°C. This high temperature is primarily due to (1) gravitational force, starting early in Earth's history, that increased the Earth's density, which in turn results in interior warming;[1] (2) heat produced by the disintegration of radioactive nuclides including uranium, thorium, and potassium (that presently reside mostly in the continental crust); and (3) residual meteoric impact heat.

In general, geothermal heat is too diffuse as it emanates from the Earth's surface to be utilized on a practical scale.[2] Exceptions are regions that contain hot water reservoirs. These are associated with spreading centers, convergent plate boundaries, transform plate boundaries, hot spots, and rifts. Furthermore, these areas are commonly related to volcanism, and their associated hot water is termed hydrothermal (water-heat).

The first instance of geothermal-generated electricity was on July 4, 1904, at Larderello, Italy. There, volcanically heated steam was (and still is) channeled into turbines that produce electricity. Today, globally, there are more than 700 electrical generation projects in 76 countries. Electrical production in 2014 exceeded 12,000 megawatts (MW), with another 30,000 MW under development. Geothermal electricity production is expected to continue growing at a rate of 4 to 5 percent annually.

Geothermal resources are not strictly renewable resources—power from them declines over time. This decline is due to decreasing reservoir pressures because of fluid withdrawal. The flow rate in one plant in the California Geysers field, for example, declined between 7 and 24 percent annually between 1982 and 1990. Furthermore, the thermal efficiency of geothermal electricity-generation plants is low, at 10 to 23 percent.

Country	Megawatts (MW) of Electricity
United States	3,093
Philippines	1,904
Indonesia	1,197
Mexico	958
Italy	843
New Zealand	628
Iceland	575
Japan	536
El Salvador	204
Kenya	167

(from the International Geothermal Energy Association, 2014)

SB TABLE 1: Ten leading countries producing geothermal electricity in 2010.

Electricity is produced from three types of geothermal systems: dry steam (vapor-dominated), flash steam (liquid-dominated), and binary cycle. Dry steam systems, such as at the Geysers in California and Larderello, Italy, require 150°C or hotter steam. These systems are preferred because steam is sent directly into turbines, with about 20 percent of the fluid returned back into the aquifer. The balance of the steam is vented into the atmosphere.

In flash steam systems, the most common type used in electrical generation, hot water (generally 180°C or higher) under high pressure at depth flashes to steam as pressure drops on the rising liquid. The steam is then directed into a turbine. Examples include the fields in Wairakei, New Zealand, Salton Sea, California, and Cerro Prieto, Mexico.

Binary cycle systems, first used in 1967 in the U.S.S.R. (now Russia), utilize water as cool as 57°C. In the binary cycle, another fluid with a much lower boiling point encounters warm to hot water. The fluid warms, flashes to "steam," which then enters the turbines.

Examples of active binary systems include those in Long Valley Caldera in California and Steamboat Hills in Nevada. They use isobutylene to extract heat from 170 to 175°C water. This type of system has a great future for producing more electricity because of the greater availability of water of appropriate temperature.

Another way geothermal electricity is generated is by pumping water into hot ground. There it heats, and is extracted as steam, that then enters turbines.

In addition to electrical generation, geothermal waters are used in spas, space heating of buildings including greenhouses, and in fish farms. For example Klamath Falls, Oregon, with a population of 20,000, has over 600 geothermal wells. Water from these wells heats homes, schools, and hospitals, as well as sidewalks.

NOTES

1 Note that when materials are squeezed or pressure is increased on them, they increase in temperature. That results in a gravitational temperature increase.

[2] Ground source heat pumps used to heat buildings extract natural warmth by circulating heat-exchanging fluids in pipes through warm ground.

REFERENCES

Arnorsson, 2000: Geological Survey of Japan, 1999; Goff and Janik, 2000; Matek, 2014.

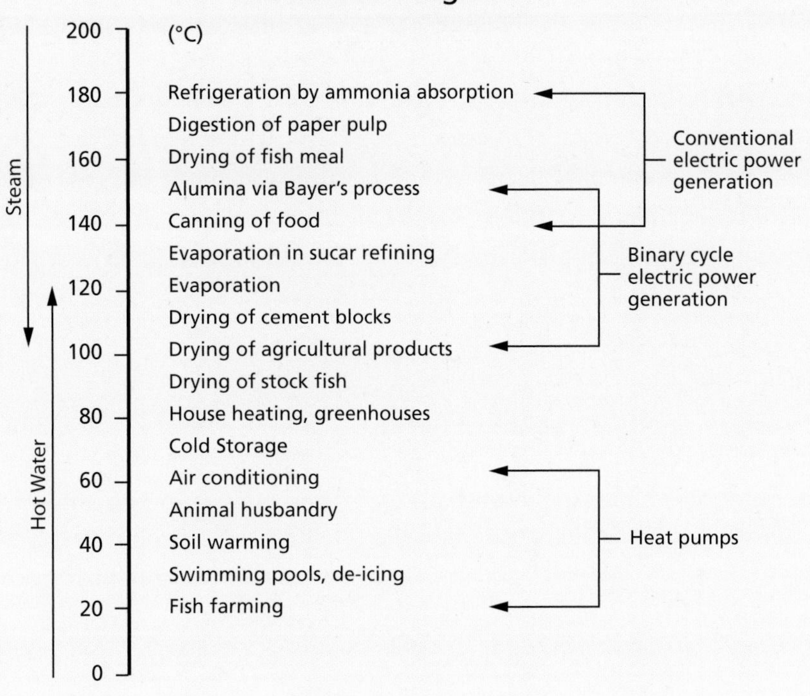

SB FIGURE 1: Negative environmental impacts of geothermal energy production include the atmospheric release of carbon dioxide, hydrogen sulfide, methane, and ammonia into the air from some systems. Groundwater may be contaminated with salt brines, and, for example, boron, lead, and arsenic. Landslides, ground subsidence (such as has happened at Wairakei, New Zealand), and the triggering of earthquakes also occur. Still, the carbon dioxide release is generally much less per energy unit produced—7 percent or less—than that from natural gas-fired plants (modified from Arnorsson, 2000).

City, Utah, the many deposits that dot south-central Arizona, southern British Columbia in Canada, and the huge, open pit copper deposits in the Andes Mountains of Chile. Bajo de la Alumbrera in the roots of the Farallon Negro volcanic complex in Argentina, mentioned earlier in this chapter, is also a porphyry-type deposit.

Water, key to eruptions, is also vital to porphyry sulfide mineral deposit formation and host rock alteration (see Water sidebar, page 27). During a volcano's life, many batches of water-bearing magma invade chambers located hundreds to thousands of meters below the cone. Some of these chambers erupt lavas, some don't. With time, the remaining magma solidifies. During this process, the pressure of water and other volatiles (such as carbon dioxide) builds up near the top of the slowly crystallizing igneous mass.

Because only limited amounts of water can be accommodated within the igneous minerals amphibole, and biotite and muscovite mica minerals, the remaining volatiles concentrate at the top of magma chambers. This hydrothermal solution becomes increasingly pressurized with magma crystallization. The resulting fluid over pressuring creates fractures that allow fluid to migrate into the surrounding country rocks. These hydrothermal solutions carrying dissolved silica, sulfur, and metals ultimately form epithermal veins.

In many systems, however, fluid pressure eventually exceeds the strength of the crystallizing magma itself. The result is brecciation (fracturing) of the hot, crystallizing rock into a myriad of large and tiny pieces. Brecciation may be so pervasive that pieces of unbroken rock are rarely more than fist-sized. Hydrothermal solutions then flow though the shattered mass, in some places up to kilometers across and hundreds of meters top to bottom. Tiny specks and veinlets of ore (chalcopyrite and bornite with gold, silver, and molybdenite) and waste minerals (abundant pyrite, quartz, etc.) precipitate within this shattered mass of rock.

Porphyry copper deposits vary in size, grade, and metals present. Mines may contain many billion tons of ore with grades approaching one percent copper.

But where did the iron come from that makes up the bulk of the sulfide minerals—pyrite? Iron is contained within the igneous minerals biotite ($K(Mg,Fe)_3(AlSi_3O_{10})(OH)_2$) and hornblende ($NaCa_2(Mg, Fe, Al)_5(Al, Si)_8O_{22}(OH)_2$), prior to mineralization. (These are black minerals visible in the granite steps at many libraries and post offices). Sulfur, in the invading hydrothermal solution, then combines with iron in these minerals to form pyrite (FeS_2). If copper is present, chalcopyrite ($CuFeS_2$) and/or bornite (Cu_5FeS_4) form at the expense of the iron-bearing minerals.

Do vein and porphyry-type mineral deposits underlie all arc-type volcanoes such as Mount Rainier and Mount St. Helens? They probably don't. Erosional processes, such as occurred at the Farallon Negro volcanic complex, have to first strip away the upper parts of volcanoes—up to hundreds to a few thousand meters (thousands of feet)—for us to know. And each volcanic system is different.

The oldest rocks in the Farallon Negro volcanic complex are 9.7 million years old: the youngest activity was 6.1 million years ago. Rocks include both intrusive and shallowly emplaced basaltic andesite, andesite, and dacite. The mineralizing/mineralized body was a 6.8-million-year-old dacite porphyry that formed the 800 million tonne[6] copper-and-gold-bearing Bajo de la Alumbrera ore deposit.

Other Mesothermal Ore Deposit Types

Other ore deposit types intimately associated with volcanic systems include porphyry molybdenum, porphyry tin, diatreme and skarn (see Figure 10.1). Climax, Colorado, and Potosi, Bolivia, are classic, well-described examples of molybdenum and tin deposits, respectively.

Diatremes are pipe-like bodies up to a mile or more across that contain rounded to angular volcanic and non-volcanic clasts and, in some places, base and precious metals (see Figure 3.14, page 50). Many if not most diatremes not only vented, but carried, fragments both up and down. Evidence includes wood found 300 m (1,000 feet) or more below the surface in some systems. This vertically directed milling or abrasion also rounded the enclosed rock fragments.

Skarn (Swedish for mixed-up rock) deposits form where magma encounters carbonate rocks. Limestone ($CaCO_3$), chemically very reactive, dissolves as hydrothermal solutions derived from cooling intrusions course through the surrounding rock. The dissolving limestone is simultaneously replaced with various other silicate, oxide, and sulfide minerals.

Skarn deposit types include: (1) iron-magnetite (Fe_3O_4) with or without gold and silver, (2) copper sulfides such as chalcopyrite ($CuFeS_2$) and bornite (Cu_5FeS_4) with gold and silver, and (3) lead-zinc-copper +/- arsenic-bearing deposits (galena (PbS), sphalerite (ZnS), and copper sulfides).

Note on Figure 10.1 that some gold-arsenic (As), antimony (Sb), and mercury (Hg) replacement-type deposits and gold-silver veins commonly form around the volcanic center.

Ore Deposits in Submarine Volcanic Environments

Some of the highest grade, and largest base metal —(copper, lead, and zinc) bearing ore deposits also containing precious metals are generated by submarine volcanic activity (see Figure 10.3). These form in both basalt-dominated spreading centers and their associated seamounts, and in felsic (silica-rich) dominated volcanic rocks. The latter type deposits are created in arc-related, convergent or subduction zone submarine environments.

Both types, called volcanogenic massive sulfide-type deposits (VMS), contain, by definition, at least 50 percent by volume sulfide minerals. Both types form on and just below the seafloor, generally at water depths exceeding 500 m (1,650 feet), and in rocks ranging in age from at least 2.5 billion years to the present. Both deposit types develop above subvolcanic intrusions within their respective volcanic sequences. Furthermore, these deposits are surrounded, and in some places capped, by volcanic rocks hydrothermally altered by the mineralization process.

The first observation of active seafloor mineralization was in the early 1960s when warm, metalliferous brines were encountered in the trough in the center of the Red Sea. This axial valley is in a new spreading center that is separating the Arabian Peninsula from northeastern Africa (see Figure 5.8, page 86).

Heavy, warm, saline brines are continuously leaking into the trough, precipitating muds rich in iron, zinc, and copper. The metal deposits are worth billions of dollars, and so efforts are in progress to recover them economically.

If hosted by basaltic rocks, VMS deposits are called Cyprus type. Kuroko-type deposits develop in submarine arc settings in more felsic (higher in silica) rocks. Pyrite is usually the most abundant sulfide mineral in both deposit types. Cyprus deposits contain trace but valuable amounts of gold in addition to copper and zinc. Kuroko deposits contain copper, lead, and zinc together with some silver, gold, and uranium. These deposits have been mined where uplifted, thrusted, or obducted onto land, and subsequently exposed on or near the surface by erosion.

These deposit types are also termed volcanogenic exhalative because much of the material deposits directly on the seafloor from very saline, high-temperature (375°C and higher), metal- and sulfur-bearing water.

CYPRUS-TYPE ZINC-COPPER DEPOSITS

Cyprus-type deposits, named for their initial description on said island in the Mediterranean Sea, have been mined for thousands of years. They form in and around spreading centers where cold, bottom sea water (slightly alkaline pH and with a salinity of about 3.5 percent sodium chloride) enters cracks in the basaltic seafloor. Sea water then descends for hundreds to more than 1,000 meters into the fractured rock. Water warms during this descent, changing dramatically in composition upon interaction with the lavas it courses through. With salinity and temperature increasing, and the solution pH dropping, this former sea water becomes a hot (temperatures of up to +400°C), salty, acidic solution capable of leaching and transporting trace amounts of base and precious metals from these rocks.

Seawater containing sulfur[7] enters basaltic rocks in the oxidized form, sulfate ($SO_4^{=}$). Sulfate is subsequently reduced (changed) to hydrogen sulfide (H_2S) during its journey. This evolved brine eventually emanates onto the cold seafloor in high-temperature submarine springs. Because of the sudden water temperature drop, hydrogen sulfide immediately combines with iron, base, and precious metals within this solution to form very fine-grained, sulfur-bearing ore and waste minerals (see Strange Life Forms sidebar, page 18). Together with non-economic (waste) minerals such as pyrite, quartz, and clays, sulfide ore minerals precipitate in massive sulfide deposits.

The thermally driven circulation of seawater, called convection, is familiar because this process operates when warming a pot of water (and also drives plate tectonics). Warmer water at the bottom of the pot rises only to be replaced by downward moving, heavier, denser, cooler water. This circular process eventually results in vigorous boiling.

Cyprus deposits are lens-shaped in cross section, with their upper parts composed of thin sulfide layers. With depth, this mass is increasingly

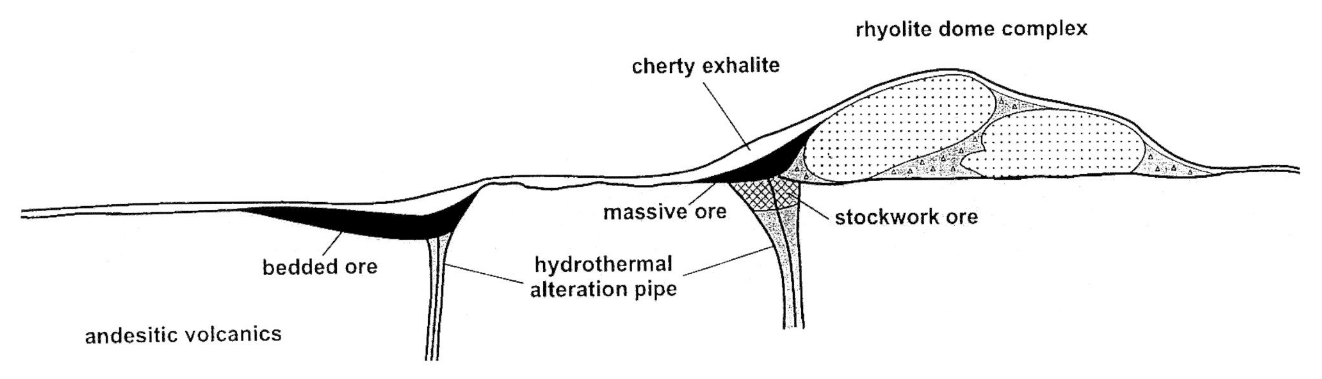

FIGURE 10.3: *Seafloor VMS-forming Hydrothermal Systems (Kuroko). Submarine volcanogenic massive sulfide deposits (VMS) showing the bedded nature of the VMS, their feeder zones located below containing stock work or vein-type mineralization. The Kuroko type is depicted here, which contains zinc, lead, and copper with lesser amounts of silver, uranium, and other metals. These bedded deposits are commonly overlain by a cherty or silica-rich exhalite layer that develops during the mineralization process and venting on to the sea bottom. Unless covered by volcanic rocks shortly after formation, these mineral deposits which may form thousands of meters below sea level will be destroyed by weathering and erosion on the ocean floor (from White and Herrington, 2000).*

crosscut by veins until just the mineralizing feeder "plumbing system" is encountered beneath the massive sulfide portion of the deposit. The lower part of the deposit, called the stringer or stockwork zone, is composed of intimately crosscutting (anastomosing) quartz-metal-sulfide-bearing veins that may continue for another several hundred meters (thousand feet) below.

Between the veins is highly altered basalt—silica and pyrite have replaced the original lava that has also lost almost all semblance of the original lava flow characteristics.

Metal grades and deposit size vary. Typical ancient Canadian deposits average 1.95 percent copper, 4.23 percent zinc, 0.8 ppm gold, and 19 ppm silver. Deposits range from a few thousand tonnes[6] up to 20 million tonnes, with the mean of 142 Canadian deposits being 5.3 million tonnes.

This type of metal occurrence is apparently common in spreading centers of all ages. However, until submersibles explored active spreading centers in the mid 1970s, little was known about deposit formation. Now we know that these occurrences develop contemporaneously with their enclosing host volcanic rocks. Furthermore, unless deposits are quickly buried by lava and/or sediments, they are rapidly destroyed.

Cyprus-type deposits are also found in New Brunswick and Newfoundland in Canada, Japan, the Philippines, and the Ural Mountains of Russia. These ore-hosting rock assemblages commonly consist of oceanic crustal and uppermost mantle rocks. The assemblages, named ophiolites, were tectonically tilted and subsequently thrusted or obducted onto land by plate tectonic forces.

KUROKO-TYPE LEAD–ZINC–COPPER DEPOSITS

Kuroko massive sulfide-type deposits[8] form in destructional submarine volcanic geologic environments. While hosted by different rock types, felsic (silica-rich) volcanic rocks (including flows, domes, and pyroclastic deposits) predominate (see Figure 10.3).

The Japanese deposits are in domes, pyroclastic, and sedimentary rocks of Miocene-age (23.7 to 5.3 million years ago). These occurrences developed in water as much as 4,000 m deep. Dacitic domes rose along deep-seated, basement faults within large submarine calderas. Following the formation of clusters of deposits between approximately 16 and 11 million years ago, the Hokuroku region of present-day northeastern Japan witnessed rapid uplift accompanied by surface erosion.

The idealized Kuroko deposit is a lenticular body showing a complex pattern of vertical mineral zoning from surface-deposited exhalative black sphalerite-rich (ZnS) kuroko ore into underlying yellow (oko and ryukako) chalcopyrite ($CuFeS_2$) and pyrite-rich (FeS_2) ore, respectively.

Most Kuroko deposits show a complex genesis in time and space of mineralizing episodes, resulting in solution and then redeposition of minerals. Some deposits were subsequently dismembered by violent submarine volcanic explosions.

Both Kuroko and Cyprus deposits show seafloor-deposited, layered ores overlying a feeder or stockwork zone of siliceous, crosscutting quartz-pyrite-chalcopyrite veins. A stockwork, with diminishing metal grades, commonly extends well below the mineable portion of the deposit.

Kuroko-type deposits occur in many parts of the world, in addition to Japan, in rocks ranging from 2.5 billion to 15 million years ago (Miocene). Ontario and Newfoundland in Canada, Australia including Tasmania, the Philippines, Fiji, Turkey, California, Alaska, and the Appalachian Mountains contain deposits.

The largest known Japanese deposit, Motoyama, contained 15 million tonnes[6] averaging 2.2 percent copper, 4.5 percent zinc, and 0.8 percent lead, plus 25 to 620 ppm silver and 0.3 to 2.1 ppm gold credits.

Copper–Nickel–Platinum (PGE) Group Sulfide Deposits Associated with Flood Basalts

Continental flood basalt provinces vary in size from the "small" 17- to 6-million-year-old, 200,000 km^2 Columbia River basalt plateau in the northwestern United States to the gargantuan (1.5 million km^2) Siberian province in north-central Russia[9] (see Figure 7.2, page 111). In addition to it's size, the Siberian Traps or Tungus mantle-derived plateau basalt, and similar rock assemblages in China and Canada,[9] contain huge copper-nickel, and PGE-bearing massive, vein and disseminated sulfide deposits.

One of the biggest Siberian deposits, Norilsk, with more than 80 percent by volume sulfides, measures 1 km by 3 km, averaging 20 m thick. At least 900 million tonnes averaging 3.72 percent nickel and 3.78 percent copper, with substantial amounts of platinum and palladium, are located within and along the basal contact of a mafic-ultramafic differentiated intrusion,[10] and the sedimentary rocks below.

While all basalt and ultramafic rocks contain these metals, the metals usually occur in trace amounts within olivine and pyroxene minerals. Apparently, within early basaltic intrusions at Norilsk and in the Talnakh and Oktyabrskoye deposits, however, metals partitioned into a high-temperature sulfide melt. The melt, being much heavier than the enclosing silicate liquid, settled to the bottom of the silicate melt and subsequently crystallized into layers, pods, veins, and disseminations. These concentrations consist of the iron sulfide pyrrhotite ($Fe_{1-x}S$) together with small amounts of pentlandite ($(Fe,Ni)_9S_8$) and chalcopyrite ($CuFeS_2$), plus trace amounts of other exotic minerals containing gold and platinum group metals.[11]

Unusual here were the huge amounts of sulfur needed to form a sulfide melt that scavenged these metals before they could enter the accepting olivine and pyroxene mineral structures. And while basalt contains up to 2,000 ppm sulfur (average 400 ppm), much more was needed to saturate the silicate melt in order to form a sulfide liquid. The most likely sulfur source was sedimentary rocks through which the basaltic magmas rose.

Nickel Sulfide Deposits Associated with Komatites (Ultramafic Lavas)

Massive nickel sulfide deposits are also found in ultramafic lavas called komatites[9] (see Table 3.1, page 31, for rock chemical compositions). These low-silica, high-magnesium lavas erupted at very high temperatures (1,300°C to 1,400°C), well above the temperature of basalt crystallization. They are characterized by olivine crystals that settled to the base of the komatiitic lava flows. Most komatites formed early in Earth's history when the Earth's interior was considerably hotter. That situation allowed these high-temperature magmas to reach the Earth's surface.

The most famous nickel sulfide deposits are those at Kambalda in western Australia. Sulfides there occur in small, layered deposits between and at the base of komatiitic lavas. The bulk of the sulfide is pyrrhotite ($Fe_{1-x}S$). It occurs with small amounts of nickel-bearing pentlandite $(Fe,Ni)_9S_8$ plus minor amounts of copper, cobalt, chromium, and platinum group minerals.

Mineralized rock consists of massive sulfide lenses, sulfides surrounding olivine grains, and sulfides disseminated within the lavas. The sulfide bodies are tabular to ribbon-shaped, generally less than 10 meters (+30 feet) thick, and vary from hundreds of meters to more than 5 km long.

The Kambalda mining camp consists of many small but high-grade deposits that contained a total of at least 25 million tons, averaging 3.5 percent nickel. Other deposits occur in Zimbabwe, Africa, and in Canada.

Diamonds from the Deep—Kimberlite Deposits

Diamonds, once used exclusively in jewelry because of their beauty, hardness, and rarity, have become one of our most important industrial minerals due to their great hardness. On an absolute hardness scale, diamond is 37; the next hardest mineral, sapphire, has a value of 9. The price of petroleum and metals, for example, would be much higher without diamond-tipped drill bits able to bore through thick, hard rock layers before needing replacement. The abrasion and polishing industries would be at a major disadvantage without diamonds.

Some of the world's largest and richest diamond fields are in stream and beach gravel placers.[12] These diamonds were released by rock weathering processes from diamond-bearing volcanic pipes called diatremes. Diatremes are composed of kimberlite and lamproite rocks.[13] Running water then transports diamonds to their new homes in stream and beach placer deposits.[14]

Diamond-bearing pipes are found in southern Africa, Russia, northern Canada, parts of the United States, Brazil, western Australia, and China (see Figure 3.14, page 50, for a cross-sectional view of a diatreme).

Diamonds form at great depths—200 to 300 km (124 to 190 mi)—in the mantle from carbon under pressure so high that graphite or pure carbon (as found in pencils), cannot exist because of its more open molecular structure. Based upon the nature of the included micro mineral grains, diamonds form within the asthenosphere and the lithospheric mantle. Diamonds apparently also develop in subduction zones where the subducting slab of oceanic crust penetrates deeply into the mantle under extremely high pressure, but relatively cool temperatures for that depth.

Diamonds are picked up by rising, volatile-rich kimberlite and lamproite magmas that bring them to the surface. Diamonds have been rising within pipes since Precambrian time, and the youngest known pipes are Mesozoic in age. There is no reason to believe the process is not continuing.

Diatreme pipe surface exposures are usually circular, between 50 to 500 m (165 to 1,640 feet) in diameter, and carrot-shaped top to bottom. They generally form topographical depressions due to the chemical instability of the host rocks when exposed to surface weathering. The depressions commonly contain lakes. In northern Canada, for example, small, circular lakes present excellent exploration targets.

Conclusions

This chapter has highlighted some of the varied types of ore deposits related to volcanic and subvolcanic processes. Clearly, both intrusive and extrusive magmatic activities have been responsible throughout Earth's history for the formation of major mineral deposits. These processes will continue as long as plate tectonics and hot spots operate.

Not discussed here are the volcanic rocks themselves, including flow, tephra, pumice, travertine, and ignimbrite deposits used for abrasives and in building and road construction; these are touched on in the Volcanic Building Materials sidebar, page 53.

CHAPTER 10 – NOTES

1 Sulfur production from most volcanoes ceased with the utilization of the Frasch process. Discovered by German-born American chemist Herman Frasch (1852 to 1914) and first used successfully in 1894, the process used hot water pumped into the top of petroleum-bearing salt domes in the Gulf Coast region of the U.S. and elsewhere. Molten sulfur then formed, rose to the surface, and was captured. Now most of the world's sulfur is derived from petroleum (oil and natural gas).

2 Ore is a geologic term that is defined economically: ore is something that is mined at a profit. For example, there are many known deposits of gold in the world but far fewer *ore deposits* of gold. If gold sells for $1,000 per ounce but it costs $1,100 per ounce to recover and refine, one has a gold deposit but not a gold ore deposit. However, if the price of gold rises above $1,100 per ounce with stable mining and processing costs, the deposit may become an ore deposit. Conversely, if the price remains at $1,000 per ounce but newer technology lowers the recovery costs, the deposit might subsequently become an ore deposit.

3 The name epithermal, defined by Burbank, Nolan, and Lindgren in 1933, refers to deposits that form on or just below the surface down to about 0.5 to 1 km, and from hydrothermal (hot water) solutions ranging from 150°C to about 250°C to 300°C. These are the Western Hemisphere high-grade bonanza deposits that the Spaniards and others sought and mined.

4 Porphyritic intrusive rocks are those that have at least two sizes of crystals, with the larger size called phenocrysts. The smaller crystals might be microscopic in size. In porphyry copper-type deposits, the phenocrysts are typically feldspars: potassium feldspar (KAlSi3O8), plagioclase ($(Na,Ca)Al_2Si_2O_8$), biotite mica, and quartz (see Figure 4.2D, page 59).

5 The Bingham Canyon deposit, Utah, is a "typical" but very large porphyry type copper-molybdenum deposit hosted by granitic rocks. It is presently the largest man-made structure on Earth. Open pit mining started in 1906 on a hill, and as of 2004 the mine was 0.97 km (0.6 mi) deep and 4 km (2.5 mi) wide. It has yielded over 2 billion tonnes of ore containing 17 million tonnes of copper, 386,000 tonnes of molybdenum, 23 million ounces of gold, and 190 million ounces of silver.

6 One metric tonne is equal to 1,000 kilograms (kg). One kg equals 2,204.6 Av. pounds. There are 12 Troy ounces in the Troy pound vs 16 Avoirdupois (Av.) ounces (used for bananas, lettuce, etc.) in the Av. pound. Each Troy ounce is equal to 1.0971 Av. ounces (Troy ounce = 31.104 grams; Av. ounce = 28.35 grams).

7 Sea water averages 900 parts per million sulfur in the form of sulfate ($SO_4^{=}$).

8 Kuroko means black ore in Japanese and is due to the abundance of dark colored (iron-rich) sphalerite (ZnS) that occurs together with galena (PbS).

9 See Chapter 7 for information about flood basalts.

10 Mantle-derived intrusions may differentiate or separate, mineralogically, during crystallization into a more magnesium- and iron-rich, silica-poor lower ultramafic or ultrabasic unit and an overlying basic unit. The basic unit is basaltic in composition, whereas the lower unit has at least 18 percent MgO, high FeO, and 45 percent or less SiO_2. See Table 3.1, page 31, for rock compositions. The mantle and komatite rocks listed in Table 3.1 are ultramafic in composition and are composed of at least 90 percent olivine and pyroxene minerals.

11 The 6 platinum group metals (PGM) include iridium (Ir), osmium (Os), palladium (Pd), platinum (Pt), rhodium (Rh), and ruthenium (Ru).

12 Placer is a term for stream and beach deposits of minerals that are (1) chemically durable in the weathering environment (won't dissolve), (2) have high specific

gravity (heavy), and (3) can withstand abrasion (physically durable). Moving water separates these heavy minerals from the lower density minerals such as quartz. Common placer minerals include gold, cassiterite (tin oxide: SnO_2), ilmenite (titanium oxide: TiO_2), native copper, platinum minerals, diamonds, sapphires and rubies.

13 Kimberlite, named for Kimberley, South Africa, is a porphyritic, alkalic (rich in potassium), water- and carbon dioxide-rich, mantle-generated peridotite. Phenocrysts of olivine and potassium-rich mica phlogophite—$KMg_3AlSi_3O_{10}(F,OH)_2$—occur in a fine-grained groundmass of Mg-rich olivine, enstatite pyroxene, Cr-rich diopside, and Cr-rich garnet crystals. Lamproite is a similar K-rich diamond-bearing rock distinguished from kimberlite by having both phenocrysts and small crystals of phlogophite mica.

14 Because of their durability diamonds and gold (and other placer minerals) may be incorporated into sediments that subsequently become hardened into sedimentary rocks. Upon weathering these "recycled" diamonds then reenter rivers forming stream placers, and beach placer deposits near where rivers enter the sea.

CHAPTER 10 – REFERENCES

Boyd and Meyer, 1979; Buchanan, 1981; Eldridge et al., 1983; Franklin, 1996; Goff and Gardiner, 1994; Guilbert and Park, 1985; Halter et al., 2004; Kirkham and Sinclair, 1996; Megaw, 2006; Nystrom and Henriqez, 1994; Ohmoto and Skinner, 1983; Sillitoe, 1973, 1993, 1997; Sillitoe and Bonham, 1984; White and Herrington, 2000; Zhang et al., 2008.

CHAPTER ELEVEN

Solar System Volcanism

Introduction

We are not alone. That is, the Earth is not unique in the solar system as a home to volcanism. Signs of past or ongoing volcanism are found on our Moon, Mars, Mercury, Venus, Neptune's moon Triton, Jupiter's moon Io, and apparently Pluto too. Within our solar system, Earth's volcanism may be the most varied in style and composition, in part because of plate tectonic processes, and also because the Earth is both chemically and structurally differentiated. But maybe it is not so different.

Most terrestrial volcanism, in terms of sheer volume, is basaltic and associated with divergent or constructive plate boundaries, leaky transform faults, and hot spots. Convergent plate boundaries produce less volcanic material, but form the gamut of compositions from basalt to rhyolite. Hot-spot volcanism, found mostly within lithospheric plates but also along plate boundaries, is responsible for some of Earth's largest calderas, huge silica-rich ash flows, weird lavas such as carbonatites,[1] and vast flood basalt provinces.

Extraterrestrial volcanism includes Earth-type, silica-based lavas. Basalt is found on the four small inner stony planets and our Moon, and andesite occurs on Mars. Volcanism of different types, however, is associated with some moons surrounding the outer gaseous giant planets. These bodies with extremely low surface temperatures erupted materials such as water, nitrogen, carbon, sulfur, and hydrocarbon compounds.

The Moon

Let's first examine our Moon. It is unique in the solar system for at least two major reasons. The Moon is (1) proportionally closest in size to the planet it orbits, with a mass of 1/81 of that of the Earth,[2] and (2) the Moon appears to have an origin different than that of other planet satellites. Current thinking holds that the Moon formed after a colossal impact between the Earth and a Mars-sized body shortly after Earth formation, about 4.5 billion years ago. Following this collision, the rocky outer part of the impactor spun away to form the Moon, and the core of the impactor was incorporated into a growing Earth. In contrast, most of the moons of other planets are believed to be captured asteroids.

Information about the Moon was initially gathered strictly by visual observations. In 1609, following the lead of Dutch inventors, Galileo built a telescope and used it to view the Moon, becoming the first person to document the existence of lunar mountains and plains. In the mid-1600s, Francesco Maria Grimaldi mapped the face of the Moon; many of the crater names he bestowed are still in use today.

Robert Hooke (1635-1703) first proposed that volcanoes existed on the Moon after peering into the large craters with his telescope. In 1785, Immanuel Kant argued the craters were of non-volcanic origin due to their gigantic size, and in 1824, Franz von Gruithuisen (1774-1852) correctly suggested that meteorite strikes formed the craters. But the real information-gathering breakthrough started in 1959 with the Soviet Union's Luna 1 flyby spacecraft. This was followed by other data-gathering flybys, hard and soft landing craft, and finally, starting in 1969, by the U.S. manned Apollo 11 through 17 space missions that landed and returned to Earth with lunar rocks and other data.[3]

Lunar samples from many locations have greatly enlarged our knowledge base. Chemical analyses and radiometric age dating of rocks, remote sensing, and detailed photographic reconnaissance have helped decipher a complex lunar history involving both volcanism and impacting. And while no volcanic activity has occurred for hundreds of millions of years, the Moon's surface shows a very active volcanic past coupled with an early period of intense meteoric bombardment.

The surface of the Moon consists mostly (84 percent) of highlands or terrae. Such extensive coverage was not apparent from Earth until lunar spacecraft viewed the far side of the moon, a region never seen from Earth. The highlands appear white due to high reflectivity that is a function of rock pulverization from intense, early meteoric impacting and rock compositions that are low in iron and high in aluminum. The impacting left vast regions of the moon covered with dust and fine- and coarse-grained breccia deposits.[4]

In these back-side-of-the-Moon highlands, volcanic rocks are found in domes and patches older than 3.85 billion years that were not destroyed by subsequent meteor impacts, and then covered with younger volcanic flows. Also, small areas of ancient (4.2-billion-year-old) lava exist in highland breccia deposits. These lavas and breccias are part of early Moon history when volcanic eruptions most likely resulted in lunar resurfacing. This happened during the period of most intense meteoric impacting, or between 4.3 and 3.8 billion years ago.

But the side of the moon always seen from Earth is a different story (see Figure 7.1, page 110). This face shows both highlands and extensive dark, smooth plains called maria.[5] Maria, composed of flood basalts 4.3 to 3.1billion years in age or younger, fill huge, generally circular meteoric impact basins.[6] Interestingly, remnants of rims of ancient, immense impact craters can be easily seen from Earth using only binoculars.

FIGURE 11.1: *Photograph of the back or far side of the Moon showing the intensely cratered ancient lunar highlands. Dark areas are composed of younger maria flood basalt (NASA Apollo 16 photo).*

Lunar lavas show both similarities and differences compared with Earthly basalts. Both basalts contain pyroxene and plagioclase silicate minerals, but lunar basalts show no alteration by water. The water-bearing mineral apatite ($Ca_{10}(PO_4)_6(OH,F,Cl)_2$), however, was found in basalt collected during the 1971 Apollo 14 mission. Water was also reported in 2009 by the Indian ISRD Chandrayaan-1 satellite. The formation and retention of hydroxide (OH) and water apparently occurs in brecciated rocks in polar regions.

Lunar basalts, like Earthly basalts, are fine-grained to glassy, produced by rapid cooling. Some contain gas escape holes or vesicles. Lunar basaltic magmas, however, appear to have been extremely fluid due to very low viscosity. This low viscosity—low resistance to flow—allowed basalt to spread like motor oil at room temperature across the lunar surface.

Both high eruption temperatures and low amounts aluminum, phosphorous, the alkali elements sodium and potassium, and high iron concentrations formed these low-viscosity lavas. Low magma viscosity together with low lunar gravity allowed these lavas to flow, in some instances, hundreds of kilometers from their vents.

It is theorized, based upon basalt chemistry, that magmas were generated at depths of 150 km to 400 km within the lunar mantle. Also, volumetrically little mantle was partially melted to generate these lavas. Most maria basalts migrated quickly to the lunar surface without spending time in subsurface holding chambers, as is common in Earthly systems. And while the thin maria flows cover about 20 percent of the lunar surface, they constitute only 1 percent of the moon's thin crust.

BASALT FLOW CHARACTERISTICS

Lunar basalts are characterized by lobate flows, similar to Earthly flows. Other lunar lava features include terraces, wrinkle ridges, lobate scarps of younger over older flows, low-relief domes and cones, collapse craters, and sinuous rilles.

Rilles were either lava channels or resulted from the collapse of lava tube roofs into channels below. Cones and domes are usually only 100 to 500 m high, a few kilometers in diameter, resemble small Earthly shield volcanoes, and are spatially associated with fissures and rilles. Wrinkle ridges are concentric structures found in all maria. They apparently formed as cooling maria lava flows settled. They may also be associated with dikes and other intrusive features.

Impact basin centers contain the thickest flow sequences, perhaps only 2 to 4 km (1.2 to 2.4 mi) thick. Basin margin basalt may be only 0.5 km (0.3 mi) thick. Maria surfaces generally appear to be smooth and featureless except for subsequent impact craters.

PHASES OF LUNAR VOLCANISM

Mare volcanism is divided into three phases: early, main, and late. The oldest lavas from the early, lunar volcanic surfacing period date at 4.2 billion years. They occur as fragments in highland breccias that formed 4.3 to 3.8 billion years ago.

The final phase of heavy bombardment (3.9 to 3.8 billion years ago) resulted in the development of several, largely intact, great basins. These basins contain ejecta blankets with the same composition as that of the highland rocks. This similarity in composition implies that the light-colored highland plains resulted from debris blasted from the big multi-ring basins.

Within the highland plains, however, are dark impact craters. These contain smooth basaltic lavas older than 3.8 billion years that are similar to the old fragments within highland breccia deposits.

The main phase of mare volcanism took place following the rapid decline of meteor impacting, starting about 3.8 billion years ago. These basalt flows were subsequently hit by smaller meteorites but not totally destroyed.

FIGURE 11.2: *Photograph of the ancient lunar Hipparchus impact crater. Note the vague but still visible outline of the maria-floored crater rim except for the more prominent right wall. Also note the conditions of younger impact craters, some pristine, which are used to tell relative impacting ages (NASA Apollo 16 photo).*

The chemistry of these flows evidently varied with time. High titanium basalts erupted between 3.8 and 3.6 billion years ago, whereas lower titanium-bearing basalts flowed into basins between 3.6 and at least 3.1 billion years ago. Some of the lower titanium basalts are also relatively high in aluminum (see Table 2.1, page 15).

Late-phase lunar volcanism commenced about 3.1 billion years ago and may have ended as recently as about 800 million years ago. This age is conjecture because the youngest known volcanic activity has yet to be dated. Because we still have relatively few lunar age dates, much of our knowledge is based on relationships between known ages, crater density, crater crosscutting relationships, and subsequent extrusive activity. For example, some maria have much lower crater densities than the 3.1-billion-year-old basalt at the Apollo 12 landing site. Estimates place the age of these younger Imbrium flows at 1.5 to 2 billion years.

But volcanism as young as 1 billion years to 800 million years may exist. The youngest lunar lava flow yet discovered covers ejecta from one of the youngest craters, the Lichtenberg Crater. This 20-km (12-mi) wide structure is estimated to be less than 2 billion years old.

GLASS BEADS

Apollo 15 and 17 landings retrieved emerald green, black, and orange beads that are chemically homogenous and of basaltic composition. They are believed to have formed in fire-fountaining vents, as occur today in Hawaii, but over 3.7 billion years ago. More than 20 varieties of beads have been encountered at all landing sites, but no lavas with the same compositions have been discovered.

These beads differ from impact-formed glass beads in that they have no attached rock or mineral particles. Surfaces of the beads contain vesicles and small amounts of lead, zinc, and chlorine. Green beads are magnesium rich and titanium poor, whereas the black and orange beads are rich in titanium.

Why might these beads be important? It is believed they were derived from magma that was generated some 400 km (250 mi) beneath the surface from partial melting of an olivine-rich mantle. The magma rose rapidly without being contaminated through overlying rocks. The beads, then, may represent our best samples of the deep lunar mantle.

Mars

Mars, the fourth planet from the sun, has been observed since before recorded history due to its red color and changing position in the night sky. It was described by the Babylonians, 3,600 years ago, as having a looping motion across the night sky. Mars to the Greeks personified the God of War, Ares. The Romans named this planet Mars for their god of war.

In 1609, Johannes Kepler discovered that the looping motion was due to Mars orbiting the Sun. In 1877, Italian astronomer Giovanni Schiaparelli called the thin dark lines he observed *canali*. This led American Percival Lowell a few years later to postulate that the canali were actually canals and therefore proof of an advanced Martian civilization. While Mars does not have civilizations or canals, its red color is due to the ubiquitous presence of iron oxide-coated (Fe_2O_3 or rust) sediments veneering its surface. This color resulted from the interaction long ago of water and oxygen on surface rocks. The Mariner 9 mission arrived in Mars orbit in late 1971. Mariner 9 waited a month in orbit until an intense global dust storm cleared, and then began mapping about 85 percent of the Martian surface.[3] Among the first features to emerge were several enormous shield volcanoes, demonstrating that the planet had at one time experienced extensive volcanism.

Mars has been a favorite space probe destination due to its relative nearness to Earth, surface visibility, and the possibility of present or past life. The United States and Soviet Union sent the most flyby and orbiting probes.[3] The United States has also sent landers and then surface rovers, starting with the Viking 2 mission in 1976. The Mars Pathfinder landing on July 4, 1997, was joined two days later by the Sojourner, the first roving robot on another planet. We observed again the red, stony,

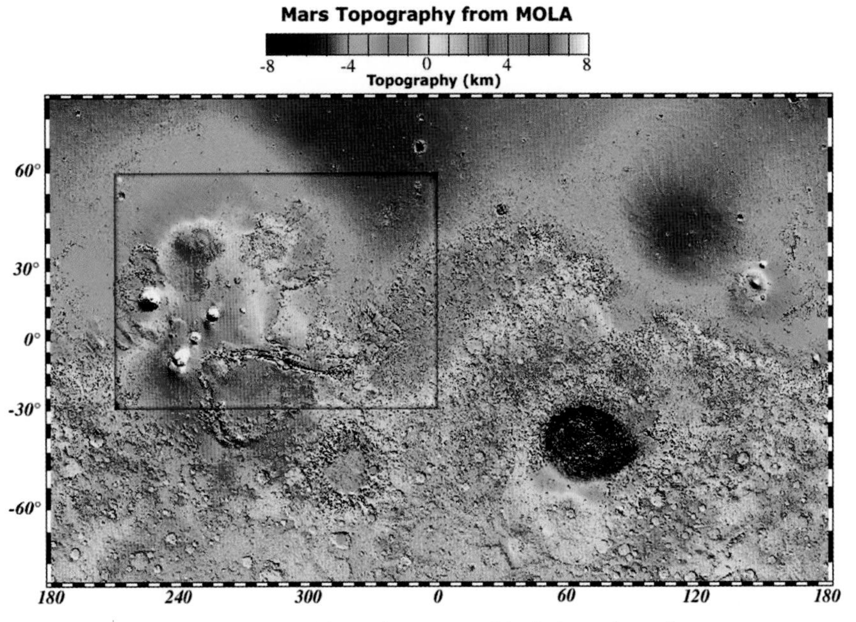

FIGURE 11.3: *False color topographical view of a region (north is up) containing the Tharis bulge located in the Martian Northern Hemisphere. The tallest volcano in the solar system, Olympus Mons, is on the left side with the giant volcanoes Ascraeus Mons, Pavonis Mons, and Arsia Mons oriented northeast to southwest. The largest volumetrically known volcano in the solar system, Alba Mons, is located farthest north. The Syria Planum region to the east of Arsis Mons volcano is bordered on the north by the Noctis Labyrinthus terrain intensely dissected by fractures and grabens (down-dropped tectonic fault blocks) (NASA Mars Orbiter Laser Altimeter (MOLA).*

FIGURE 11.4: *Detailed view of the Tharis Bulge region denoted by the box outlined in Figure 11.3. Alba Mons is 6 to 7 km (19,680 to 22,965 feet) high with a basal diameter of 1,200 km (745 mi). The summit caldera is 120 km (75 mi) wide. Alba Mons is located in a major fracture zone near the border of the northern lowlands. Note the very subtle concentric pattern of fractures and grabens trending north into the volcano, encircling it, and diverging to the northeast north of the volcano (NASA MOLA false color map).*

lifeless Martian surface that resembles Death Valley, California.[7]

More Martian missions have followed in the search for water and signs of life and to investigate Mars' atmosphere and surface geology. In 2004, two rovers, MER A "Spirit" and MER B "Opportunity," started investigating the Martian surface. They were followed in 2008 by the Phoenix Lander that set down near the north pole and the MSL "Curiosity" rover in August of 2012 sent to investigate Martian climate and geology.

MARTIAN EPOCHS

Based on the aerial density of impact craters, researchers have delineated three major periods or epochs of Martian history. The oldest, the Noachian (4.1 to 3.7 billion years ago), is followed by the Hesperian (3.7 to 1.8 billion years ago). The Amazonian epoch is the youngest (1.8 billion years ago to the present). But without radiometric age dating, we do not know the exact ages of these epochs. Researchers determined Martian epochal ages from 11 igneous Martian meteorites found on Earth years ago. They have age dates between 1.3 billion years and 165 million years. A twelfth Martian meteorite has an age of 4.5 billion years (see Martian Meteorites sidebar, page 153). In all, as of March, 2014, 132 Martian meteorites have been found on Earth and have been grouped into four categories.

The densely cratered highlands in the south are of Noachian age, with meteorite ALH84001 (aged 4.5 billion years) possibly being representative of these surface materials. Isolated areas of volcanic activity apparently occurred during early Martian history.

During Hesperian time (3.7 to 1.8 billion years?), massive eruptions of fluid lavas created volcanic plains that now cover vast areas of Mars. The lavas are, most likely, basalt and basaltic andesite accompanied by lots of tephra. These eruptions built the volcanic centers on the rim of the Hellas impact basin and in the dark region of Syrtis Major.

Late Hesperian and Amazonian time (1.8 billion years to the present) witnessed voluminous volcano- and plains-building eruptions. These are especially concentrated around the Elysium and Tharsis regions. The lack of intense cratering in the Tharsis regions suggests that volcanism here is even younger.

SURFACE FEATURES

The Martian surface has two major terrain types: Cratered highlands located primarily in the Southern Hemisphere and low-lying northern plains. A 1- to 3-km-high escarpment separates them. Most of the volcanic landforms are within the northern plains province.

Viking orbiters with thermal infrared scanners indicated the bright Martian surface is coated with dust, whereas the darker regions are veneered with sand and coarser particles. Thermal scanners, however, did not find any heat-producing anomalies, or hot spots, on the Martian surface.[8]

Radar signals beamed from Earth and reflected from portions of the Martian surface displayed a strong scattering behavior indicative of surface roughness or particles 10 cm (4 inches) or greater across. The region west of Tharsis Montes, however, showed very low radar reflectance, suggesting the surface material was either thick enough to absorb signals or is of unusual relief.

In the 1970s, Viking 1 and 2 Orbital Landers investigated the Martian surface in detail. They found dark rocks, red dust, and pink sky, but no signs of life. The next surface investigation, begun on July 4, 1997, used the Mars Pathfinder Lander. The robot rover transmitted images and surface rock and soil chemical data back to Earth.

Other successful Martian probes were Mars Exploration Rovers "Spirit" and "Opportunity," launched on June 10 and July 8, 2003, respectively. They determined rock and soil compositions, geologic processes including sedimentation associated with water, and whether conditions might have once been conducive for life. The 2005-launched Mars Reconnaissance Orbiter went into orbit on March 10, 2006, in order to characterize Martian climate and to identify possible sites of hydrothermal activity, water-related landforms, and future landing sites.

A newer Martian excursion, the Phoenix Mars Lander, touched down successfully on May 25, 2008. The Phoenix investigated the surface environment, including climate and weather, surface-shaping processes, and the biological potential of the northern region. In addition to discovering water ice beneath the surface, the rover determined the mineralogy, chemistry, and biological potential of surface rocks. It also detected snow falling from clouds. The latest Mars surface rover, MSL Curiosity, became operational on August 6, 2012. It is examining past and present habitability, climate, and geology. The MAVEN orbiter became operational on September 25, 2014. Its mission is to investigate the Martian atmosphere.

These Martian probes have shown that Mars has a great variety of volcanic features. The density variation of impact craters on volcanic features suggests a long volcanic history. This has been substantiated by radiometric age dating of Martian meteorites found on Earth (see Martian Meteorites sidebar, page 153). The probes have also shown that Mars contains hundreds of volcanoes and other volcanic features, most much larger than Earthly counterparts.

Volcanic features cataloged to date include very large and small shield volcanoes, lava flows (some with levees and lava tubes), vast volcanic plains, low, circular domes (some with summit depressions), scoria or cinder cones, rifts, dike-like structures, and non-cratered mesas and buttes. Also found are tuff or ignimbrite deposits and volcanoes that likely emitted the ash flows.

VOLCANOES

Four massive volcanoes were initially observed in the equatorial Tharsis region. Three other regions containing large volcanic complexes were subsequently discovered. The Viking Orbital spacecraft that investigated the Martian surface between June 19, 1976, and August 17, 1980, further refined the work of the 1971 Mariner 9. It discovered that volcanic regions are non-uniformly distributed on the planet.

While the largest concentration of volcanic centers (13) occurs on and around the 4,000-km (2,440-mi) diameter Tharsis bulge or uplift, 10 are distributed in the other three regions: Hellas, Elysium, and Syrtis Major. The Tharsis bulge is approximately 10 km (32,800 feet) higher than the average Martian elevation and twice as high as the ancient southern, cratered highland.

With a diameter of nearly 2,000 km (1,200 mi), the 6.8-km (22,000-foot) high shield volcano Alba Mons[9] has the largest known volume in the solar system (see Figures 11.3 and 11.4). The eighth tallest Martian volcano has gentle slopes of between 0.1° and 2° and a summit caldera 110 km (67 mi) across. In comparison, Earth's largest volcano, Mauna Loa on the Island of Hawaii, is taller at 34,000 feet above its seafloor base, but has a much smaller volume.[10]

The gentle slope angles of Alba Mons appear to be due to very fluid lavas that flowed down its slopes in channels, some extending 1,500 km from the summit caldera. Much lava also flowed through lava tubes. With the latter type of transport, lava, well insulated from heat loss, is able to travel great distances before solidification.

This volcano appears to have had a long and complex history. Schneeberger and Pieri (1991), for example, believe that early sheet flow eruptions were followed by massive pyroclastic depositing. Central vent eruptions then followed and were succeeded by summit collapse and caldera formation.

The immensity of Alba Mons and the other, large shield volcanoes is due in part to a lack of Martian plate tectonics. This also explains the lack of chains of volcanoes showing age progressions, compared with the Earthly Hawaiian volcanoes. The Hawaiian age progression is due to the Pacific Plate moving over a fixed hot spot (see Chapter 5). Plate tectonics also accounts for the lineation of subduction zone volcanoes. The linear trends of volcanoes that occur on Mars appear controlled by fissures or cracks.

Tyrrhenus Mons, located in the cratered southern highlands, is believed by some researchers to be an example of a volcano constructed at least in part by pyroclastic deposits. It not only has very gentle

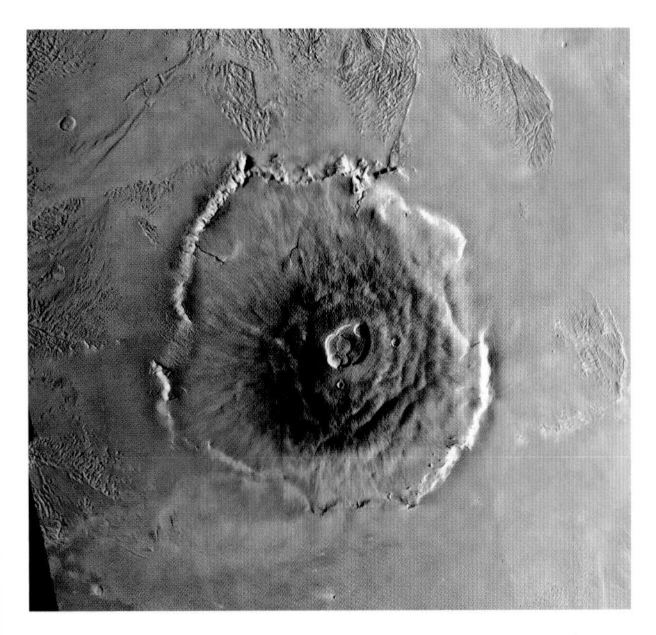

FIGURE 11.5: *Olympus Mons volcano, Mars, showing the complex summit caldera, basal scarps of unknown origin, and lava flows drapped over portions of the scarp. Olympus Mons, the highest known volcano in the solar system, is 21.1 km (69,460 feet) high and 600 km (373 mi) wide, with the basal scarp up to 6 km (19,685 feet) high (NASA photo).*

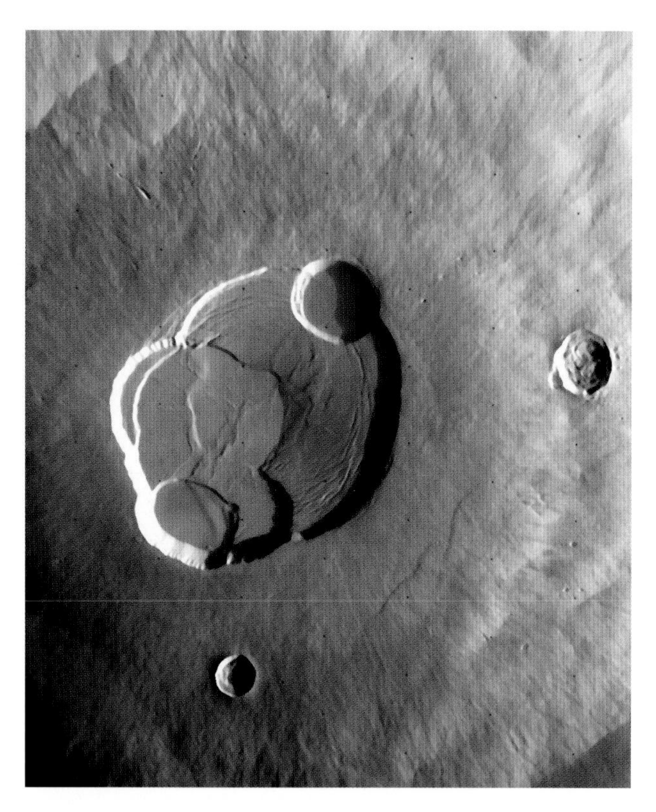

FIGURE 11.6: *Summit caldera of Olympus Mons, Mars, showing multiple collapse basins and lava flows oriented radially to the rim. The caldera is approximately 3 km (9,800 feet) deep, and 65 by 85 km (40 by 53 mi). Small meteorite craters Karzok and Pangboche, respectively, appear to the east and south of the caldera (NASA photo).*

slopes, but a radiating pattern of what appear to be erosional channels cut into ignimbrite deposits.

The tallest, most famous Martian volcano, Olympus Mons,[11] is 21.1 km (69,460 feet) high and 600 km (365 mi) across. The volume of Olympus Mons is huge, about equal to that of the entire Hawaii-Emperor volcanic chain on Earth.

Olympus Mons is capped by a 3-km (2-mi)-deep caldera complex containing six craters, together some 65 km by 85 km (40 by 53 mi) across. This volcano has slope angles of between 3° and 5°, typical of Earthly shield volcanoes. Olympus Mons is built of long, narrow flows, channel flows with levees, and lava tubes, all emanating from the summit caldera and flank vents. It is believed to have begun forming during late Hesperian time.

The volcano is surrounded by a puzzling, huge basal escarpment or cliff. Ideas for its origin include faulting, gravity sliding outward from the volcano, and some form of surface erosion.

Mapping by Mars Global Surveyor (MGS) in 1997 also revealed many small shield volcanoes, 2 to 10 km (1.2 to 6 mi) in diameter, with very gentle slopes. They are commonly associated with fracture systems and appear similar to those found on the Snake River Plain of Idaho, in Iceland, and in Hawaii. Typical Martian examples may be seen in the Tempe-Mareotis Province. Determination of their compositions by Thermal Emission Spectrometry, however, has not yet been possible due to the ubiquitous red dust that covers all rocks.

Another distinctive Martian volcanic feature is the tholus, or isolated domical hill or small mountain. Tholi usually have steeper slopes (about 8°) than mons and shield volcano constructs, and diameters of less than 200 km (125 mi). Ceraunius Tholus is interesting in that it has erosive sinuous channels carved into its flanks, possibly from late lava or pyroclastic flows.

Recent Martian missions have also revealed that Mars once had surface water, perhaps seas, and contains fluvially carved valleys and sedimentary rocks. Some of these valleys were subsequently partially filled with lava flows. A classic example can be seen in Martes Valles.

VOLCANIC PLAINS

Volcanic plains are one of the most extensive Martian features. Based on high-resolution imaging, researchers have divided them into three types: (1) those that are relatively featureless and of presently unknown origin, (2) those formed by lava flows with lobate margins, and (3) those with complex surfaces.

Plains built of flows with lobate margins and individual flow thicknesses of tens to hundreds of meters surround the Tharsis and Elysium volcanic areas, the Lunae Planum region east of the Tharsis region, and the Hesperia Planum plain northeast of the Hellas basin. Crater densities indicate that the latter two regions are of Hesperian age, whereas the former two plains are younger, or of Amazonian age.

The Medusae Fossae Formation southwest of the Tharsis region is representative of plains with complex surfaces. Some researchers believe the layered complex, topped with a more resistant cap rock that protects the underlying, more erodible materials, represents ash flow or ignimbrite units. The cap rock may be the more welded portions of ash flows. To date, however, no possible source vents have been located. But if this theory is correct, the vents may have been buried by subsequent volcanic units.

Martian Meteorites on Earth!

Almost all meteorites found on Earth (to date more than 61,000) are 4.5 to 4.6 billion years old, of igneous derivation, and are pieces of solar system asteroids. However, 132, as of March 3, 2014, came from Mars.[1] Answers to how they got to Earth, how we know that they really are from Mars, and what these rocks have told us about Mars come from Viking Landers probes.[1]

How did Martian and lunar rocks get to Earth? Meteoric impacting on the Martian surface blasted chunks of the crust into space. Because Mars is so much smaller than Earth, an exit velocity of only 5.4 km per second was required to escape Martian gravity. After liberation from Mars, these chunks of rock spent millions of years in space before encountering Earth's gravitational field and falling to the surface.

How do we know these meteorites are from Mars? We can thank the Viking Landers that relayed back to Earth between 1976 and 1980 analyses of rocks, soil, and the Martian atmosphere. These atmospheric data matched analyses of gases trapped in the interior of the meteorites.

Furthermore, through radiometric age dating we know the ages of the meteorites, and their residence time in space between leaving the Martian crust and arriving on Earth. Meteorite crystalline ages fall into three age groups varying from 165 million years to 4.5 billion years.[2] Time in space varies from less than 1 million years to about 20 million years. These meteorites, then, were derived from Mars during three or possibly four impact events over the last 20 million years.

What have we learned about Mars from the meteorites? Chemical analyses show that there are four types of Martian rocks, and they differ from Solar System meteorites. The Shergotty (S) type are basaltic, Nakhla (N) are composed of clinopyroxene ($Ca(Mg, Fe)(SiO_3)_2$), Chassigny (C) are dunite or olivine-rich ($(Mg, Fe)_2SiO_4$), and ALH84001 is composed of orthopyroxene ($(Mg,Fe)SiO_3$).

Finally, where are the best places to find meteorites on Earth? Most rocks from Mars likely end up in the oceans, but some hit land. They're scattered over the Earth's surface but are easier to find on barren ground. For this reason, searches have concentrated on deserts in North Africa and Oman and also the ends of Antarctic glaciers. Why the *ends* of glaciers? The ice covers vast "collection areas" and moves like a conveyor belt, carrying ice, rocks, and dust slowly but surely to the terminus. There, "prospectors" look for meteorites.

NOTES

1 At least 200 meteorites found on Earth came from the Moon.

2 Debate exists as to the actual crystalline ages of Martian meteorites. For example, radiometric age determinations of Shergotty (S) meteorites range from 150 to about 600 million years. However, some researchers believe these are reset ages and that the true ages vary from 4.1 to 4.3 billion years.

REFERENCES

Crumpler and Aubele, 2000; Geissler, 2000; Gladman et al., 1996; McSween, 1994; NASA - Meteorites from Mars!, 1996; Spudis, 2000; Webster, 2013: Wikipedia, 2015; Zimbelman, 2000.

MARTIAN ROCK AND SOIL COMPOSITIONS

The compositions of Martian meteorites found on Earth vary in mineralogy and chemical composition. The meteoritic material, however was probably derived from some depth during impacting, and so the chemical analyses of materials from the 1997 Pathfinder probe greatly helped us understand surface compositions. One surprise was the higher silicon dioxide content (52 to 57 percent) than that of Martian meteorite data. (Martian meteorites have 37 to 53 percent silica, or within the range of ultramafic to basaltic magma—see Table 2.1, page 15, and Table 3.1, page 31, for definitions/compositions/comparisons). The Pathfinder data are in the chemical range of basaltic andesite.

Thermal Emission Spectrometer analyses on the Mars Global Surveyor suggest that the southern portion of Mars is basaltic in composition. The northern part, including Pathfinder's landing site, is more andesitic, or higher in silicon dioxide. The higher silica content may be due to a more evolved magma (magmatic differentiation of basaltic magma). Alternatively, basalt alteration by water could also have changed the rock chemistry in the northern regions.

Venus

The Evening Star, is the brightest object in the night sky after our Moon. With an equatorial diameter of 12,104 km (7,517 mi), it is only slightly smaller than its sister planet, our Earth (12,756 km (7,921 mi). Venus has no satellite moons and a weak to nonexistent magnetic field. While known to the Romans as the Goddess of love and beauty, Venus in reality is a hellish planet with surface temperatures of 750°C (900°F) and atmospheric pressure 90 times that on Earth. The pressure is equivalent to being about 1 km deep in our ocean. And the thick atmosphere that totally clouds our view of the surface, essentially water-free, is composed primarily of carbon dioxide (97 percent), with hydrochloric and sulfuric acid droplets.

United States flyby missions[3] to Venus include Mariner 2 that reached the planet in 1962, Mariner 5 in 1967, Mariner 10 in 1977, Pioneer Venus in 1978, Galileo in 1990, Magellan in 1990, plus many sent from the USSR starting in 1965. These probes have provided information about the composition of the atmosphere and mapped the surface, using radar that is capable of penetrating thick, opaque atmospheres.

Resolution of surface features in 1978 was about 100 km. The first detailed views of the surface were done by radar in the early 1980s by a Soviet Union orbiter. Resolution then was about 10 km. This mapping revealed many volcanoes and strongly deformed highlands called Tesserae, containing the oldest rock. This Soviet Venera mission also placed a lander on the surface that showed Hawaiian-like fluid basaltic lava flows. In the early 1990s, the Magellan probe radar mapped 97 percent of the surface with a resolution of 120 m (400 feet).

We have learned that Venus has been subjected to very extensive volcanism. Furthermore, the paucity of impact craters in comparison to our Moon and Mercury suggests that volcanic activity has been both extensive and relatively recent. Impact crater density is similar to Earth—about one per million square kilometers. This means that the longevity of impact craters is probably between 300 million years and 1 billion years.

Estimates of complete volcanic resurfacing vary but may be about 700 million years. And if magmatic activity is similar to that on Earth, with about a 10-to-1 ratio of intrusive to extrusive output, Venus may produce between 3 and 46 km^3 (0.7 to 11 mi^3) per year. The Earthly value is about 20 km^3 (4.8 mi^3) per year.

Analyses of atmospheric sulfur dioxide indicates that volcanoes have been recently active. Values were higher in 1978 during the Pioneer Venus mission than found by the 1990 to 1994 Magellan probe. This suggests that volcanic activity has waned yet is still ongoing.

Russian and American Venusian probes, then, have helped us to understand the great variety, complexity, and youth of Venusian land forms. While the Venusian surface contains thousands, perhaps millions, of volcanoes and vast quantities of lavas showing both smooth and rough surfaces,[12] the major style or mechanism of volcanism differs from that on Earth. On Earth, plate tectonic forces steadily transfer internal heat to the surface. Venus appears similar to Mars in that heat exits the planet's interior primarily through hot spots. Plate tectonics similar to Earth's does not exist.

Venusian tectonics is dominated by global rift zones and broad, low dome-like structures called coronae.[13] These appear to form by magmatic upwelling and subsidence. Also, because of a lack of surface water, Venusian magmas probably contain little water. Therefore, large, explosive eruptions similar to those on Earth may not be common, or may not occur at all.

MAGMATIC FEATURES

Researchers have cataloged at least 1,738 Venusian magmatic centers with diameters of 20 km (12 mi) or more. Centers include volcanoes with diameters that vary from less than 1 km (0.6 mi) to more than 1,000 km (620 mi), volcanic fields, intrusive magmatic centers with associated lava fields, calderas, and complex ring features called coronae[13] and arachnoids.[14] The latter two phenomena are believed to result from intrusive activity. Linear patterns of magmatic centers, abundant on Earth, are uncommon on Venus.

LARGE VOLCANIC FEATURES

To date, geologists have identified 168 large, shield-type volcanoes, some with diameters exceeding several thousand kilometers. These large volcanoes characteristically have radially oriented lava flows and are generally only about 1.5 km (4,800 feet) in

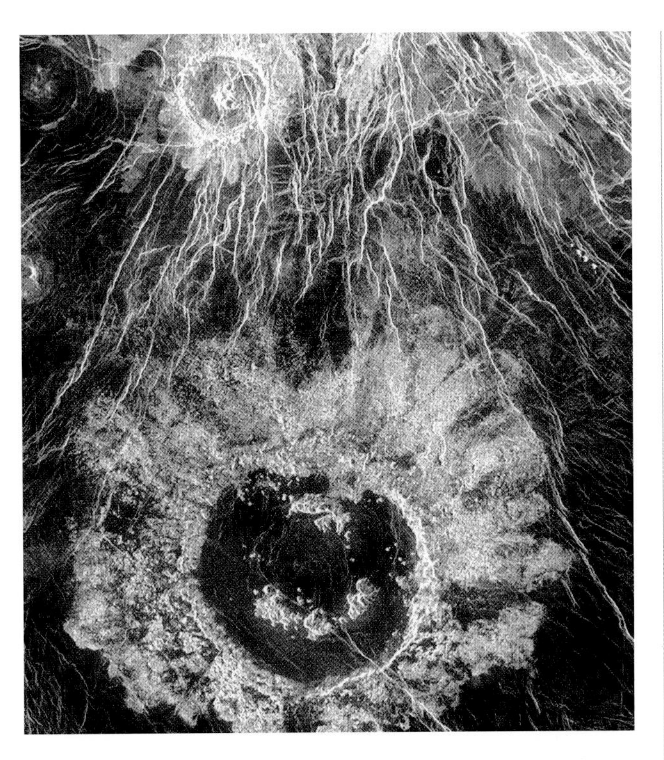

FIGURE 11.7: *Wheatley Volcano, Venus, showing a multi-ring crater 72-km (48-mi) wide crater. Note light gray, rough-surfaced ejecta around crater and superposed surface fractures (NASA Magellan probe image).*

FIGURE 11.8: *Sapas Mons, a flat topped, twin-peaked volcano on Venus, is approximately 400 km (250 mi) across at its base (NASA Magellan probe image).*

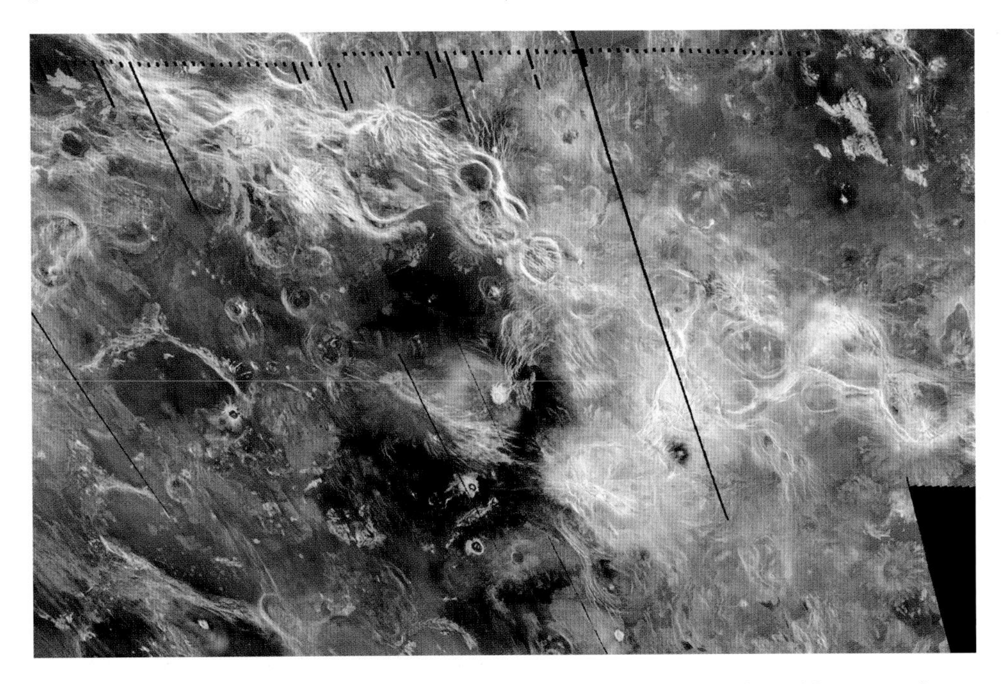

FIGURE 11.9: *Corona chain, Venus. Low, dome-like structures possibly formed by magmatic upwelling and subsidence processes (NASA Magellan probe image).*

elevation. Big volcanoes commonly occur in broad rises or swells located at the intersection of rifts or belts of fractures.

Large volcanoes may also have smaller volcanoes on their summits. Chloris Mons is a classic example, similar to Mount Etna on Sicily, and Mauna Kea on Hawaii. Both smooth- and blocky-surfaced flows are present. Other large Venusian volcanoes with complex summit regions may have radial fracture patterns, steep-sided volcanic domes, and circular structures. These features have also been observed around Venusian coronae.[13]

INTERMEDIATE-SIZED VOLCANIC FEATURES

Another 289 intermediate-sized volcanoes with diameters between 20 and 100 km (12 to 62 mi) have been mapped to date. They include volcanoes

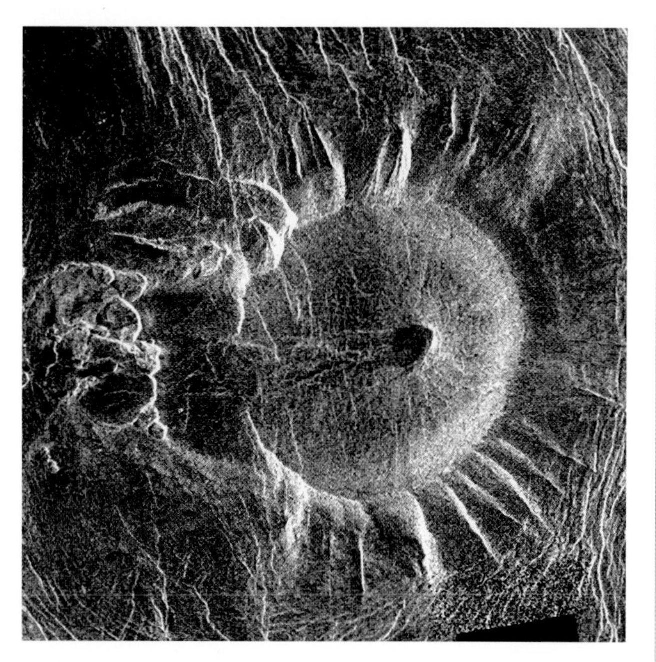

FIGURE 11.10: *Tick Crater, Venus, showing an outer rim 30 km (18.6 mi) in diameter (NASA Magellan probe image).*

FIGURE 11.11: *"Crater Farm" on Venus. Volcanoes clockwise from the upper left are Danilova, Aglaonice, and Saskia. Aglaonice crater is 65 km (40 mi) in diameter (NASA Magellan probe image).*

with radially erupted flows, similar to the larger shield type, and domes with steep, fluted or scalloped slopes. The distinctive sides of the latter two dome types may be the result of slope failure.

Domes commonly occur in clusters and may have summit pits and complex fracture patterns. Venusian domes are generally much larger than their Earthly counterparts. Diameters are commonly 20 to 80 km (12 to 50 mi). They may be due to magmatic evolution resulting in silicic compositions and intrusions, as found on Earth. (See Chapter 8 for magma generation and magma evolution).

SMALL VOLCANIC FEATURES

Small Venusian volcanic features have diameters of less than 20 km (12 mi); these may number in the millions. They tend to cluster in the 647 volcanic shield fields. Shield fields are tens to hundreds of kilometers in diameter and each contains tens to hundreds of vents. Shield fields are also found within coronae and arachnoids.

Morphologies of the small volcanic features are similar to their larger relatives. They include radially patterned volcanoes, different types of domes, and gentle-sloped shield volcanoes, the most abundant type. These shield volcanoes commonly show uniform spacing over large areas but without apparent structural control. They exist in what are termed shield plains. They overlie tessera and are one of the oldest volcanic plain features on Venus. Their closest Earthly analog may be seamounts found in our deep ocean basins.

OTHER VENUSIAN VOLCANIC FEATURES

Venus, like Earth, has circular-to-elongate calderas. Ninety seven of these fracture-rimmed collapse structures are known. Most have diameters of between 60 and 80 km (37 and 50 mi) with floors 1 to 3 km (0.6 to 1.9 mi) below their surroundings. Lava flows floor some of these structures. Other calderas developed after central vent eruptions.

Most of the Venusian surface, however, is composed of dark, low-elevation, featureless plains. Termed lava plains, they may result from flooding by smooth-surfaced basalt.

Lava flow fields, also prevalent, show distinguishable lava flows. Individual flows within the lava fields are long—100 km (62 mi) to greater than 1,000 km (620 mi). They issued from fissures or rifts, calderas, and small volcanoes. Lava surfaces vary from smooth (dark radar reflection) to blocky (bright radar reflection).

Thick, blocky, lobate flows with transverse ridges and troughs are another notable Venusian volcanic feature. Their appearance is similar to silicic, viscous flows here on Earth.

Finally, immense, linear-to-sinuous channels or valleys exist. They vary from 0.5 to 10 km (0.3 to 6.2 mi) wide and are 100 (62 mi) to 1,000 km (620 mi)

FIGURE 11.12: *Photograph taken 11,588 km (7,200 mi) above Mercury of several processes that affected the surface of our innermost planet. Lower left is an ancient, probably smooth lava-filled impact crater 230 km (143 mi) in diameter still showing a vague but discernable rim. It contains a smaller, younger impact crater 85 km (53 mi) in diameter, also with a smooth lava (?) surface. This inner crater contains a still younger, small impact crater (see two upper left inset photos).*

Note the formation of a subtle northeast-southwest-trending lobate scarp (also depicted in the third small inset photo down) that cuts through all three craters described above It may have formed from compressional tectonic stresses that crumpled Mercury's surface.

The youngest major feature in this region is the large impact crater visible along the top of the photograph. This crater, like many others seen on the Moon and Mercury, has a central uplift or mountain. This forms by crustal rebounding during impacting.

Note the three "gullies" leading radially away from the crater. They are actually chains of small, secondary impact craters formed by the ejected debris from the large crater during impact formation (see lower left photo inset) (NASA Messenger probe photo).

FIGURE 11.13: *Photograph taken from 1,800 km (1,180 mi) above Mercury of Polygnotus Crater, which is 133 km (63 mi) across at top of photo. Polygnotus was a 5th century B.C. Greek painter. The large crater to the left is Boethius, named for a 6th century Roman philosopher. Note its smooth plains veneered with basalt and bearing a north-trending fault scarp. Also note that there are central uplifts or mountains in some of the smaller craters (NASA Messenger probe photo).*

FIGURE 11.8: *Photograph taken from 10,500 km (6,520 mi) of smooth, sparsely impact-cratered Mercury surface. Image is 270 km (170 mi) across and shows a light-gray, shield-type volcano with a kidney-shaped vent near the top. The volcano has an impact crater to the west, and the volcano sits inside the rim of Caloris impact basin (note part of the rim is still visible to the right of the volcano (NASA Messenger probe photo).*

long, with one more than 6,800 km (4,250 mi) long. These channels generally occur on lava plains and volcanic centers.

Very fluid flow behavior (low viscosity) is a prerequisite for the great lengths of these flows. Crumpler and Aubele list potential magma types that possess very low viscosity.[15] They include very low silica (less than 45 percent) ultramafic rocks, lunar-type high-titanium basalts, carbonatites, and even sulfur magmas.

Crumpler and Aubele (2000) further speculate that the great length and width of the channels are not the result of lava tube flow and roof collapse. Rather, lava-filled channels may have had moving plate-like crusts that helped insulate the lava flowing below from cooling.

Mercury

Mercury, named by the Romans for the fleet-footed messenger of the gods, is the innermost and smallest planet in our solar system.[16] While larger than our Moon with an equatorial diameter of 4,880 km (3,030 mi), it is smaller than the Jovian moon Ganymede (5,262 km, 3,268 mi) and Saturn's moon Titan (5,150 km, 3,198 mi), and barely larger than Jupiter's moon Callisto (4,800 km, 2,981 mi).

Details of the lunar-like landscape of Mercury were first seen on March 29, 1974, when the Mariner 10 probe flew to within 705 km (438 mi) of the surface.[3] The probe confirmed what was suspected: Mercury has essentially no atmosphere, and has not witnessed volcanism for perhaps 2 billion years or more. Mercury has, however, a weak magnetic field, indicating it has at least a partially molten iron-bearing core.

Much of the surface of Mercury, therefore, is old and heavily cratered. The surface also has intercrater plains and smooth plains. The intercrater plains have fewer impact craters than the oldest surface. Crater diameters are generally less than 15 km (9.3 mi). Smooth plains have still fewer impact craters, and so are the youngest Mercurial feature.

Lavas, probably basaltic, erupted onto the Mercurial surface following intense early meteoric bombardment. These formed the intercrater plains. As the planet cooled, it contracted and so developed long, lobate fault scarps, often several kilometers high. This fracturing event was succeeded by the final volcanic outpourings that formed the smooth plains.

Since the end of volcanism, Mercury has been subjected to additional meteoric impacting. These impacts have spread dust and debris over the planet's surface.

Cryovolcanism

Flyby probes to some of the outer giant gaseous planets have revealed moons with recurring volcanism. But volcanism includes cryovolcanism, the eruption of liquids and gases onto very cold surfaces where they instantaneously freeze. While active cryovolcanism was detected on Neptune's Triton moon by Voyager 2 in 1979, it is believed to occur on other satellites too. Most moons are covered with water ice. Jupiter's moon Io, covered in sulfur, is the exception.

But cryovolcanic activity can occur only if fluids (liquids and gases) are generated beneath a moon's rigid, cold crust. Furthermore, these fluids must be able to reach the surface. Fluid-heating mechanisms include radiogenic decay of unstable, long-lived isotopes; frictional heating caused by daily tidal-induced changes in satellite shape by their nearby neighbors; gravitational accretion of particles; and solar radiation.

Spectroscopic and theoretical studies indicate that cryomagma may be composed of various fluids, including water, methane, ammonia, carbon monoxide and dioxide, nitrogen, hydrogen sulfide, and sulfur dioxide.

Whether cryovolcanism occurs depends on the size of the moon and its composition, orbital path, and location in the solar system. The amount of silicate determines the amount of radiogenic heat generated. A moon's orbital path around its gaseous giant host determines the amount of tidal heating. Location within the solar system helped determine composition during the moon's formation,[17] and larger moons retain more heat than smaller bodies. The paucity of impact craters on many water ice-covered moons implies that cryovolcanic activity has been recent.

One major cryovolcanic puzzle is what could cause water to erupt through a less dense icy crust above. Liquid water is denser than ice, so under stable conditions, it would not move upward through a water-ice crust. How then do we explain water eruptions?

One mechanism is overpressure. If some of the water beneath a solid crust freezes, it will expand, forcing the remaining liquid upward. The expansion could also fracture the surface crust above, allowing liquid water to escape under pressure.

Liquid water might erupt if it includes other ingredients. Adding ammonia, for example, reduces water's density, making eruption more likely.

Finally, subsurface heating, whether by tidal forces or other means, may play a role.

Satellite Volcanism

Solar system flyby probes, starting with Voyager 1 on March 8, 1979, witnessed active volcanism for the first time in our solar system other than on Earth. This was the startling observation of Prometheus Volcano emitting sulfur on Jupiter's colorful moon Io. Voyager 2 subsequently saw a cryovolcanic eruption on Neptune's moon Triton (diameter 2,706 km (1,680 mi).

NEPTUNE'S MOON TRITON

Triton (diameter 2,706 km, 1,680 mi) is covered by frozen nitrogen and methane, with lesser amounts of carbon monoxide and carbon dioxide. Its surface temperature is 38°K, or well below the freezing temperature of nitrogen. Yet the observed volcanic eruption or geysers sent nitrogen 8 km above the frozen surface. It is speculated that these nitrogen eruptions are due to solar effects such as sublimation (the nitrogen changes from solid to gas without passing through a liquid state).[15]

MOONS OF URANUS

Voyager 2 also observed the moons of Uranus: Miranda (484 km diameter), Ariel (1,160 km), Titania (1,600 km), Umbriel (1,190 km), and Oberon (1,550 km). Umbriel and Oberon have high crater densities and so have long been volcanically inactive. Miranda and Ariel, however, show extensive tectonic disruption followed by some ice resurfacing. Titania, the largest moon of Uranus, underwent large-scale tectonism that produced down-dropped blocks or grabens (valleys) 20 to 75 km wide and up to 1,500 km long.

MOONS OF SATURN

Information about the seven moons of Saturn is scant. Enceladus (diameter of 502 km, 312 mi), covered with nearly pure water ice, is believed to have had recent cryovolcanism. Titan, Saturn's largest moon (5,150 km, 3,198 mi) has a dense atmosphere and so we know little about it.

JUPITER'S MOONS

Jupiter has four moons, starting with the inner, silicate-rich Io (diameter of 3,630 km, 2,254 mi) and Europa (3,138 km, 1,948 mi), then outer ice- and silicate-rich Ganymede (5,262 km, 3,268 mi), and finally Callisto (4,800 km, 2,980 mi).

Both heavily cratered Callisto and Ganymede show tectonic rifting and faulting but appear to have had little, if any, cryovolcanism. Europa and Io, however, are very different. Europa, with few impact craters, has an icy crust, believed to overlay a sea. The surface is a complex of icy blocks and slabs similar to those seen, for example, in refrozen Arctic Ocean ice pack. The surface ice of Europa may be a product of eruptions of melt from below and subsequent surface refreezing. The prominent, intricate ridge patterns may result from convection within the icy crust. This sea below is floored by a silicate core that probably produces radiogenic-derived heat.

In this model, surface crustal ice is denser than the subsurface portion, resulting in solid state convection of ice. Alternatively, resurfacing of Europa may result from icy cryovolcanism venting from between the ridge structures.

MOON IO

Io, the innermost Jovian moon, was discovered together with her three sister moons by Galileo (Galilei) on January 7, 1610. This moon was subsequently named Io for the maiden loved by Zeus (Jupiter).

Together with observations from Earth, Voyager 1 and 2 and the Galileo spacecraft have discovered that Io is the most volcanically active body in our solar system. It boasts higher heat flow than from Earth, volcanoes that tower 16 km (53,000 feet) above its surface, calderas, and lava flows with temperatures of up to 1527°C.[18]

Io has more than 70 active volcanic centers with at least 15 eruptions observed by the year 2000. This continuous volcanic activity results in constant resurfacing by both sulfur and silicate compounds. Furthermore, gravity and magnetic studies revealed that Io has differentiated into a rock shell

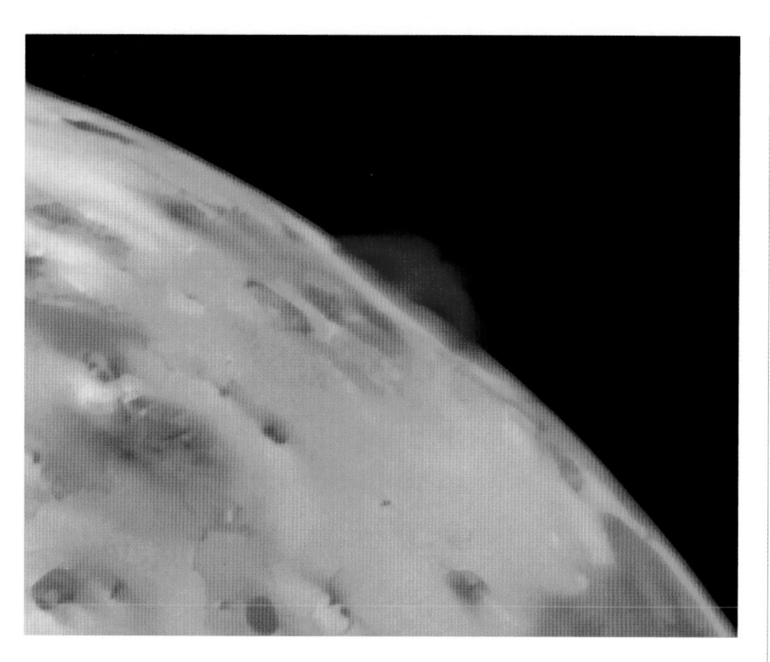

FIGURE 11.15: *The first observation of an volcanic eruption in our solar system other than on Earth, of volcano Prometheus on Jupiter's moon Io. Photo taken 3/04/1979 by NASA Voyager 1 about 800,000 km (500,000 mi) from Io.*

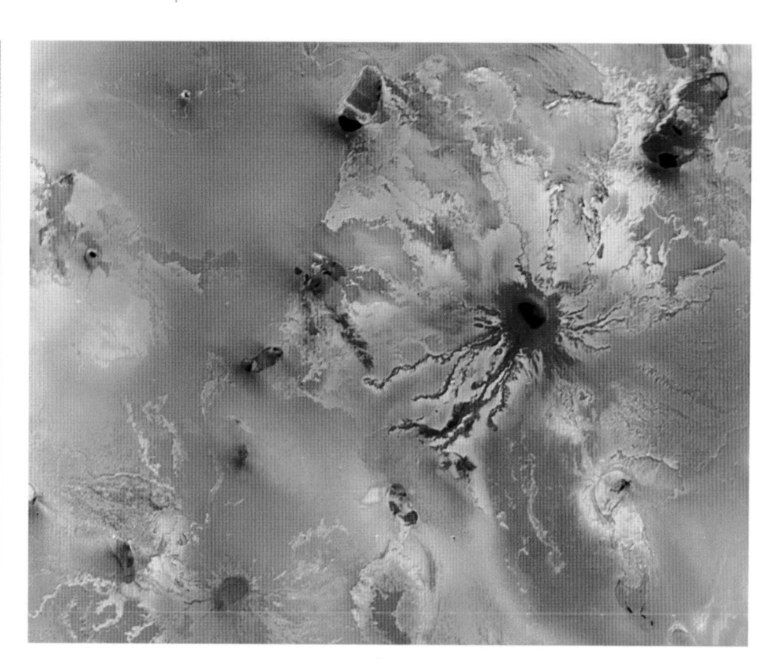

FIGURE 11.16: *Ra Patera, a large shield volcano on Io, is approximately 300 km (186 mi) in diameter (NASA Galileo photo).*

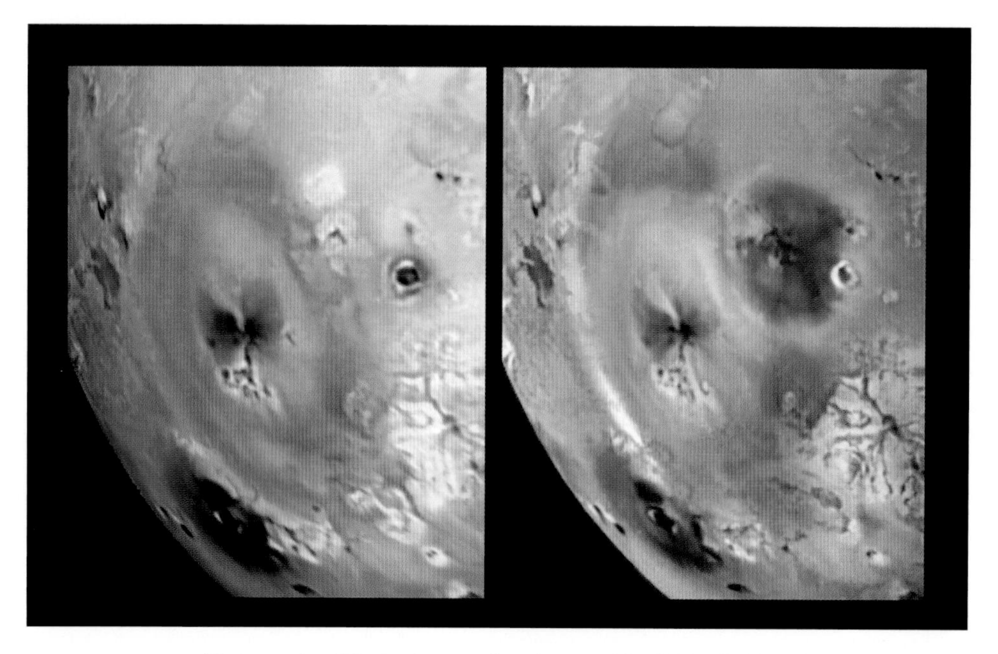

FIGURE 11.17: *Photographs of Jupiter's moon Io taken on 4/04/1997 (left) and on 9/19/1997 (right) showing the results of a new eruption (dark area 400 km (240 mi) in diameter) surrounding Pillar Patera Volcano. Pyroclastic silicate material is suspected. Eruption height was 120 km (75 mi).*

The red halo is believed to be sulfur deposited as S3 and S4 "snowflakes" from an earlier eruption. These red sulfur molecules convert to the more stable yellow S8 molecules in time. Note how younger, dark gray pyroclastic material has covered part of the older, red sulfur halo (NASA Galileo photo taken from 487,000 km (302,000 mi) away from Io).

or mantle that overlies a iron-nickel-sulfide core.

In addition, Io spews plumes of frozen sulfurous materials to heights of 500 km (310 mi) above its surface. Such great heights are apparently achieved for two reasons: Io has about 1/6 Earth's gravitational pull and its atmospheric pressure is extremely low. In comparison, should Old Faithful Geyser in Yellowstone National Park erupt on Io, the top of the geyser would reach 37 km (23 mi), rather than its usual 50 m (155 feet).

Sulfurous plumes are created by liquid sulfur dioxide contacting hot material in Io's crust. Boiling coupled with rapid expansion results in the upward high-velocity jetting of sulfur dioxide gas. This gas freezes into "snow flakes" that settle back to Io's surface.

Red colors seen around vents are due to the formation of trisulfur (S_3) and tetrasulfur (S_4) "snowflake" molecules. These eventually convert to the more stable yellow octasulfur (S_8) molecules that color most of Io's surface.

But why is such a small celestial body so volcanically active? Because Io lies between giant Jupiter and the moons Europa and Ganymede, extreme tidal forces appear to drive volcanism. These bodies exert enormous pulls on the crust of Io, resulting in solid crustal bulges of as much as 100 m (330 feet). This crustal tidal bulge constantly moves around Io, flexing its crust.[19]

CHAPTER 11 – NOTES

1. Carbonatite is rare igneous rock associated with hotspot volcanism on continental lithosphere. (The most famous example is Ol Doinyo Lengai Volcano located in Tanzania, Africa.) Lavas are composed of between 55 and 95 percent calcite ($CaCO_3$) and potassium carbonate (K_2CO_3), commonly with small amounts of apatite ($CaPO_4$), calcium-magnesium-rich pyroxene, and garnet. Some carbonatites contain radioactive element- and/or rare earth-bearing minerals. Sodium-rich (approximately 30 percent Na_20) carbonatites also exist on the Earth.
2. The Moon has a equatorial diameter of 3,476 km (2,158 mi) and a density of 3.344 grams per cubic centimeter, compared with the Earth's diameter of 12,756 km (7,921 mi) and a density of 5.52 g/cm^3.
3. See Wikipedia – "List of Solar System probes" on the internet for an exhaustive list of both successful and unsuccessful U.S. and other nations' space probes.
4. Breccia is composed of angular fragments of pre-existing rocks that have been broken. Breccias can form in many ways. Lunar breccias probably formed primarily from impacting and less so from breakage during landslides into craters. Constant and extreme heating and cooling also breaks rocks.
5. Maria (Latin, plural for seas) are dark areas of the moon composed of flood basalt flows. The singular is mare.
6. In his 1949 book *The Face of the Moon,* Ralph Baldwin first suggested that maria were composed of basalt and not sediments based on the many basalt flow characteristics he was able to view from Earth.
7. The 1997 Mars Pathfinder Lander was preceded by the Viking 1 and 2 landers of July 1976 and September 1976, respectively.
8. Maybe Mars is not thermally "dead." Twice during spring 2012, mysterious plumes of gas, 966 km (600 mi) in diameter were seen (from Earth) rising more than 241 km (150 mi) above the Martian surface. The origin and composition of the plumes are presently unknown.
9. Alba Mons original name was Alba Patera. A patera is an irregular or complex caldera-type volcano with scalloped edges and gentle flank slopes. They appear to be older than shield-type volcanoes, such as Olympus Mons, and may have witnessed explosive eruptions producing pyroclastic deposits.
10. Mauna Loa is taller than Alba Mons, with an total height from its seafloor base of 10,365 m (34,000 feet), and greater slope angles of between 1° and 4°.
11. "Mons" is a large, isolated mountain.
12. Radar showing a bright surface indicates roughness, whereas dark reflections indicate smooth surfaces.
13. Coronae (crowns or garlands) are circular or elliptical structures consisting of concentric ridges or fractures. They may be domes, plateaus, or depressions. They may have a peripheral trough or moat, and abundant volcanic and tectonic land forms. To date, 208 have been identified. They vary in size from 200 to 1,000 km in diameter; large coronae are less common. Coronae are believed to be associated with mantle plumes or hot spots.
14. Arachnoids (cobweb-like) are similar to coronae. They have circular fracture patterns or ridges, but contain more strongly developed radial fractures or ridges. These radial features are several radii longer than the diameter of the arachnoid. Arachnoids contain lava flows and small shield volcanoes and are generally less than 200 km in diameter. To date, 265 have been catalogued. They may be intrusions.
15. Viscosities (resistance to flow) equivalent to that of water are believed necessary for generating the long flows. The viscosity of Earthly basalt, for example, is from 10 to 1,000 times greater.
16. The inner four planets are rocky and relatively gas-poor in comparison to the four outer gaseous giant planets (Jupiter, Saturn, Neptune, and Uranus). This aspect of the inner planets is due to their nearness to the Sun, which effectively boiled off most gases early in solar system formation.
17. The great expansion that takes place with a change from the solid to gaseous state would cause gaseous cryovolcanic eruptions.
18. This temperature is more than 200°C higher than Earthly basaltic lava flows. Most of Io's surface is about -150°C.
19. To understand how heat is generated by solid crustal bulging, consider what happens when a thin piece of metal is flexed. Notice how hot the metal becomes at the point of flexure. Eventually the strip of hot metal will break.

CHAPTER 11 – REFERENCES

Bussey and Spudis, 2004; Crisp, 2004; Crumpler and Aubele, 2000; Drake, 2011; Forget et al., 2008; Gaddis, 2004; Geissler, 2000; Lopes, 2004; Lopes-Gautier, 2000; Lopes and Gregg, 2004; Mouginis-Mark and Robinson, 1992; Murchie et al., 2011; Neal and Taylor, 1992; Patrick and Howe, 1994; Prockter, 2004; Spudis, 2000; Stofan, 2004; Zimbelman, 2000.

CHAPTER TWELVE

Eruption Forecasting

Introduction

Volcanologists are working to accurately predict eruptions, just as seismologists are striving to predict when, where, and how large the next earthquake will be. This task is becoming increasingly important as more people live and work on volcano slopes. But what are the societal consequences of accurately forecasting these naturally occurring, devastating events?

In 1975, Chinese authorities, noting increasing seismic activity and odd animal behavior, evacuated more than 1 million people from Haicheng City just days before half the city was destroyed by an intense 7.3 Richter Scale magnitude earthquake. This action by a totalitarian state probably saved tens of thousands of lives. Those of us who do not live in China, however, may imagine less compliant human reactions to predicted catastrophic events.

Analyses of responses before, during, and after Hurricane Katrina that devastated New Orleans in 2005 is informative. Even though officials expected Katrina to make landfall as a Category 5 storm, the largest possible, thousands of people remained in the city. Many lives were lost when the city flooded. Many others who stayed had to be rescued at great expense and risk to all involved.

The same scenario—a general reluctance to leave—would no doubt happen should a major eruption (or earthquake) be forecast near a large U.S. city, such as Portland, Oregon. Where would the people go? How would they be housed and fed? Who would stay behind to pump gas, fight fires, prevent looting, and maintain civil order? What about those who decided at the last moment to leave by way of congested roads? Would riots ensue as tempers and frustrations flared?

And what if the cataclysmic event did not occur as predicted? What if it hit a day earlier or two days, or a week, or a month later? Imagine the lawsuits! Imagine the liability and damaged credibility of the scientist(s) who made the dire prediction in good faith, using the best available science.

Scientists are striving for ever more accurate prediction tools, but foretelling the future is likely to remain a gamble. Even with highly accurate disaster forecasts, the human response and behavior will likely be problematic, perhaps even negating the value of the advanced warning.

Volcanologists have made great strides in forecasting eruptions using geological, geochemical, and geophysical techniques. But this difficult task is compounded by the various pre-eruption characteristics of the different volcano types, the individualized nature of each volcano, and the sheer number of potentially active volcanoes in the world (between 1,300 and 1,500). Also problematic is the wide interval between the end of dormancy and the climax of an eruption. This interval can vary from days to decades and span an array of signs that an eruption may (or may not) be imminent.

Simkin and Siebert (2000) compiled data on 252 eruptions and found that in 42 percent of these the climactic phase began during the first day, many within the first hour. In 52 percent of the cases, the climactic eruptive event occurred within the first week. Of the 60 or so known eruptions per year, about three (5 percent) cause deaths.

Based on past volcanic activity, we can estimate the frequency of future eruptions. Those with a Volcanic Explosive Index (VEI) of 2 occur every few weeks. Several eruptions can be expected per year having a VEI of 3, whereas an eruption the size of Mount St. Helens with a VEI of 5 occurs on average every 10 years. A Krakatau-sized eruption (VEI of 6), producing more than 10 km^3 of ejecta, is expected about every 100 years. Finally, the largest eruptions such as those from Toba and Yellowstone with VEIs of 8 may happen about every 100,000 years. Fortunately for us, the larger the eruption and destructive force, the less frequent the occurrence.

Eruption Predictive Techniques

Kilauea and Mauna Loa volcanoes on the island of Hawaii are good places to learn about eruption prediction methodologies. The most active volcano in the world, Kilauea, in particular, has provided much useful data because it erupts frequently, announces seismically (through earthquakes) what will happen and where, deviates little in both precursory and eruptive behavior, and is both accessible for study and well monitored.

Kilauea also has a relatively simple magmatic plumbing system: semi-horizontal layered basalt flows and intrusions. In contrast, subduction zone volcanoes rest on faulted, metamorphosed, and intruded heterogeneous assemblages of basement rocks. These volcanoes generally erupt infrequently but explosively due to their more silicic magmas. Our knowledge base, therefore, is limited and brimming with complex variables.

Volcano Shape and Eruptive History

What predictive techniques are used and how effective are they? Scientists first consider the physical condition of the volcano and its historical record derived from studying and dating volcano-sedimentary deposits around it.

Prior to 1980, Mount St. Helens was a perfectly shaped, little-eroded cone. Known as the Fujiyama of North America, Mount St. Helens also had an unusually low tree line for its latitude. Both of these characteristics were due to very recent, post ice age growth, well documented by radiometrically dated (carbon 14) volcanic deposits around its base. Eruptions were actually witnessed between 1831 and 1857, and were documented in a painting by Paul Kane in 1847 of a flank dome eruption. The painting presently hangs in the Royal Ontario Museum in Toronto, Canada.

Given these data and an eruption recurrence interval (ERI) of approximately every 225 years over the last 4,500 years, no geologist would have bet against an eruption in the not too distant future. In fact, U.S. Geological Survey geologists Dwight Crandell and Donal Mullineaux wrote in 1978 that "The volcano's behavior pattern suggests that the current quiet interval will not last as long as a thousand years; instead an eruption is more likely to occur within the next 100 years, perhaps even before the end of this century."

In the spring of 1991, volcanologists, because of seismicity and the nature of violently erupted pyroclastic deposits surrounding Mount Pinatubo in the Philippines, decided the region should be evacuated. Great loss of life was thereby averted during the subsequent cataclysmic explosive eruption of June 15 (see the Famous Calderas sidebar, page 104).

Volcano shape and eruptive history delineated from deposits surrounding the mountain, then, aid in forecasting possible sizes and types of future eruptions, but not when they will begin. Unfortunately, not all volcanic eruptions originate from well-shaped, little-eroded, relatively recently active volcanoes like Mount St. Helens. Many, such as El Chichón in 1982 in Mexico and Mount Pinatubo, had not erupted for hundreds or thousands of years. These volcanoes with long repose periods between eruptions pose additional problems. This is especially true if the volcano is highly eroded, as was Mount Pinatubo. Furthermore, volcanoes with long repose periods also commonly have very large, explosive eruptions.

In light of the above, there are some "common sense" ways to mitigate volcanic hazards for populated regions. They include (1) the delineation of potentially hazardous zones, (2) avoidance of these zones except for appropriate uses, such as for animal grazing, (3) automated warning systems integrated with local and state civil defense systems, (4) the installation, if feasible, of diversion barriers for lava and lahar flows, (5) remedial actions during eruptions to thwart or lessen deaths and damage, and (6) safe places to house and feed evacuees.

Hazard zone maps of the distributions and ages of eruptive products have proved very useful in saving lives and limiting structural damage. Hazard zone mapping also applies to potentially dangerous areas subjected to coastal storms, river flooding, and earthquakes. Such maps are useful in determining appropriate land uses (see Mount Shasta sidebar, page 66, for hazard zone maps).

Warning systems are vital for survival, as was evidenced by the relatively few deaths (about 350) that accompanied the 1991 Mount Pinatubo eruption. With adequate warning and evacuations, tens of thousands of lives were saved, including U.S. servicemen stationed at Clark Air Force Base on the east slope of the volcano. Compare this with the outcome of the Nevado del Ruiz eruption in Colombia in 1985. Even though a large eruption was predicted and lahars had devastated the Amero region in 1845, a communication break down prevented the Red Cross evacuation order from reaching Amero. Tragically, three quarters of the population—more than 23,000 people—perished.

Nothing short of evacuation to higher ground would have saved the folks of Amero, but in other locations physical barriers and small dams have saved people and structures from lava and lahars. Barriers have been used, for example, in Hawaii and on Mount Etna on Sicily.

Finally, other methods have been used to stop or divert lava flows in order to protect people and structures. Just prior to World War II, bombs were dropped on a lava flow issuing from Mauna Loa near Hilo, Hawaii. In the 1970s, water was sprayed on lava flows during an eruption on the island of Heimaey off the Icelandic coast. This action prevented the lava flow from closing the harbor.

Eruption Predictive Methods

VISUAL CHANGES FROM "NORMAL" VOLCANO BEHAVIOR

Before modern monitoring techniques, folks living around volcanoes became apprehensive when they saw changes in the behavior of their volcanic neighbor. This remains a key part of successful eruption prediction—the detection of any change from the "normal" or background state of a volcano. Changes to watch for include new or increased steam venting, higher temperatures at the vents, warming ground coupled with snow and ice melting or vegetative changes, and seismicity. (Smoke or ash emissions are actual eruptive phases.)

Higher temperatures and increasing amounts of steam emanating from vents and fumaroles may also be due to factors other than intruding magma. For example, rainfall may trigger greater steam output through the porous volcanic summit complex. Also, changes in volcano plumbing systems may result in increasing (or decreasing) steam venting and fumarole temperatures. When the temperature of a crater lake rises, however, it is most likely due to magmatic heating.

INSTRUMENTS USED FOR PREDICTION

Methods for determining where and when an eruption will occur include seismicity studies, ground deformation monitoring, and gravity, electrical, and magnetic field measurements. Researchers are also increasingly relying on ground- and satellite-based thermal and geochemical monitoring of volcanoes, vents and fumaroles.

Seismic Monitoring

The first indication that Mount St. Helens was awaking from dormancy was an earthquake with a Richter scale magnitude of 4.2 detected on March 20, 1980. Its depth (hypocenter or point of origin) was 4 km (2.5 mi) under the mountain. The epicenter (geographic location) and all subsequent quakes were restricted to a 3-km (1.9-mi) diameter region directly below the volcano above. In retrospect, seismic activity actually started under the volcano on March 16. It increased with time, reaching about 50 events per day with magnitudes of 3 or greater. Subsequent earthquake hypocenters were almost entirely at a depth of 2.5 km (1.5 mi) or less under the mountain. By May 18, more than 10,000 earthquakes had been detected.

Volcanologists developed and refined our understanding of the relationship between seismicity and eruptions in part by studying Hawaiian volcanoes. There, seismic activity begins usually 60 to 70 days before an eruption with hypocenters at depths of about 70 km. Magma rises approximately 1 km per day, so it issues forth about two months after the initial, deep earthquake activity.

Volcanically generated seismic waves are differentiated into three types: high-frequency Short Period (SP or A waves), low-frequency Long Period (LP or B waves), and harmonic tremors.[1] Each earthquake type provides researchers with important data (Figure 12.1).

Short-period (SP) quakes of a few hundredths of a second duration are believed to be caused by sudden, rock-breaking events associated with rising magma. They commonly become more frequent during pre-eruptive phases and typically occur in swarms. Swarms can be of very long duration because of the time magma takes to rise. This type of seismic activity is not always followed by an eruption.

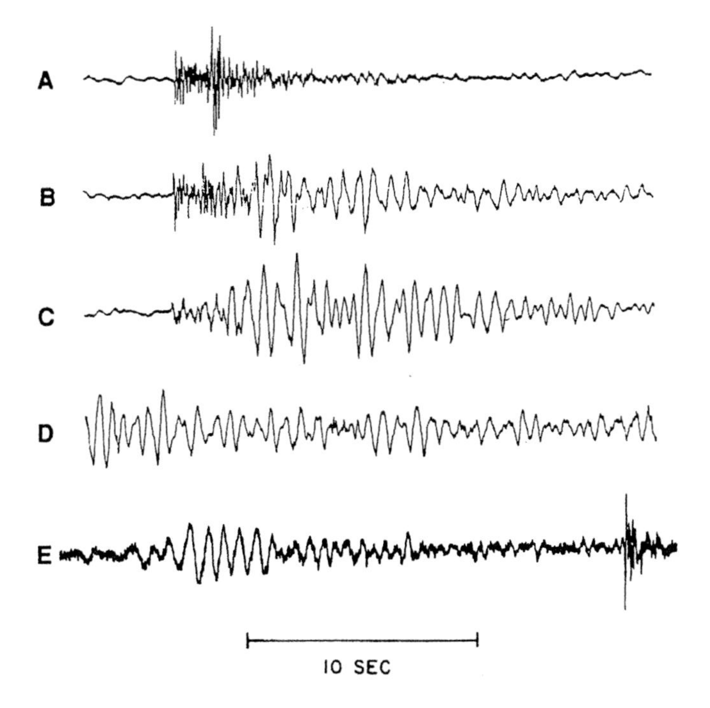

FIGURE 12.1: *Various Seismic Wave Forms. (A) is a high-frequency, volcano-tectonic earthquake detected 6.8 km below Redoubt volcano in Alaska, (B) is a hybrid or mixed-frequency earthquake with a hypocenter 0.6 km (2,000 feet) below Redoubt Volcano, (C) is a low-frequency or long-period (LP) event 0.4 km (1,300 feet) below Redoubt Volcano (probably due to fluid pressurization or magma movement), (D) is a volcanic harmonic tremor, or continuous signal with duration of minutes to days, and (E) is the trace of an explosion quake followed by a prominent air-wave arrival event on the left associated with Pavlof Volcano (from McNutt, 1986, 2000, and Power et al., 1994).*

Long-period (LP), low-frequency events last 0.2 to 2 seconds and have frequencies of between 0.5 and 10 cycles per second. They are believed to be caused by interactions between magma, other fluid, or gas and surrounding rock. They can be of shallow or deep origin.

While deeper LP seismicity is theorized to result from magma movement, shallow LP events are likely due to the resonance of fluid-filled cavities. LP waves may evolve from SP seismicity as gas bubbles form in the rising magma, which increases magma viscosity and in turn lowers acoustic velocity.

Rocks also vibrate as magmas or gases rise toward the surface in volcanic plumbing systems. These harmonic tremors or "hum" commonly begin prior to a volcanic outburst and continue during the eruption. Such tremors occur sporadically, have a continuous signal when active, and are the most common type of seismicity related to volcanoes.

At Mount Pinatubo in 1991, for example, during the week before the cataclysmic eruption, monitors detected at least 400 LP events daily at a depth of 10 km (6.2 mi). Long-Period waves were also detected prior to the 1989 Redoubt Volcano eruption in Alaska.

U.S. Geological Survey volcanologist Bernard Chouet reasons that there is a significant difference between deep and shallow LP waves. Shallow LP waves (depths of about 1 km or less) are due to an interaction of magma and groundwater that causes vibrations or "hum." When this happens, Chouet theorizes, an eruption is imminent. Furthermore, Chouet believes that the faster the system pressurizes, the sooner an eruption will occur.

Observations also indicate that seismic activity (known as "noise") and ground deformation such as cone swelling or inflation tend to increase shortly before an eruption. The pre-eruptive increase in seismic noise has been quantified by Pennsylvania State University volcanologist Barry Voight. He postulated that the odds are high that a volcano will erupt after seismicity exponentially accelerates.

An analysis of seismic records show that this method would have predicted eruptions at Mount St. Helens in the 1980s, Redoubt in 1989, Galeras in Colombia on January 14,1993 (which killed 7 volcanologists), Izu Oshima in 1990 in Japan, Soufriere Hills on the island of Montserrat in 1995, and the 1998 eruption of Colima in Mexico. While this predictive method utilizing LP waves commonly herald impending eruptions, in some situations magma never surfaces.

Yet an increase in LP waves does not always signal impending eruptions—in some situations an eruption is long delayed and in others magma never surfaces. The Rabaul Caldera in Papua New Guinea became restless in 1983, but the volcano did not erupt until 1994. The dramatic increase in seismicity due to magma intrusion in the Long Valley Caldera in California in 1997, though ongoing, has yet to result in an eruption. That not all earthquake swarms portend eruptions was determined statistically by Benoit and McNutt (1996) who analyzed 327 swarms between 1979 and 1989. They discovered that while 191 (58 percent) were followed by an eruption, 136 (42 percent) were not.

Ground Deformation

One of the oldest methods to determine that magma was moving into the upper parts of a volcano was with tiltmeters. Used for years in Hawaii, they were also employed on Mount St. Helens in

1980. With magma injection, cones may swell, and tiltmeters record the magnitude and rate of bulging (Kilauea, for example, bulges 1 to 2 m (3 to 6 feet) before erupting). Estimates can then be made as to depths and amounts of new magma below. But a tiltmeter was not necessary to discover the bulge on the northwest flank of Mount St. Helens prior to the May 18 eruption. By the 18th, it was approximately 2 km (1.2 mi) in diameter and attained a maximum uplift of 100 m (330 feet).

The advent of Earth-orbiting satellites has revolutionized ground deformation studies and monitoring of volcano thermal emissions. Using satellite interferometry, volcanologists can determine surface deformation to within as little as a few millimeters. These measurements enable volcanologists to determine when magma has moved to within 5 to 10 km or less of the Earth's surface.

Volcanologists have used satellite interferometry together with the Global Positioning System (GPS),[2] which precisely measures latitude, longitude, and elevation, to delineate volcanic bulges in many parts of the world. The systems are so sensitive that they can detect a 1-mm change of elevation over a horizontal distance of 1 km. In 1996, for example, scientists discovered a 10-cm-high, 15-km (9-mile)-wide uplift on the flanks of the Three Sisters volcanoes in the Cascade Range of west-central Oregon west of Bend. This "slight" bulge is believed to be the result of approximately 21 million m^3 (742 million $feet^3$) of magma rising into a magma chamber 6 to 7 km below the surface. It is also theorized that rising magma pressurizes the volcanic plumbing system above, and that the longer the land rises, the more likely an eruption will take place.

More recently, Chang et al., 2007, noted that rapid uplift started in mid-2004 within the Yellowstone Caldera in Wyoming. GPS and interferometric measurements show rates of uplift of up to 7 centimeters per year. The scientists theorize that a 1,200-km^2 (465-mi^2) intrusion exists at a depth of 10 km (6.2 mi) and that the injection rate is 0.1 km^3 per year. The uplift, however, could also be due to surface pressurization by magmatic fluids, and not by new, intruding magma.

Another promising satellite-based volcano monitoring technique is Synthetic Aperture Radar (SAR). With SAR, large areas can be imaged and subsequently compared with images of the same region taken as little as one month previously. Technical developments may improve SAR's usefulness for volcano monitoring, but challenges remain when the scanning area includes steep topography, snow cover, or vegetation.

Gravity, Electrical, and Magnetic Field Measurements

Other techniques under development for forecasting eruptions include monitoring changes in the Earth's magnetic, electrical, and gravity fields in and around volcanoes. Theoretically, these field parameters near or under volcanoes should change prior to eruptions. To date, field changes have been observed in the volcanic regions of Japan, Hawaii, New Zealand, and Kamchatka, Russia. The most useful method so far is microgravity monitoring, which measures both gravity and elevation changes or ground deformation.

Geochemical Monitoring

Magmas, commonly saturated in gases such as water, carbon dioxide, and sulfur dioxide, produce systems of exsolved, or free gas phases. Because gases are important eruption drivers, monitoring their amounts, changing quantities, and component ratios has become an important eruption prediction tool.

Starting back in the 1960s, geologists, such as Professor Dick Stoiber at Dartmouth College, began collecting gas samples from Central American volcano fumaroles. It was theorized that increasing amounts of carbon dioxide, chlorine, and sulfur dioxide, and especially rising ratios of sulfur dioxide to carbon dioxide, signaled that magma was moving closer to the surface, thereby increasing the odds of an eruption.[3]

Recently, volcanologist William Rose of Michigan Technological University theorized that "both an increase in gas emissions and a decrease in gas emissions are bad" Why? Emissions of gases, and especially sulfur dioxide, decreased significantly from Galeras volcano in Colombia, South America, just prior to a killer eruption in 1993. This change was not due to an aborted eruption, but because magma plugged the eruptive vent. Pressure then built up within the conduit, resulting in an explosion just as a party of volcanologists was examining fumaroles within the crater. Seven volcanologists were killed.

The amount of sulfur dioxide gas released, then, decreased prior to the eruption because of vent plugging, and not because of decreased magma degassing. Bernard Chouet, who investigated the seismic record prior to the eruption, discovered that there had been shallow LP seismic waves signaling that an eruption might follow.

Volcanoes, even the same kind, have differences and thus may have different pre-eruption gas signatures. For example, recent studies of gases emitted from Mount Etna on Sicily show that increasing carbon dioxide/sulfur dioxide ratios accompany the refilling of shallow conduits, which in turn leads to pressurization and the triggering of eruptions.

Since 1971, volcanologists have remotely monitored sulfur dioxide emitted during eruptions with the Correlation Spectrometer (COSPEC)[4] (see Chapter 9 for more details). One problem with basing eruption predictions on sulfur dioxide emissions is that this gas is very soluble in water. Low emission rates can be deceiving and may actually be due to gas dissolving in water rather than there being low amounts of sulfur dioxide in the system.

Other predictive techniques include determining changes in concentrations of the radionuclides lead-210, polonium-210, bismuth-210, and radon-226 in volcanic gas plumes. These radioactive or unstable isotopes have different half-lives[5] and also different volatilities in magmas. Researchers have monitored these isotopes at Mount Etna to better understand short-term changes within the magmatic reservoirs. Determinations of these nuclides, together with sulfur compounds, it is theorized, will allow researchers to characterize volcanic activity and determine the volume of the degassing magma.

Summary

Volcano monitoring techniques will continue to improve, and new ones will be developed with time. This is especially true with satellite-mounted, and remote-positioned instruments. Non-monitored volcanoes will be added to the "watch" list, and no doubt new predictive techniques will be developed. But the development and operational costs associated with present and to-be-discovered methods will probably be high. And so for volcanoes located in remote parts of Africa, the Kamchatka region of the Russian far east, Indonesia, the Philippines, and Alaska, automated satellite monitoring systems will become more prevalent.

While we don't worry as much about most of those remote, sparsely populated regions, eruption predictive techniques must be improved in the volcanically active, more heavily populated parts of the world. These include many cities in Indonesia, Central and South America, the Philippines, the Naples region of Italy, and parts of Africa such as Goma in the Democratic Republic of the Congo.

Ideally, we will use ground-based and especially satellite-based techniques employing continuous, real-time data collection and monitoring. This type of information-gathering will enable scientists to obtain much needed baseline, background data instantaneously. With those data, it will be easier and perhaps cheaper to identify indicators of imminent eruptions.

Where Do We Go from Here?

When faced with volcanic, flood, or earthquake hazards, the best strategy is to avoid dangerous sites and to prepare for emergencies. Yet around the globe people continue to build structures in harm's way, whether on river and coastal floodplains, on landslide-prone hills, and on dormant volcano slopes. Ironically, while humans have the unique capacity to learn from past experiences and plan accordingly, rational judgment is clouded by innate optimism that destructive events won't be repeated soon (or to us), ignorance,[6] and the almost willful forgetting of bad past experiences.

Risk perception is also related to religious, socioeconomic, and political views. For example, people choose to live on the slopes of Mount Etna in Italy because of various and good reasons. They include established property ownership, productivity, lifestyles, climate, scenic beauty, and family history. Add to these factors a growing population and increasing land values, with the best lands already in use, and we have the recipe for annually increasing loss of life and property by naturally occurring, destructive forces. Rather than "conquering nature" with our increasingly sophicated technology, as many of us were taught in school, we are becoming more vulnerable to the "whims" of "Mother Nature." This is especially true as world

Cause	Deaths (percent)	events (percent)
Pyroclastic Flow	29	15
Indirect (1)	23	5
Tsunami	21	5
Lahar (mudflow)	15	17
Pyroclastic Fall	2	21
Avalanche	2	3
Gas	14	
Flood	1	2
Earthquake	<1	2
Lightning	<1	1
Unknown Causes	7	20

Modified from Simkin et al., 2001; Abbott, 2004
1 = subsequent deaths due to famine, disease, etc.

TABLE 12.1: *Volcanic-associated deaths (275,000) related to 530 volcanic events.*

population grows, the Earth's atmosphere warms, and crowding results in people living in more hazardous places.

Clearly, the perceived risks of natural hazards does not match reality. Most people receive information about risky phenomena through communication with others and the media, not by direct experience. And therein lies a major problem: seeing and experiencing makes believing easier. This has been demonstrated by research showing that experience leads to a perception of the high risk of natural hazards.

If we wish to improve avoidance responses to hazards, we must improve forecasting of impending natural disasters, enhance communication between scientists and local authorities, and ensure rapid warning responses from those authorities to the general population. Increased education of both officials and the affected population by scientists about the hazards associated with the phenomena is also vitally important. And because most people don't witness destructive processes, visuals are our next best educational tool.

One video, entitled "Understanding Volcanic Hazards" by Maurice and Katia Krafft, was shown to the local population prior to the 1991 Pinatubo eruption. It proved especially effective in persuading officials and people living on and around the volcano to evacuate. A total of 80,000 souls left the immediate region, including those at Clark Air Force Base. The video, prepared for the International Association of Volcanology, contained striking footage of lahars, pyroclastic flows, and the effects of ash fall-crushed buildings during other eruptions.

Local and federal governments are also remiss to allow people to live in these potentially dangerous places. Whether on floodplains, faults, or volcanic slopes, governments should only allow select, non-dangerous land uses there. For example, logging and cattle grazing should be allowed on the slopes of Mount Shasta, but not recreational homes. Parks and grazing lands should be encouraged in flood-prone areas, not homes.

In the final analysis, the use of these lands for inappropriate or dangerous purposes comes back to "bite" us all through the unnecessary loss of life, property, and taxpayer dollars required to insure and rebuild, once again, in these locations unsuitable for living and working.

CHAPTER 12 – NOTES

1 A period (P) is the duration in time of one cycle in a repeating sequence of cycles. Frequency (f) is the number of occurrences of an event per unit of time such as 60 revolutions per minute)

2 The Global Positioning System (GPS) was developed in 1973 and became fully operational by 1995.

3 Chlorine (Cl), sulfur dioxide (SO_2), and carbon dioxide (CO_2) dissolve in silicate magmas. As magmas rise toward the surface, pressure drops in the magmatic liquid. Under diminishing pressure, carbon dioxide gas is the first to exsolve from the magma. Monitoring gases, then, will commonly detect increasing carbon dioxide concentrations in fumarolic gases before chlorine and sulfur dioxide concentrations rise.

4 COSPEC was originally designed and used in the early 1970s for remotely monitoring the amount of sulfur dioxide emitted from industrial smoke stacks.

5 Half-life refers to the time it takes for radioactive or unstable isotopes (species of an element) to decay into a different nuclide or isotope "daughter" products of half of the original amount. For example, the half-life of the isotope uranium-238, from which lead-210, bismuth-210, and polonium-210 are derived is about 4.5 billion years. At the end of that time, only 0.5 pound of uranium-238 remains from the original pound. If we wait another 4.5 billion years (or 9 billion years from the start), only 0.25 of a pound of uranium-238 remains, and so on. The rest of the starting uranium-238 will have decayed into "daughter" products, and finally into lead (Pb). The remaining amount of the starting, radioactive isotope halves after each half-life as the amount of the starting isotope approaches but never quite reaches zero.

Half-lives vary from fractions of a second for the most unstable nuclides or radioactive isotopes to billions of years for the most stable of the radioactive isotopes such as uranium-238 and rubidium-87. The radon-222 half-life is 3.82 days, bismuth-210 is 5 days, polonium-210 is 4.6 months, and lead-210 is 22 years.

6 A good example of human ignorance is the concept of the "100-year flood," clearly not understood by most people living in or near floodplains. The concept means that there is a one-in-a-hundred chance of a flood of that

enormous volume per year. Too many people assume this means that such a flood can occur only once every 100 years—that once such a flood has occurred, it will be another 100 years before another one. This is exactly wrong. In reality, there could be back-to-back years of 100-year floods.

CHAPTER 12 – REFERENCES

Abbott, 2004; Aiuppa et al., 2007; Andres and Schmid, 2001; Baxter, 1990; Benoit and McNutt, 1996; Burlini et al., 2007; Cashman and Hoblitt, 2004; Chang et al., 2007; Chouet, 2003; Christiansen and Peterson, 1981; Crandell and Mullineaux, 1978; Decker and Decker, 2006; Endo et al., 1981; Forget et al., 2007; Francis and Rothery, 2000; Freundt et al., 2006; Gaillard and Dibben, 2007; Kerr, 2001, 2002, 2003; Lange and Avent, 1973; Le Cloarec and Pennisi, 2001; McNutt, 1986; McNutt, 2000; McNutt et al., 2000; Murray et al., 2000; Power et al., 1994; Rymer and Williams-Jones, 2003; Sharpton and Ward, 1990; Simkin and Siebert, 2000; Simkin et al, 2001; Stone, 2003; Tilling, 2000; Tilling and Lipman, 1993; Turner et al., 2008; Utada et al., 2007; Voight, 2003; VRPAB, 2008; Walter and Amelung, 2007; Witham, 2005; Zimmer and Erzinger, 2003.

References

Abbott, P. L., 2004. *Natural Disasters*, Boston: McGraw Hill, 4th ed.

Aiuppa, A., Moretti, R., Federico, C., Giudice, G., Gurrieri, S., Liuzzo, M., Papale, P., Shinohara, H., and Valenza, M., 2007. "Forecasting Etna eruptions by real-time observation of volcanic gas composition," *Geology*, v. 35, no. 12, p. 1115-1118.

Anderson, D. L., 2005. "Scoring hotspots: the plume and plate paradigms," in Foulger, G.R. et al., eds., *Plates, plumes, and paradigms*, Geological Society of America Special Paper 388, p. 31-54.

Anderson, D. L., and Natland, J. H., 2005. "A brief history of the plume hypothesis and its competitors: concept and controversy," in Foulger, G.R., et al., eds, *Plates, plumes and paradigms*, Geological Society of America Special Paper 388, p. 119-145.

Anderson, D. L., and Schramm, K. A., 2005. "Global hotspot maps," in Foulger, G.R., et al., eds., *Plates, plumes, and paradigms*, Geological Society of America Special Paper 388, p. 19-29.

Anderson, D. L., Tanimoto, T., and Zhang, Y., 1992. "Plate tectonics and hotspots: the third dimension," *Science*, v. 256, p. 1645-1651.

Andres, R. L., and Schmid, J. W., 2001. "The effects of volcanic ash on COSPEC measurements," *Journal of Volcanology and Geothermal Research*, v. 108, p. 237-244.

Arnorsson, S., 2000. "Exploitation of geothermal resources," in Sigurdsson, H., ed., *Encyclopedia of Volcanoes*, p. 1243-1258.

Ashley, R. P., and Silberman, M. L., 1976. "Direct dating of mineralization at Goldfield, NV by potassium-argon and fisson track methods," *Economic Geology*, v. 71, p. 904-924.

Asimow, P. D., 2000. "Melting the mantle," in Sigurdsson, H., ed., *Encyclopedia of Volcanoes*, p. 55-68.

Bacon, C. R., 1983. "Eruptive history of Mount Mazama and Crater Lake caldera, Cascade range, U.S.A.," *Journal of Volcanology and Geothermal Research*, v. 18, p. 57-115.

Bacon, R. A., Gardner, J. V., Mayer, L. A., Buktenica, M. W., Dartnell, P., Ramsey, D. W., and Robinson, J. E., 2002. "Morphology, volcanism, and mass wasting in Crater Lake, Oregon," *Geological Society of America Bulletin*, v. 114, p. 675-692.

Bacon, C. R., and Lanphere, M. A., 2006. "Eruptive history and geochronology of Mount Mazama and the Crater Lake region, Oregon," *Geological Society of America Bulletin*, v. 118, p. 1331-1359.

Bailey, R. A., 1987. "Long Valley Caldera, Eastern California," *Geological Society of America Field Guide - Cordilleran Section*, p. 163-168.

Bailey, R. A., Dalrymple, G. B., and Lanphere, M. A., 1976. "Volcanism, structure, and geochronology of Long Valley caldera," *Journal of Geophysical Research*, v. 81, p. 725-744.

Bailey, R. A., Miller, C. D., and Sieh, K., 1989. "Field guide to Long Valley caldera and Mono-Inyo craters volcanic chain, eastern California," ICC Field Trip T313 in *Volcanism and Plutonism of western North America*, v. 1., 28th International Geological Congress, American Geophysical Union, p. 1-36.

Baker, E. M., Kirwin, D. J., and Taylor, R. G., 1986. *Hydrothermal breccia pipes*, Contributions of the Economic Geology Unit, Geology Department James Cook University of North Queensland.

Ball, P., 2008. "They really were the Dark Ages," *Nature News*, published on line.

Baxter, P. J., 1990. "Medical effects of volcanic eruptions," *Bulletin of Volcanology*, v. 52, p. 532-544.

Beget, J. F., 2000. "Volcanic tsunamis," in Sigurdsson, H., ed., *Encyclopedia of Volcanoes*, p. 1005-1013.

Benoit, J. P., and McNutt, S. R., 1996. "Global volcanic earthquake swarm database and preliminary analyses of volcanic earthquake swarm duration," *Annali de Geofisca*, v. 39, p. 221-229.

Bindeman, I. N., 2006. "The secrets of supervolcanoes," *Scientific American*, v. 294, no. 6, p. 37-43.

Bindeman, I. N., and Valley, J. W., 2001. "Low 18O rhyolites from Yellowstone: magmatic evolution based on analyses of zircons and individual phenocrysts," *Journal of Petrology*, v. 42, p. 1491-1517.

Blay, C., and Siemers, R., 2004. *Kauai's Geologic History*, TEOK Investigations.

Bluth, G. J. S., Schnetzler, C. C., Krueger, A. J., and Walter, L. S., 1993. "The contribution of explosive volcanism to global atmospheric sulphur dioxide concentrations," *Nature*, v. 366, p. 327-330.

Bowen, N. L., 1928. *The Evolution of Igneous Rocks*, New York: Dover.

Bowring, S. A., 2014. "Closing the gap," *Nature Geoscience*, v. 7, p. 169-170.

Bowring, S. A., and Williams, I. S., 1999. "Priscoan (4.00 - 4.03 Ga) orthogneisses from northern Canada," *Contributions to Mineralogy and Petrology*, v. 134, p. 2-16.

Boyd, F. R., and Meyer, H. O. A., 1979. "Kimberlites, diatremes and diamonds," (I) "Their Geology, Petrology, and Geochemistry"; (II) "The Mantle sample: Inclusions in Kimberlites and Other Volcanics," *American Geophysical Union.*

Branney, M. J., and Kokelaan, B. P., 2002. *Pyroclastic density currents and the sedimentation of ignimbrites*, Geological Society Society of London, Memoir 27, p. 152.

Brantley, S. R. (Ed.), 1990. "The eruption of Redoubt Volcano, Alaska, December 14, 1989 - August 31, 1990," *U.S. Geological survey Circular 1031.*

Bryan, S. E., Riley, T. R., Jerram, D. A., Stephens, C. J., and Leat, P. T., 2002. "Silicic volcanism: an undervalued component of large igneous provinces and volcanic rifted margins," in Menzies, M. A., Klemperer, S. L., Ebinger, C. J., and Baker, J., eds., *Volcanic Rifted Margins,* Geological Society of America Special Paper 362, Boulder, Colorado, p. 97-118.

Bryan, T. S., 1990. *Geysers: What they are and how they work*, Roberts Rinehart.

Bryan, T. S., 2005. *The Geysers of Yellowstone*, Mountain Press, p. 64.

Buchanan, L. J., 1981. "Precious metal deposits associated with volcanic environments in the southwest," in Dickinson, W. R., and Payne, W. D., eds., *Relations of tectonics to ore deposits in the southern Cordillera*, Arizona Geological Society, Digest, v. 14, p. 237-262.

Bullard, F. M., 1962. *Volcanoes*, University of Texas Press.

Burke, K. C., and Wilson J. T., 1976. "Hot spots on the Earth's surface," *Scientific American*, Freeman, v. 236, San Francisco, p. 31-42.

Burlini, L., Vinciguerra, S., DiToro, G., DeNatale, G., Meredith, P., and Burg, J. P., 2007, "Seismicity preceding volcanic eruptions: new experimental insights," *Geology*, v. 35, p. 183-186.

Bussey, B., and Spudis, P. D., 2004. "Small spacecraft exploration of the moon," *Acta Astronautica*, v. 55, p. 637-641.

Carey, S., and Bursik, M., 2000. "Volcanic plumes," in Sigurdsson, H., ed., *Encylopedia of Volcanoes*, p. 527-544.

Carey, S., and Sigurdsson, H., 1986. "The 1982 eruptions of El Chichón volcano, Mexico (2): Observations and numerical modeling of tephra-fall distribution," *Bulletin of Volcanology*, v. 8, p. 127-141.

Carey, S. and Sigurdsson, H., 1989. "The intensity of plinian eruptions," *Bulletin of Volcanology*, v. 51, p. 28-40.

Carey S., Sigurdsson, H., Mandeville, C., and Bronto, S., 2000. "Volcanic hazards from pyroclastic flow discharge into the sea: examples from the 1883 eruption of Krakatau, Indonesia," in McCoy, F. W., and Heiken, G., eds., *Volcanic Hazards and Disasters in Human Antiquity*, Geological Society of America Special Paper 345, p. 1-14.

Carrigan, C. R., and Grubbins, D., 1979. "The source of the Earth's magnetic field," *Scientific American*, v. 240, #2, p. 118-130.

Cas, R. A. F., and Wright, J. V., 1987. *Volcanic Successions*, London: Allen and Unwin.

Casadevall, T. J. (ed), 1994. "Volcanic ash and aviation safety - proceedings of the first international symposium on volcanic ash and aviation safety," *U.S. Geological Survey Bulletin 2047*.

Cashman, R. V., and Hoblitt, R. P., 2004. "Magmatic precursors to the May 18, 1980 eruption of Mount St. Helens," *Geology*, v. 32, p. 141-144.

Chang, W., Smith, R. B., Wicks, C., Farrell, J. M., and Puskas, C. M., 2007. "Accelerated uplift and magmatic intrusion of the Yellowstone caldera, 2004-2006," *Science*, v. 318, p. 952-956.

Chapin, C. E., and Elston, W. E., eds., 1979. *Ash-Flow Tuffs*, The Geological Society of America Special Paper 180.

Chernicoff, S., Fox, H. A., and Tanner, L. H., 2002. *Earth Geologic Principles and History*, Boston: Houghton Mifflin.

Chesner, C. A., Rose, W. I., Deino, A., Drake, R., and Westgate, J. A., 1991. "Eruptive history of Earth's largest Quaternary calder (Toba, Indonesia) clarified," *Geology*, v. 19, p. 200-203.

Chouet, B. A., 2003. "Stalking nature's most dangerous beasts," *Science*, v. 299, p. 215-2030.

Christiansen, R. L., 1979. "Cooling units and composite sheets in relation to caldera structure," Chapin C. E. and Elston. W. E., eds., *Geological Society of America Special Paper 180*, p. 29-42.

Christiansen, R. L., 1984. "Yellowstone magmatic evolution: its bearing on understanding large-volume explosive volcanism," in *Explosive Volcanism: Inception, Evolution, and Hazards*, Washington, DC: National Academy of Sciences, p. 84-95.

Christiansen, R. L., Foulger, G. R., and Evans, J. R., 2002. "Upper-mantle origin of the Yellowstone hotspot," *Geological Society of America Bulletin*, v. 114, p. 1245-1256.

Christiansen, R. L., and Peterson, D. W., 1981. "Chronology of the 1980 activity," in Lipman, P. W., and Mullineaux, D. R., eds., *The 1980 eruptions of Mount St. Helens*, Washington, DC: U.S. Geological Survey Professional Paper 1250, p. 17-30.

Clague, D. A. and Dalrymple, G. B., 1987. "The Hawaiian-Emperor volcanic chain, Part I, Geologic evolution," in Decker, R. W., Wright, T. I., and Stauffer, P. H., eds., *Volcanism in Hawaii*, U.S. Geological survey Professional Paper 1350, v. 1, p. 5-54.

Clague, D. A., and Dalrymple, G. B., 1989. "Tectonics, geochronology and origin of the Hawaiian-Emperor volcanic chain: in Winterer," E. L., et al., eds., *The Geology of North America, v. 1, The eastern Pacific Ocean and Hawaii*, The Geological Society of America, Boulder, CO, p. 188-217.

Clouard, V., and Bonneville, A., 2001. "How many Pacific hotspots are fed by deep-mantle plumes?" *Geology*, v. 29, p. 695-698.

Clouard, V., and Bonneville, A., 2005. "Ages of sea mounts, islands and figaus on the Pacific plate," in Foulger, G. R., et al., eds., *Plates, plumes and paradigms*, Geological Society of America Special Paper 388. p. 71-90.

Coffin, M. F., and Eldholm, O., 1993. "Large igneous provinces:" *Scientific American*, v. 268, no. 10, p. 42-49.

Coffin, M. F., and Eldholm, O., 1994. "Large igneous provinces: crystal structure, dimensions and external consequences," *Reviews in Geophysics*, v. 32, p. 1-36.

Corliss, J. B., Dymond, J., Gordon, I. I., Edmond, J. M., von Herzen, R. P., Ballard, R. D., Green, K., Williams, D., Bainbridge, A., Crane, K., and van Andel, T. J. H., 1979. "Submarine thermal springs on the Galapagos rift," *Science*, v. 203, p. 1073-1082.

Courtillot, V., Davaille, A., Besse, J., and Stock, J., 2003. "Three distinct types of hotspots in the

Earth's mantle," *Earth and Planetary Science Letters*, v. 205, p. 295-308.

Courtillot, V., Jaupart, C., Manighetti, P., and Besse, J., 1999. "On causal links between flood basalts and con-tinental breakup," *Earth and Planetary Science Letters*, v. 166, p. 177-195.

Crandell, D. R., 1989. "Gigantic debris avalanche of Pleistocene age from ancestral Mount Shasta volcano, California, and debris-avalanche hazard zonation," *U.S. Geological Survey Bulletin 1861*.

Crandell, D. R., Miller, C. D., Glicken, H. X., Christiansen, R. L., and Newhall, C. G., 1984. "Catastrophic debris avalanche from ancestral Mount Shasta volcano, California," *Geology*, v. 12, p. 143-146.

Crandell, D. R., and Mullineaux, D. R., 1978. "Potential hazards from future eruptions of Mt. St. Helens volcano, Washington," *U.S. Geological Survey Bulletin*, 1383-C.

Crandell, D. R., and Nichols, D. R., 1989. *Volcanic Hazards at Mt. Shasta*, U.S. Geological Survey unnumbered Series General Information.

Crisp, J. A., 2004. "Volcanoes on Mars: a view from the surface," in Lopes and Gregg, eds., *Volcanic worlds: Exploring the Solar System's Volcanoes*, p. 111-126.

Critias in Plato the Collected Dialoques including the Letters, 1961. Hamilton, E., and Cairns, H., eds., Bollinggen Series LXXI, Princeton University Press, p. 1743.

Crumpler, L. S., and Aubele, J. C., 2000. "Volcanism on Venus," in Sigurdsson, H., ed., *Encyclopedia of Volcanoes*, p. 727-769.

Dartevelle, S., Ernst, G. G. J., Stix, J., and Bernard, A., 2002. "Origin of the Mount Pinatubo climatic eruption cloud: Implications for volcanic hazards and atmospheric impacts," *Geology*, v. 30, p. 663-666.

Decker, R., and Decker, B., 1981. "The eruption of Mount St. Helens," *Scientific American*, v. 244, p. 68-80.

Decker R. and Decker, B., 2006. *Volcanoes*, 4th ed. New York: Freeman.

Delmelle, P., and Stix, J., 2000. "Volcanic gases," in Sigurdsson, H., ed., *Encyclopedia of Volcanoes*, p. 803-815.

de Rita, D., and Giampaolo, C., 2006. "A case study – Ancient Rome was built with volcanic stone from the Roman land," in Heiken, G., ed., *Tuffs – Their Uses, Hydrology and Resources*, Geological Society of America Special Paper 408, p. 127-131.

Dragovich, J. D., Pringle, P. T., and Walsh, T. J., 1994. "Extent and geometery of the Mid-Holocene Osceola mudflow in the Puget Lowland - implications for Holocene sedimentation and paleography," *Washington Geology*, v. 22, no. 3, p. 3-26.

Drake, N., 2011. "Venus unveiled," *Science News*, v. 180, no. 12, p. 26-27.

Druitt, T. H., Edwards, L., Mellors, R. M., Pyle, D. M., Sparks, R. S. J., Lanphere, M., Davies, M., and Barriero, B., 1999. *Santorini Volcano*," London: Geological Society Memoir No. 19.

Druitt, T. H., and Francaviglia, V., 1992. "Caldera formation on Santorini and physciography of the islands in the late Bronze Age," *Bulletin of Volcanology*, v. 54, p. 484-493.

Dvorak, J., and Gasparini, P., 1990. "Historical activity at Camp Flegrei caldera, southern Italy," *Earthquakes and Volcanoes*, v. 22, no. 6, p. 256-267.

Eldridge, C. S., Barton, P. B., Jr., and Ohmoto, H., 1983, "Mineral textures and their bearing on formation of the Kuroko orebodies," in, Ohmoto, H. and Skinner, B.J., eds., *The Kuroko Volcanogenic Massive Sulfide Deposits, Economic Geology Monograph 5*, Economic Geology Pub. Company, p. 241-281.

Endo, E. T., Malone, S. D., Noson, L. L. and Weaver, C. S., 1981. "Locations, magnitudes, and statistics of the March 20-May 18 earthquake sequence," in Lipman, P. W., and Mullineaux, D. R., eds., *The 1980 eruptions of Mount St. Helens, Washington*, U.S. Geological Survey Professional Paper 1250, p. 93-107.

Ernst, R. E., and Buchan, K. L., eds, 2001. *Mantle Plumes: Their Identification Through Time*, Geological Society of America Special Paper 352.

Fisher, R. V., Heiken, G., and Hulen, J. B., 1997. *Volcanoes Crucibles of Change*, Princeton, NJ: Princeton University Press.

Fisher, R. V., Smith, A. L., and Roobob, M. J., 1980. "Destruction of St. Pierre, Martinique, by ash-cloud surges, May 8 and 20, 1902," *Geology*, v. 8, p. 472-476.

Fisher, R. V., and Schmincke, H. U., 1984. *Pyroclastic Rocks*, Berlin, New York: Springer Verlag.

Fiske, R. S., Hopson, C. A., and Waters, A. C., 1963. *Geology of Mount Rainier National Park Washington*, U.S. Geological survey Professional Paper 444.

Forget, F. Spiga, A., Dolla, B., Vinalie, S., Melchiorri, R., Drossant, P., and Gendrin, A., 2007. "Remote sensing of surface processes," *Journal of Geophysical Research*, v. 112.

Foulger, G. R., Natland, J. H., Presnall, D. C., and Anderson, D. L., eds., 2005. *Plates, Plumes and Paradigms*, Geological Society of America Special Paper 388.

Francis, P., 1983. "Giant volcanic calderas," *Scientific American*, v. 248, p. 60-70.

Francis, P., and Oppenheimer, C., 2004. *Volcanoes*, Oxford University Press, 2nd edition.

Francis, P., and Rothery, D., 2000. "Remote Sensing of active volcanoes," *Annual Reviews of Earth and Planetary Sciences*, v. 28, p. 81-106.

Francis, P., and Self, S., 1983. "The eruption of Krakatau," *Scientific American*, v. 249, no. 5, p. 172-187.

Franklin, B., 1784. "Meterological impingnations and conjectures," *Memoirs of the Literary and Philosophical Society of Manchester*, v. 2, p. 373-377.

Franklin, J. M., 1996. "Volcanic-associated massive sulphide base metals," in *Geology of Canadian Mineral Deposit Types*, eds Eckstrand, O. R., Sinclair, W. D., and Thorpe, R. I., Geological Survey of Canada, no. 8, p. 158-183.

Freundt, A., Kutterolf, S., Schmincke, H. U., Hansteen, T., Wehrmann, H., Perez, W., Strauch, W., and Navarro, M., 2006. *Volcanic hazards in Nicaragua: past, present and future*, Geological Society of America Special Paper 412, p. 141-165.

Funiciello, R., Heiken, G., Levich, R., Obenholzner, J., and Petrov, V., 2006. "Construction in regions with tuff deposits," in Heiken, G., ed. *Tuffs – Their Uses, Hydrology and Resources*, Geological Society of America Special Paper 408, p. 119-126.

Gaddis, L., 2004. "The face of the Moon: lunar volcanoes and volcanic deposits, in Lopes and Gregg," eds., *Volcanic Worlds*, p. 81-96.

Gaillard, J. C., and Dibben, C. J. L., 2007. "Volcanic risk perception and beyond," *Journal of Volcanology and Geothermal Research*, v. 172, p. 163-169.

Garcia, M., 2002. "Giant landslides northeast of O'ahu: when, why and how?" in Takahashi, E., and four others, *Hawaiian Volcanoes deep underwater perspectives*, Geophysical Monograph 128, p. 221-229.

"Geological Survey of Japan," 1996, 1999. *Geothermal Research Notes*.

Geissler, P., 2000. "Cryovolcanism in the outer Solar System," in Sigurdsson, H., ed., *Encyclopedia of Volcanoes*, p. 785-800.

Geist, D., and Richards, M., 1993. "Origin of the Columbia plateau and Snake River plain: deflection of the Yellowstone plume," *Geology*, v. 21, p. 789-792.

Gilluly, J., Waters, A. C., and Woodford, A. O., 1968. *Principles of Geology*, 3rd edition, San Francisco: Freeman and Company.

Gladman, B. J., Burns, J. A., Duncan, M., Lee, P., and Levison, H. F., 1996. "The exchange of impact ejecta between terrestrial planets," *Science*, v. 271, p. 1387-1392.

Glicken, H., 1990. "The rockslide-debris avalanche of the May 18, 1980 eruption of Mount St. Helens - 10th anniversary perspectives," *Geoscience Canada*, v. 17, p. 150-153.

Goff, F., and Gardner, J. N., 1994. "Evolution of a mineralized geothermal system, Valles caldera, New Mexico," *Economic Geology*, v. 89, p. 1803-1832.

Goff, F., and Janik, C. J., 2000. "Geothermal systems," in Sigurdsson, H., ed., *Encyclopedia of Volcanoes*, p. 817-834.

Graham, C. M., Wilde, S. A., Valley, J. W., and Peck, W. H., 2001. "Evidence from detrital zircons for the existence of continental crust and oceans on the Earth 4.4 Gyr ago," *Nature*, v. 409, p. 175-178.

Grieve, R. A. F., 1990. "Impact cratering on Earth," *Scientific American*, v. 262, p. 66-73.

Grove, T. L., 2000. "Origin of Magmas," in Sigurdsson, H., ed., *Encyclopedia of Volcanoes*, p. 133-147.

Guilbert, J. M., and Park, C. F. Jr., 1985. *The Geology of Ore Deposits*, New York: Freeman.

Gudmundsson, A., 2000. "Dynamics of volcanic systems in Iceland: example of tectonism and volcanism at juxtaposed hot spot and mid-ocean ridge systems," *Annual Reviews of Earth and Planetary Science*, v. 28, p. 107-140.

Gudmundsson, A. et al., (12 coauthors), 2012. Ash generation and distribution from the April–May 2010 eruptions of Eyjafjallajokull, Iceland, *Nature Scientific Report 2,* Article #572.

Guilbert, J. M., and Park, C. F., Jr., 1985. *The Geology of Ore Deposits,* Freeman, New York, p. 985.

Halmer, M. M., Schmincke, H. U., and Graf, H. F., 2002. "The annual volcanic gas input into the atmosphere, in particular into the stratosphere: a global data set for the past 100 years," *Journal of Volcanology and Geothermal Research*, v. 115, p. 511-528.

Halter, W. E., Bain, N., Becker, K., Heinrich, C. A., Landtwing, M., VonQuadt, A., Clark, A. H., Sasso, A. M., Bissig, T., and Tosdal, R. M., 2004. "From andesitic volcanism to the formation of a porphyry Cu-Au mineralizing magma chamber: the Farallon Negro Volcanic Complex, northwestern Argentina," *Journal of Volcanology and Geothermal Research*, v. 136, p. 1-30.

Hammond, P. E., 1998. "Tertiary andesitic lava-flow complexes (stratovolcanoes) in the southern Cascade Range of Washington – observations on tectonic processes within the Cascade arc," *Washington Geology*, v. 26, no. 1, p. 20- 28.

Hamblin, W. K., and Christiansen, E. H., 2004. *Earth's Dynamic Systems*, Prentice-Hall, 10th ed.

Harris, S. L., 1988. *Fire Mountains of the West*, Missoula, MT: Mountain Press.

Harris, S. L., 1990. *Agents of Chaos*, Missoula, MT: Mountain Press.

Heiken, G., 1979. "Pyroclastic flow deposits," *American Scientist*, v. 67, p. 564-571.

Heiken, G., 2006. "Introduction," in Heiken, G., ed., *Tuffs - their uses, hydrology and resources*, Geological Society of America Special Paper 408, p. 3-4.

Hildreth, W., 2004. "Volcanological perspectives on Long Valley, Mammoth Mountain, and Mono Craters: several contiguous but discrete systems," *Journal of Volcanology and Geothermal Research*, v. 136, p. 169-198.

Hildreth, W., Fierstein, J., and Lanphere, M., 2003. "Eruptive history and geochronology of the Mount Baker volcanic field, Washington," *Geological Society of America Bulletin*, v. 115, p. 729-764.

Hildreth, W., and Mahood, G. A., 1986. "Ring-fracture eruption of the Bishop tuff," *Geological Society of America Bulletin*, v. 97, p. 396-403.

Hill, D. P., and Prejean, S., 2005. "Magmatic unrest beneath Mammoth Mountain, California," *Journal of Volcanology and Geothermal Research*, v. 146, p. 257-283.

Hon, K., Gansecki, C., and Kauahikaua, J., 2003. "The transition from A'a to Pahoehoe crust on flows emplaced during the Pu'u O'o-Kupaianaha eruption," in Heliker, C., Swanson, D. A., and Takahashi, T. J., eds., *The Pu'u O'o-Kupaianaha eruption of Kilauea volcano, Hawai'i: the first 20 years*, U.S. Geological survey Professional Paper 1676, p. 89-103.

Hooper, P. R., 1997. "The Columbia River flood basalt province: current status," in Mahoney, J. J. and Coffin, M. F., eds., *Large Igneous Provinces: Continental, Oceanic and Planetary Flood Volcanism, American*, Geophysical Union Monograph 100, p. 1-27.

Hooper, P. R., 2000. "Flood basalt provinces," in Sigurdsson, H., ed.: *Encyclopedia of Volcanoes*, p. 345-359.

Hopson, C. A., and Melson, W. G., 1990. "Compositional trends and eruptive cycles at Mount St. Helens," *Geoscience Canada*, v. 17, p. 131-141.

Huffman, A. R., 1990. "An endogenous mechanism for extinction," *Geotimes*, v. 35, no. 8, p. 16-17.

Hughes, J. M., Stoiber, R. E., and Carr, M. J., 1980. "Segmentation of the Cascade volcanic chain," *Geology*, v. 8, p. 15-17.

Humphreys, E. D., Dueken, K. G., Schutt, D. L., and Smith, R. B., 2000. "Beneath Yellowstone: Evaluating plume and non plume models using telesismic data," *GSA Today*, v. 10, p. 1-7.

Humphreys, W. J., 1913. "Volcanic dust and other factors in the production of climatic changes, and their possible relation to ice ages," *Bulletin Metrological Weather Observations*; v. 6, p. 1-34.

Hyndman, D. W., 1985. *Petrology of Igneous and Metamorphic Rocks*, 2nd ed. McGraw-Hill.

Iverson, R. M., Schilling, S. P., and Vallance, J. W., 1998. "Objective delineation of lahar-inundation hazard zones," *Geological Society of America Bulletin*, v. 110, p. 972-984.

Jeanloz, R., 2000. "Mantle of the Earth," Sigurdsson, H., ed.: *Encyclopedia of Volcanoes*, p. 41-54.

Jerram, D. A., 2002. *Volcanology and facies architecture of flood basalts*, Geological Society of America Special Paper 362, p. 119-132.

Kelley, D. S., Baross, J. A., and Delaney, J. R., 2002. "Volcanoes, fluids, and life at mid-ocean ridge spreading centers," *Annual Reviews of Earth and Planetary Sciences*, v. 30, p. 385-491.

Kennett, J. P., and Thunell, R. C., 1975. "Global increase in Quaternary explosive volcanism," *Science*, v. 187, p. 497-503.

Kerr, R. A., 2001. "Oregon's rising, an eruption to follow?" *Science*, v. 292, p. 12-13.

Kerr, R. A., 2002. "Of ocean weather and volcanoes," *Science*, v. 295, p. 260-261.

Kerr, R. A., 2003. "High-tech fingers on Earth's erratic pulse," *Science*, v. 299, p. 2016-2020.

Kieffer, S. W., 1981. "Fluid dynamics of the May 18 blast at Mount St. Helens, in St Helens," in Lipman, P. W., and Mullineaux, D. R., eds., *The 1980 eruptions of Mount St. Helens, Washington*, U.S. Geological survey Professional Paper 1250, p. 379-400.

Kirkham, R. V., and Sinclair, W. D., 1996. "Porphyry copper, gold, molybdenum, tungsten, tin and silver deposits," in Eckstrand, O. R., Sinclair, W. D., and Thorpe, R. I., eds., *Geology of Canadian mineral deposit types*, Geological Survey of Canada, Geology of Canada no. 8, p. 421-446.

Klein, E. M., and Langmuir, C. H., 1989. "Local versus global variations in ocean ridge basalt compositions," *A reply: Journal of Geophysical Research Atmospheres*, v. 94, p. 4241-4252.

Koppers, A. A. P., 2011. "Mantle plumes perserve," *Nature Geoscience*, v. 4, p. 816-817.

Lacroix, A., 1904. *La Montagne Pelee et ses eruptions*, Paris: Masson et Cie.

Ladbury, R., 1996. "Model sheds light on a tragedy and a new type of eruption," *Physics Today*, v. 219, no. 5, p. 20-22.

LaMarche, V. C., Jr., and Hirschboeck, K. K., 1984. "Frost rings in trees as records of major eruptions," *Nature*, v. 307, p. 121-126.

Lamb, H. H., 1970. "Volcanic dust in the atmosphere; with a chronology and assessment of its meteorological significance," *Philosophical Transactions Royal Society of London*, V. A266, p. 495-533.

Lambert, I. B., and Sato, T., 1974. "The Kuroko and asso-ciated ore deposits of Japan: a review of their features and metallogenesis," *Economic Geology*, v. 69, p. 1215-1236.

Lange, I. M., 2002. *Ice Age Mammals of North America – a Guide to the Big, the Hairy and the Bizarre*, Missoula, MT: Mountain Press.

Lange, I. M., and Avent, J. C.,1973. "Ground-based thermal infrared surveys as an aid in predicting volcanic eruptions in the Cascade Range," *Science*, v. 183, p. 279-281.

Latter, J. H., 1981. "Tsunamis of Volcanic origin: Summary of causes, with particular reference to Krakatoa, 1983": *Bulletin of Volcanology*, v. 44, no. 3, p. 467-490.

Lawver, L. A., and Muller, R. D., 1994. "Iceland hotspot track," *Geology*, v. 22, p. 311-314.

Lay, T., 1994. "The fate of descending slabs," *Annual Review in Earth and Planetary Sciences*, v. 22, p. 33-61.

LeCloarec, M. F., and Pennisi, M., 2001. "Radionuclides and sulfur content in Mount Etna plume in 1983-1995: new constraints on magma feeding system," *Journal of Volcanology and Geothermal Research,* v. 108, p. 141-155.

Lee, M. Y., Chen, C. -H., Wei, K. -Y., Lizuka, Y., and Carey, S., 2004. "First Toba super eruption revival," *Geology*, v. 32, p. 61-64.

Lindgren, W., 1933. *Mineral Deposits*, 4th ed., New York: McGraw-Hill.

Lipman, P. W., 1975. *Evolution of the Platoro caldera complex and related volcanic rocks, southeastern San Juan Mountains, Colorado*, U.S. Geological Survey Professional Paper 852, 128 p.

Lipman, P. W., 1997. "Subsidence of ash-flow calderas: relation to caldera size and magma chamber geometry," *Bulletin of Volcanology*, v. 59, p. 198-218.

Lipman, P. W., 2000. "Calderas," in Sigurdsson, H., ed., *Encyclopedia of Volcanoes*, p. 643-662.

Lockridge, P. A., 1989. "Volcanoes and tsunamis," *Earth Science*, p. 24-25.

Lopes, R., 2004. "Io, a world of great volcanoes," in Lopes, M. C., and Gregg, K. P., eds. *Volcanic Worlds*, Chichester, U.K.: Praxis Publishing.

Lopes, R. M. C., and Gregg, T. K. P., eds., 2004. *Volcanic Worlds*, Heidelberg: Springer-Verlag.

Lopes-Gautier, R., 2000. "Volcanism on Io," in Sigurdsson, H., ed.: *Encyclopedia of Volcanoes*, p. 709-726.

Luhr, J. F., Carmichael, I. S. E., and Varekamp, J. C., 1984. "The eruptions of El Chichón Volcano, Chiapas, Mexico: mineralogy and petrology of the anhydrite-bearing pumices," *Journal of Volcanology and Geothermal Research*, v. 23, p. 69-108.

Lutgens, F. K., and Tarbuck, E. J., 2006. *Essentials of Geology*, 9th ed., Upper Saddle River, NJ: Pearson Prentice Hall.

MacDonald, K. C., and Fox, P., 1990. "The mid-ocean ridge," *Scientific American*, v. 262, no. 6, p. 72-79.

MacDougall, J. D.,1988. *Continental Flood Basalts and MORB: A Brief Discussion of Similarities and Differences in Their Petrogensis, in Continental Flood Basalts*, Kluwer Academic Press, p. 331-341.

Malfait, W. J., Seifert, R., Petitgirard, S., Perrillat, J. P., Mezouar, M., Ota, T., Nakamura, E., Lerch, P., and Sanchez-Valle, C., 2014. "Supervolcano eruptions driven by melt buoyancy in large silicic magma chambers," *Nature Geoscience*, v. 7, p. 122-125.

Malone, S. D., Endo, E. T., Weaver, C. S., and Ramey, J. W., 1981. *Seismic monitoring for eruption prediction in 1980 eruption of Mt. St. Helens*, in Lipman, P. W., and Mullineau, eds., U.S. Geological Survey Professional Paper 1250.

Malone, S. D., 1990. "Mount St. Helens, the 1980 re-awakening and continuing seismic activity," *Geoscience Canada*, v. 17, p. 146-150.

Mandeville, C. W., Carey, S., and Sigurdsson, H., 1996. "Magma mixing, fractional crystallization and volatile degassing during the 1883 eruption of Krakatau volcano, Indonesia," *Journal of Volcanology and Geothermal Research*, v. 74, p. 243-274.

Marsh, B. D., 1979. "Island-arc volcanism," *American Scientist*, v. 67, no. 2, p. 161-173.

Marshak, S., 2004. *Essentials of Geology*, New York: Norton & Company.

Matek, B., 2014. *Annual U.S. and Global Geothermal Power Production Report.*

McBirney, A. R., 1968. "Petrochemistry of Cascade andesite volcanoes, in Andesite Conference Guidebook," *Oregon Department of Geological and Mineralogical Industries Bulletin*, no. 62, p. 101-107.

McCormick, M. P., Thomason, L. W., and Trepte, C. R., 1995. "Atmospheric effects of Mt. Pinatubo eruption," *Nature*, v. 373, p. 399-404.

McCoy, F. W. and Heiken, G., 2000. *The Late-Bronze age explosive eruption of Thera (Santorini), Greece: regional and local effects*, Geological Society of America Special Paper 345, p. 43-70.

McGee, K. A., Doukas, M. P., Kessler, R., and Gerlach, T. M., 1997. *Impacts of volcanic gases on climate, the environment, and people*, U.S. Geological Survey Open-File Report 97-262.

McNutt, S. R.,1986. "Observations and analysis of b-type earthquakes, explosions, and volcanic tremor at Pavlof volcano, Alaska," *Bulletin of the Seismological Society of America*, v. 76, p. 153-175.

McNutt, S. R., 2000. "Seismic monitoring," in Sigurdsson, H., ed.: *Encyclopedia of Volcanoes*, p. 1095-1119.

McNutt, S. R., Rymer, H., and Stix, J., 2000. "Synthesis of volcano monitoring," in Sigurdsson, H., ed.: *Encylopedia of Volcanoes*, p. 1165-1183.

McSween, H. Y., Jr., 1994. "What we have learned about Mars from SNC meteorites," *Meteoritics*, v. 29, p. 757-779.

Megaw, P. K., 2006. "Exploration of low sulfidation epithermal vein systems," Mag Silver Corporation, www.magsilver.com/i/pdf/Epithermal_Vein_Story-Jan232006.pdf.

Miller, C. D., 1980. *Potential hazards from future eruptions in the vicinity of Mount Shasta Volcano, northern California*, U.S. Geological Survey Bulletin 1503.

Miller, C. D., 1985. "Holocene eruptions at the Inyo volcanic chain, California: Implications for possible eruptions in Long Valley caldera," *Geology*, v. 13, p. 14-17.

Miller, C. D., 1989. *Potential hazards from future volcanic eruptions in California*, U.S. Geological Survey Bulletin 1847.

Miller, T. P., and Casadevall, T. J., 2000. "Volcanic ash hazards to aviation," in, Sigurdsson, H., ed., *Encyclopedia of Volcanoes*, p. 915-930.

Miller, C. D., Mullineaux, D. R., Crandell, D. R., and Bailey, R. A., 1982. *Potential hazards from future volcanic eruptions in the Long Valley-Mono Lake area, east-central California and southwest Nevada-a preliminary assessment*, U.S. Geological Survey Circular 877.

Mills, M. J., 2000. "Volcanic aerosol and global atmospheric effects," in Sigurdsson, H., ed.: *Encylopedia of Volcanoes*, p. 931-943.

Molnar, P., and Tapponnier, P., 1980. "The collision between India and Eurasia," in Bolt, B. A., ed., *Earthquakes and Volcanoes, Readings from Scientific American*, Freeman, p. 62-73.

Moore, J. G., Bryan, W. B., Beeson, M. H., and Normark, W. R., 1995. "Giant blocks in the South Kona landslide, Hawaii," *Geology*, v. 23, p. 125-128.

Moore, J. G., Clague, D. A., Holcomb, R. T., Lipman, P. W., Normark, W. E., and Torresan, M. E., 1989. "Prodigious submarine landslides on the Hawaiian Ridge," *Journal of Geophysical Research*, v. 94, p. 17, 465-17, 484.

Moore, J. G., and Clague, D. A., 1992. "Volcano growth and evolution of the island of Hawaii," *Geological Society of America Bulletin*, v. 104, p. 1471-1484.

Moore, J. G., and Normark, W. R., 1994. "Giant Hawaiian landslides," *Annual Reviews of Earth and Planetary Sciences*, v. 22, p. 119-144.

Moos, D., and Zoback, M. D., 1993. "State of stress in the Long Valley caldera, California," *Geology*, v. 21, p. 837-840.

Morgan, W. J., 1971. "Convective plumes in the lower mantle," *Nature*, v. 230, p. 42-43.

Morgan, W. J., 1972. "Deep mantle convection plumes and plate motions," *American Association of Petroleum Geology Bulletin*, v. 56, p. 203-213.

Mouginis-Mark, P. J., and Robinson, M. S., 1992. "Evolution of the Olympus Mons Caldera, Mars," *Bulletin of Volcanology*, v. 54, p. 347-360.

Mullineaux, D. R., 1974. *Pumice and other pyroclastic deposits in Mount Rainier Natioal Park, Washington*, U.S. Geological Survey Bulletin 1326.

Mullineaux, D. R., and Crandell, D. R., 1981. "The eruptive history of Mount St. Helens," in Lipman, P. W. and Mullineaux, D. R., eds., *The 1980 Eruptions of Mount St. Helens, Washington*, U.S. Geological Survey Professional Paper 1250, p. 3-15.

Murchie, S. L., Vervack, R. J., Jr., Anderson, B. J., 2011. "Journey to the Innermost Planet," *Scientific American*, v. 304, no. 3, p. 34-39.

Murphy, J. B., Oppliger, G. L., Brimhall, G. H., Jr., and Hynes, A., 1999. "Mantle plumes and mountains," *American Scientist*, v. 87, p. 146-153.

Murray, J. B., Rymer, H., and Locke, C. A., 2000. "Ground deformation, gravity, and magnetics," in Sigurdsson, H., ed., *Encyclopedia of Volcanoes*, p. 1121-1140.

NASA, 1996. *Meteorites from Mars!*, Earth Science and Solar System Exploration Division, NASA Lyndon B. Johnson Space Center.

Neal, C. R., and Taylor, L. A., 1992. "Petrogenesis of mare basalts: a record of lunar volcanism," *Geochimica et Cosmochimica Acta*, v. 56, p. 2177-2211.

Neukum, G. and 11 other authors, 2004. "Recent and episodic volcanic and glacial activity on Mars revealed by the High Resolution Stereo Camera." *Nature*, v. 432, p. 971-979.

Newhall, C. G., and Dzurisin, D., 1988. *Historical unrest at large calderas of the world*, U.S. Geological Survey Bulletin 1855, v. 1 and v. 2.

Newhall, C. G., and Self, S., 1982. "The Volcanic Explosivity Index (VEI): an estimate of explosive magnitude for historical volcanism," *Journal of Geophysical Research*, v. 87, p. 1231-1238.

Normark, W. R., Morton, J. L., Koski, R. A., and Clague, D. A., 1983. "Active hydrothermal vents and sulfide deposits on the southern Juan de Fuca Ridge," *Geology*, v. 11, p. 158-163.

Nystrom, J. O., and Henriquez, F., 1994. "Magmatic features of iron ores of the Kiruna type in Chile and Sweden: ore textures and magmatic chemistry," *Economic Geology*, v. 89, p. 820-839.

O'Connor, J. M., Jokat, W., le Roex, A. P., Class, C., Wijbrans, J. R., Kebling, S., Kulper, K. F., and Nebel, O., 2012. "Hotspot trails in the South Atlantic controlled by plume and plate tectonic processes," *Nature Geoscience*, v. 5, p. 735-738.

Ohmoto, H., and Skinner, B. J., (eds.), 1983. *The Kuroko and Related Volcanogenic Massive Sulfide deposits*, Economic Geology Monograph 5.

Pallister, J. S., Hoblitt, R. P., Crandell, D. R., and Mullineaux, D. R., 1992. "Mount St. Helens a decade after the 1980 eruptions: magmatic models, chemical cycles, and a revised hazards assessment," *Bulletin of Volcanology*, v. 54, p. 126-146.

Panteleyev, A., 1986. Ore deposits #10. "A Canadian Cordilleran model for epithermal gold-silver deposits," *Geoscience Canada*, v. 13, no. 2, p. 101-111.

Patrick, R. R., and Howe, R. C., 1994. "Volcanism on terrestrial planets," *Journal of Geological Education*, v. 42, p. 225-238.

Perez, W., and Freundt, A., 2006. "The youngest explosive basaltic eruptions from Masaya caldera (Nicaragua): Stratigraphy and hazard assessment," in Rose, W. I. et al. eds., *Volcanic Hazards in Central America*, Geological Society of America Special Paper 412, p. 189-207.

Perfit, M. R., and Davidson, J. P., 2000. "Plate Tectonics and volcanism," in Sigurdsson, H., ed., *Encyclopedia of Volcanoes*, p. 89-113.

Perkins, M. E., and Nash, B. P., 2002. "Explosive silicic volcanism of the Yellowstone hotspot: the ash fall tuff record," *Geological Society of America Bulletin*, v. 114, p. 367-381.

Peterson, D. W., and Tilling, R. I., 1980. "Transition of basaltic lava from pahoehoe to aa, Kilauea, Hawaii: Field observations and key factors," *Journal of Volcanology and Geothermal Research*, v. 5, p. 271-193.

Peterson, U., Noble, D. C., Arenas, M. J., and Goodel, P. G., 1977. "Geology of the Julcani mining district, Peru," *Economic Geology*, v. 72, p. 931-949.

Pichler, H., and Friedrich, W. L., 1980. "Mechanism of the Minoan eruption of Santorini," in Doumas, C., ed., *Thera and the Aegean World*, v. 2, p. 15-30.

Pinatubo Volcano Observatory Team, 1991. "Lessons from a major eruption: Mt. Pinatubo, Philippines," *Transactions of the American Geophysical Union (EOS)*, v. 72, no. 49, p. 545-551.

Ping, C. L., 2000. "Volcanic soils," in Sigurdsson, H., ed., *Encyclopedia of Volcanoes*, p. 1259-1270.

Pitt, A. M., 1989. Map showing earthquake epicenters (1964-1981) in Yellowstone National Park and vicinity, Wyoming, Idaho and Montana: U.S. Geological Survey Miscellaneous Field Studies Map MF-2022.

Poli, S., and Schmidt, M. W., 2002. "Petrology of subducted slabs," *Annual Reviews of Earth and Planetary Sciences*, v. 30, p. 207-235.

Power, J. A. et al., 1994. "Seismic evolution of the 1989–1990 eruption sequence of Redoubt

Volcano, Alaska," *Journal of Volcanology and Geothermal Research,* v. 62, p. 69-94.

Prockter, L., 2004. "Ice volcanism on Jupiter's moons and beyond," in eds. Lopes and Gregg, *Volcanic Worlds*, p. 145-177.

Pyle, D. M., 2000. "Sizes of volcanic eruptions," in Sigurdsson, H., ed., *Encyclopedia of Volcanoes*, p. 263-269.

Rampino, M. R., and Ambrose, S. H., 2000. "Volcanic winter in the Garden of Eden: the Toba super eruption and the Late Pleistocene human population crash," in McCoy, F. W., and Heiken, G., eds., *Volcanic hazards and disasters in human antiquity*, Geological Society of America Special Paper 345, p. 71-82.

Rampino, M. R., and Self, S., 1982. "Historic eruptions of Tambora (1815), Krakatau (1883), and Agung (1963), their stratospheric aerosols, and climatic impact," *Quaternary Research*, v. 18, p. 127-143.

Rampino, M. R., and Self, S., 2000. "Volcanism and biotic extinctions," in Sigurdsson, H., ed., *Encyclopedia of Volcanoes*, p. 1083-1091.

Rampino, M. R., and Self, S., and Stothers, R. B., 1988. "Volcanic winters," *Annual Reviews of Earth and Planetary Science*, v. 16, p. 73-99.

Rampino, M. R., and Stothers, R. B., 1988. "Flood basalt volcanism during the past 250 million years," *Science*, v. 241, p. 663-668.

Raymond, L. A., 2002. *Petrology: the study of igneous, sedimentary and metamorphic rocks*, McGraw Hill, 2nd ed.

Reid, M. E., 2004. "Massive collapse of volcano edifices triggered by hydrothermal pressurization," *Geology*, v. 32, p. 373-376.

Reidel, S. P., and Tolan, T. L., 1992. "Eruption and emplacement of flood basalt: an example from the large-volume Teepee Butte Member, Columbia River Basalt Group," *Geological Society of America Bulletin*, v. 104, p. 1650-1671.

Reidel, S. P., Tolan, T. L., and Beeson, M. H., 1994. "Factors that influenced the eruptive and emplacement histories of flood basalt flows: a field guide to selected vents and flows of the Columbia River Basalt Group," in Swanson, D. A., and Haugerud, R. A., eds., *Geologic Field Trips in the Pacific Northwest*, 1994 Geological Society of America annual Meeting, Seattle, WA, p. 1B-1-18.

Robock, A., 2000. "Volcanic eruptions and climate: Reviews in Geophysics," *American Geophysical Union*, v. 38, p. 191-219.

Rogers, N., and Hawkesworth, C., 2000. "Composition of magmas," in Sigurdsson, H., ed., *Encyclopedia of Volcanoes*, p 115-131.

Rose, W. I., and Chesner, C. A., 1987. "Dispersal of ash in the great Toba eruption, 75 ka," *Geology*, v. 15, p. 913-917.

Ross, C. S., and Smith, R. I., 1961. "Ash-flow tuffs: their origin, geologic relations and identification," *U.S. Geological Survey Professional Paper 366.*

Rutherford, M. J., and Gardiner, J. E., 2000. Rates of magma ascent," in Sigurdsson, H., ed., *Encyclopedia of Volcanoes*, p. 207-217.

Rymer, H., and Williams-Jones, G., 2003. *Introduction: Journal of Volcanology and Geothermal Research*, v. 123, p. vii-viii.

Sanderson, K., 2009. "Volcanoes stirred by climate change," *Nature*, www.nature.com/news/2009.

Santar, B. D., et al. (and 12 coauthors), 2014. "Volcanic contribution to decadal changes in troposphic temperature," *Nature Geoscience Letters*, v. 7, p. 1-10.

Sarasin, P., and Sarasin, F., 1901. *Verhandlungen der Naturforschenden Gesellschaft, in Basel*, v. 13.

Sarna-Wojciciki, A. M., Shipley, S., Waitt, R. B., Jr., Dzurisin, D., and Wood, S. H., 1981. "Areal distribution, thickness, mass, volume, and grain size of air-fall ash from the six major eruptions of 1980," in Lipman, P.J. and Mullineaux, D.R., eds., *The 1980 eruptions of Mount St. Helens, Washington*, U.S. Geological Survey Professional Paper 1250, p. 577-600.

Schmincke, H. -U., 2004. *Volcanism*, Berlin: Springer.

Seach, J., 2015. "Volcanic generated Tsunami": volcanolive.com.

Segall, P. and Anderson, K., 2014. "Look up for magma insights," *Nature Geoscience*, v. 7, p. 168-171.

Self, S., and Rampino, M. R., 1981. "The 1883 eruption of Krakatau," *Nature*, v. 294, no. 5843, p. 699-704.

Self, S., Rampino, M. R., Newton, M. S., and Wolff, J. A., 1984. "Volcanological study of the great Tambora eruption of 1815," *Geology*, v. 12, p. 659-663.

Self, S., Thordarson, T., Keszthelyi, L., Walker, G. P. L., Hon, K., Murphy, M. T., Long, P., and Finnemore, S., 1996. "A new model for the emplacement of Columbia River basalts as large, inflated pahoehoe lava flow fields," *Geophysical Research Letters*, v. 23, p. 2689-2692.

Self, S., Thordarson, T., and Keszthelyi, L., 1997. "Emplacement of continental flood basalt lava flows: in large igneous provinces," *Geophysical Monograph 100*, p. 381-410.

Sharp, W. D., and Clague, D. A., 2002. *An older, slower Hawaiian - Emperor bend*, EOS, Transactions American Geophysical Union, v. 83, no. 47, p. F1282.

Sharpton, V. I., and Ward, P. D., eds. 1990. Global catastrophes in Earth history: an interdisciplinary conference on impacts, volcanism, and mass morality, Geological Society of America Special Paper 247, p. 620.

Sheridan, M. F., 1979. "Emplacement of pyroclastic flows," a review, in Chapin, C. E. and Elsron, W. E. eds., Geological Society of America Special Paper 180, p. 125-136.

Sherrod, D. R., 1989. "Klamath Basin: 28th International Geological Congress Field trip T312: *South Cascades arc Volcanoes, California and Southern Oregon*, v. 1. in, ed. Muffler, L. S.: Volcanism and Plutonism of western North America.

Sherrod, D. R., Nishimitsu, Y., and Tagami, T., 2003. "New K-Ar ages and the geological evidence against rejuvenated-stage volcanism at Haleakala, East Maui, a postshield-stage volcano of the Hawaiian island chain," *Geological Society of America Bulletin*, v. 115, p. 683-694.

Sheth, H. C., 2005. "From Deccan to Reunion: no trace of a mantle plume," in Foulger, G. R. et al., eds., *Plates, plumes, and paradigms*, Geological Society of America Special Paper 388, p. 477-501.

Shirey, S. B., Kamber, B. S., Whitehouse, M. J., Mueller, P. A., and Basu, A. R., 2008. "A review of the isotopic and trace element evidence for mantle and crustal processes in the Hadean and Archean: implications for the onset of plate tectonic subduction," in Condie, K., and Pease, V., eds., *When did Plate Tectonics Begin on Earth?* Geological Society of America Special Paper 440, p. 1- 29.

Siebert, L., Alvarado, G. E., Vallance, J. W., and van Wyk de Vries, B., 2006. "Large-volume volcanic edifice failures in Central America and associated hazards," in Rose, W. I., et al., eds. *Volcanic hazards in Central America*, Geological Society of America Special Paper 412, p. 1-26.

Sigurdsson, H., 1990. *Assessment of the atmospheric impact of volcanic eruptions*, Geological Society of America Special Paper 247, p. 99-110.

Sigurdsson, H., 2000a. "Introduction," in Sigurdsson, H., ed., *Encyclopedia of Volcanoes*, p. 1-13.

Sigurdsson, H., 2000b. "Volcanic episodes and rates of volcanism," in Sigurdsson, H., ed., *Encyclopedia of Volcanoes*, p. 271-279.

Sillitoe, R. H., 1973. "The tops and bottoms of porphyry copper deposits," *Economic Geology*, v. 68, p. 799-815.

Sillitoe, R. H., 1993. "Epithermal models: genetic types, geometrical controls and shallow features, in Kirkham, R. V. et al., eds., *Mineral Deposit Modeling,* Geological Association of Canada Special Publication 40.

Sillitoe, R. H., 1997. "Characteristics and controls of the largest porphyry copper-gold and epithermal gold deposits in the circum-Pacific region," *Australian Journal of Earth Sciences,* v. 44, p. 373-388.

Sillitoe, R. H. and Bonham, H. F., Jr., 1984. "Volcanic landforms and ore deposits," *Economic Geology*, v. 79, p. 1286-1298.

Sillitoe, R. H., Grauberger, G. L., and Elliott, J. E., 1985. "A diatreme-hosted gold deposit at Montana Tunnels, Montana," *Economic Geology,* v. 80, p. 1707-1721.

Simkin, T., and Siebert, L., 2000. "Earth's volcanoes and eruptions: an overview," in Sigurdsson, H., ed.: *Encyclopedia of Volcanoes*, p. 249-261.

Simkin, T., Siebert, L., and Blong, R., 2001. "Volcano fatalities - lessions from the historical record," *Science*, v. 291, p. 255.

Sisson, T. W., Vallance, J. W., and Pringle, P. T., 2001. "Progress made in understanding Mount Rainier's hazards," *EOS*, v. 82, no. 9, p. 113-120.

Skinner, B. J. and Porter, S .C., 1995. *The Dynamic Earth*, 3rd ed.

Smith, R. L., 1960a. "Ash flows," *Geological Society of America Bulletin*, v. 71, p. 795-842.

Smith, R. L., 1960b. *Zones and zonal variations in welded ash flows*, U.S. Geological Survey Professional Paper 354-F, p. F149-F159.

Smith, R. L., 1979. *Ash-flow magmatism*, Geological Society of America Special Paper 180, p. 5-27.

Sisson, F. W., Vallance, J. W., and Pringle, P. T., 2001. "Progress made in understanding Mount Rainier hazards," *EOS*, Transactions, American Geophysical Union, v. 82, no. 9, p. 113, 118-120.

Sorey, M. L., McConnell, V. S., and Roeloffs, E., 2003. "Summary of recent research in Long Valley caldera, California," *Journal of Volcanology and Geothermal Research*, v. 127, p. 165-173.

Sparks, R. S. J., and Walker, G. P. L., 1973. "The ground surge deposit: a third type of pyroclastic rock: Nature," *Physical Science*, v. 241, p. 62-64.

Spera, F. J., 2000. "Physical properties of magmas," in Sigurdsson, H., ed. *Encyclopedia of Volcanoes*, p. 171-190.

Spudis, P. D., 2000. "Volcanism on the Moon," in Sigurdsson, H., ed.: *Encyclopedia of Volcanoes*, p. 697-708.

Stevens, T. A., and Eaton, G. P., 1975. "Geologic environment of ore deposition in the Creede district, San Juan Mountains, Colorado," *Economic Geology*, v. 70, p. 1023-1037.

Stofan, E., 2004. "Earth's evil twin: the volcanic world of Venus," in Lopes, R. M. C. and Gregg, T. P. K., eds., *Volcanic Worlds*, p. 61-80.

Stoffers, P., 2006. "Submarine volcanism and high-temperature hydrothermal venting on the Tonga arc, southwest Pacific," *Geology*, v. 34, p. 453-456.

Stone, R., 2003. "Stalking nature's most dangerous beasts," *Science*, v. 299, p. 2016-2030.

Storey, B. C., 1995. "The role of the mantle plumes in continental breakup: case histories from Gondwana," *Nature*, v. 377, p. 301-308.

Stothers, R. B., 1984a. "The great Tambora eruption in 1815 and its aftermath," *Science*, v. 224, p. 1191-1198.

Strothers, R. B., 1984b. "Mystery cloud of AD 536," *Nature*, v. 307, p. 344-345.

Swanson, D. A., Cameron, K. A., Evarts, R. C., Pringle, P. T., and Vance, J. R., 1989. "IGC Field Trip T106Cenozoic volcanism in the Cascade Range and Columbia Plateau, southern Washington and northernmost Oregon," in 28th International Geological Congress.

Swanson, D. A., 1990. "A decade of dome growth at Mount St. Helens, 1980-90," *Geoscience Canada*, v. 17, p. 154-157.

Tanton, L. T. E., Grove, T. L., and Donnelly-Nolan, J., 2001. "Hot, shallow melting under the Cascades volcanic arc," *Geology*, v. 29, p. 631-634.

Thordarson T., and Self, S., 1996. "Sulfur, chlorine and fluorine degassing and atmospheric loading by the Roza eruption, Columbia River Basalt Group, Washington, USA," *Journal of Volcanology and Geothermal Research*, v. 74, p. 49-73.

Tilling, R. I., no date. *Eruptions of Mount St. Helens: past, present and future,* U.S. Geological Survey document, p. 46.

Tilling, R. I., 2000. "What have volcanologists learned from the May 18, 1980, eruption of Mount St. Helens?" *Geotimes*, v. 45, no. 5, p. 14-19.

Tilling, R. I., and Lipman, P. W., 1993. "Lessons in reducing volcanic risk," *Nature*, v. 364, p. 277-280.

Toksoz, M. N., 1975. "The subduction of the lithosphere, in Volcanoes and the Earth's Interior," *Scientific American*, San Francisco: Freeman, p. 6-16.

Tunnicliffe, V., 1992. "Hydrothermal-vent communities of the deep sea," *American Scientist*, v. 80, p. 336-349.

Turner, M. B., Cronin, S. J., Bebbington, M. S., and Platz, 2008. "Developing probabilistic eruption forecasts for dormant volcanoes: a case study from Mt. Taranaki, New Zealand," *Bulletin of Volcanology*, v. 70, on line

Utada, H., Takahashi, Y., Koyama, T., Morita, Y., Koyama, T., and Kagiyama, T., 2007. "ACTIVE system for monitoring volcanic activity: A case study on Izu-Oshima Volcano, Central Japan," *Journal of Volcanology and Geothermal Research*, v. 164, p. 217-242.

Verbeek, R. V. M., 1885. Krakakau, Batavia, Government Press.

Vink, G., Morgan, W. J., and Vogt, P. R., 1985. "The Earth's Hot Spots," *Scientific American*, v. 252, no. 4, p. 50- 59.

Voight, B., 2003. "Stalking nature's most dangerous beasts," *Science*, v. 299, p. 2015-2030.

"Volcanic risk perception and beyond (VRPAB)," 2008. *Journal of Volcanology and Geothermal Research*, v. 172, p. 163-169.

Wallace P. J., 2001. "Volcanic SO2 emissions and the abundance and distribution of evolved gas in magma bodies," *Journal of Volcanology and Geothermal Research*, v. 108, p. 85-106.

Wallace, P., and Anderson, A. T., Jr., 2000. "Volatiles in magmas," in Sigurdsson, H., ed.: *Encyclopedia of Volcanoes*, p. 149-170.

Wallace, S. R., Muncaster, N. K., Jonson, D. C., MacKenzie, W. B., Bookstrom, A. A., and Surface, V. E., 1968. "Multiple intrusion and mineralization at Climax, Colorado," in Ridge, J. D., ed., *Ore Deposits of the United States*, v. 2, p. 605-640.

Walter, T. R., and Amelung, F., 2007. "Volcanic eruptions following M>9 megathrust earthquakes: implications for the Sumatra-Andaman volcanoes," *Geology*, v. 35, p. 539-542.

Wang, S., and Liu, M., 2006. "Moving hotspots or reorganized plates?" *Geology*, v. 34, p. 465-468.

Wark, D. A., Hildreth, W., Spear, F. S., Cherniak, D. J., and Watson, E. B., 2007. "Pre-eruption recharge of the Bishop magma system," *Geology*, v. 35, p. 235-238.

Watts, A. B. and Masson, D. G., 1995. "A giant landslide on the north flank of Tenerife, Canary Islands," *Journal of Geophysical Research*, v. 100, p. 24, 487-24, 498.

Webster, G., 2013. "NASA Rover confirms Mars origin of some meteorites," NASA, October.

Wexler, H., 1952. "Volcanoes and world climate," *Scientific American*, v. 1898, no. 4, p. 3-5.

White, N. C., and Herrington, R. J., 2000. "Mineral deposits associated with volcanism," in Sigurdsson, H., ed.: *Encylopedia of Volcanoes*, p. 897-912.

White, J. D. L., and Ross, P. S., 2011. "Maar-diatreme volcanoes: A review," *Journal of Volcanology and Geothermal Research*, v. 201, p. 1-29.

Wiesner, M. G., Wang, Y., and Zheng, L., 1995. "Fallout of volcanic ash to the deep South China Sea induced by the 1991 eruption of Mount Pinatubo (Philippines)," *Geology*, v. 23, p. 885-888.

Wignall, P. B., 2001. "Large igneous provinces and mass extinctions," *Earth Science Reviews*, v. 53, p. 1-33.

Wilde, S. A., Valley, J. W., Peck, W. H., and Graham, C. M., 2001. "Evidence from detrital zircons for the existence of continental crust and oceans on Earth 4.4 Ga ago," *Nature*, v. 409, p. 175-178.

Williams, H., 1942. *The Geology of Crater Lake National Park*, Carnegie Institution of Washington, Pub. 540, p. 162.

Williams, H., 1951. "Volcanoes," *Scientific American*, v. 187, no. 11, p. 3-11.

Wilson, J. T., 1963. "A possible origin of the Hawaiian Islands," *Canadian Journal of Physics*, v. 41, p. 863-870.

Windley, B. F., 1984. *The Evolving Continents*, Chichester, UK: Wiley.

Witham, C. S., 2005. "Volcanic disasters and incidents: a new database," *Journal of Volcanology and Geothermal Research*, v. 148, p. 191-233.

Yamaguchi, D. K., 1985. "Tree-ring evidence for a two-year interval between recent prehistoric explosive eruptions of Mount St. Helens," *Geology*, v. 13, p. 554-557.

Yamaguchi, D. K., Pringle, P. W., and Lawrence, D. B., 1995. "Field sketches of late - 1840s eruptions of Mount St. Helens, Washington," *Washington Geology*, v. 23, no. 2, p. 3-8.

Yang Lee, M., Hwa Chen, C., Yen Wei, K., Lizuka, Y., and Carey, S., 2004. "First Toba super eruption revival," *Geology*, v. 32, p. 61-64.

Young, K., 1975. *Geology: The Paradox of Earth and Man*, Houghton Mifflin.

Zhang, M., O'Reilly, S. Y., Wang, K. -L., Hronsky, J., and Griffin, W. L., 2008. "Flood basalts and metallogeny: the lithospheric mantle connection," *Earth-Science Reviews*, v. 86, p. 145-174.

Zimbelman, J. R., 2000. "Volcanism on Mars," in Sigurdsson, H., ed.: *Encyclopedia of Volcanoes*, p. 771-783.

Zimmer, M., and Erzinger, J., 2003, "Continous H2O, CO2, 222Rn, and temperature measurements on Merapi Volcano, Indonesia," *Journal of Volcanology and Geothermal Research*, v. 125, p. 25-38.

Index

About the Author

Ian Lange, a graduate of Dartmouth College, with a Ph.D. from the University of Washington in geology and stable isotope geochemistry, has studied volcanic rocks and mineral deposits in Peru, the continental United States, and British Columbia. He has worked for exploration companies as well as for his own personal love of the topic. Dr. Lange has also assessed mineral deposits for the U.S. Geological Survey in Alaska. In addition, he has been involved in ground-based thermal remote sensing of dormant and active Cascade and Central American volcanoes. Dr. Lange has taught geology at the university level in California, New Hampshire, and Montana.

When not studying volcanoes, volcanic rocks, and mineral deposits, Dr. Lange taught for over 30 years at the University of Montana. During his teaching career, he authored and co-authored more than 100 papers about mineral deposits and volcanic rocks in national and international geological journals and *Science*. One of Dr. Lange's interests is Pleistocene ice-age animals. His recent book, entitled *Ice Age Mammals of North America: A Guide to the Big, the Hairy, and the Bizarre,* combines biology and geology. He and his wife, Jo-Ann, travel to Mexico each January to study Mexican geology and culture.